Microwave Field-Effect Transistors— Theory, Design and Applications

ELECTRONIC & ELECTRICAL ENGINEERING RESEARCH STUDIES

ELECTRONIC DEVICES AND SYSTEMS RESEARCH STUDIES SERIES

Series Editor: **Professor C. S. Aitchison**
Department of Electronics, Chelsea College,
University of London, England

1. Microwave Field-Effect Transistors—Theory, Design and Applications
   Raymond S. Pengelly

Microwave Field-Effect Transistors— Theory, Design and Applications

Raymond S. Pengelly, M.Sc.,C.Eng.,M.I.E.E.

Group Leader, Gallium Arsenide Integrated Circuits,
Allen Clark Research Centre,
Plessey Research (Caswell) Ltd., England

RESEARCH STUDIES PRESS
A DIVISION OF JOHN WILEY & SONS LTD.
Chichester · New York · Brisbane · Toronto · Singapore

RESEARCH STUDIES PRESS

Editorial Office:
58B Station Road, Letchworth, Herts. SG6 3BE, England

British Library Cataloguing in Publication Data:

Pengelly, R. S.
  Microwave field-effect transistors.—(Electronic devices
  and systems research studies)
  1. Field-effect transistors   2. Microwave devices
  I. Title   II. Series
  621.3815'284   TK7871.95

  ISBN 0 471 10208 3

Printed in Great Britain

This book is dedicated to my parents in recognition of my education; to my wife, Alex, for her continuous encouragement during its writing and to my children, Tim, Bob and Demelza for being themselves.

Preface

The development of solid state active microwave devices can be considered as beginning with refinements of the traditional p-n junction. Examples are the varactor and parametric amplification, the amplifying and oscillating tunnel diode, and the amplifying and oscillating Impatt device. Prior to the arrival of these devices and their associated circuit techniques, microwave systems utilised thermonic devices such as the klystron and the magnetron, or at the lowest microwave frequencies miniature versions of the triode.

Thus the microwave system designer - in contradistinction to the low frequency system designer - was deprived of the system benefits of a two port device. In the mid 1960s, the performance of the silicon bipolar transistor was extended into the microwave region giving for the first time the advantages of two port techniques.

The gallium arsenide metal semiconductor field effect transistor (MESFET) has now largely replaced the bipolar transistor as the microwave two port active device and is used for linear and non linear analogue as well as digital functions at frequencies in excess of 18 GHz. The recently developed combination of lumped circuit elements on the gallium arsenide chip is a potentially powerful technique.

This monograph describes the current state of the art and is written by Ray Pengelly of Plessey Research (Caswell) Ltd. who heads arguably the most forward looking research team in Western Europe on this topic.

The work detailed in this monograph will be of great interest to microwave device, circuit and systems designers as well as people from outside this field.

C.S. Aitchison

Foreword

The field effect transistor (FET) at microwave frequencies using the
III-V compound gallium arsenide (GaAs) has been one of the most exciting devices to emerge from the solid state microwave community over
the past twenty years. This device has reached operational frequencies of well over 30 GHz with the expectation that 0.1μm gate length
devices will give acceptable performance to 50 GHz. The field effect
transistor has now become an established item in microwave systems of
today in such applications as low-noise amplifiers, mixers, oscillators, power amplifiers, switches and multipliers. Indeed many
microwave systems would not be possible at their present day performance levels if it were not for the unique solution that the FET
offers in providing a reliable, reproducible and flexible device. In
many cases costly parametric amplifiers have been replaced with compact, low cost GaAs FET units whilst bulky travelling wave tubes with
their associated large power supplies are giving way to power FET
amplifiers albeit at lower power levels of up to 100W or so. This
monograph attempts to give a comprehensive introduction to the theory,
design and application of field effect transistors with most emphasis
placed on gallium arsenide-based devices. A theoretical review of
both low-noise and power FET operation is given in Chapters 2 and 3
following an introductory chapter with examples of the agreement
between theoretical predictions and practical results. The requirements and growth of GaAs are detailed in Chapter 4 together with a
review of the fabrication methods used to produce the devices.
Chapters 5, 6 and 7 give design information and accompanying examples
from within the microwave industry on the applications of the FET to
low noise and power amplifiers, frequency conversion circuits and
oscillators respectively. Chapter 8 contains a review of the means
by which microwave transistors can be packaged hermetically whilst
retaining acceptable performance. Chapter 9 deals with the increasing use of GaAs FETs in circuit applications such as switching and
phase shifting. This chapter serves as an introduction to the use of
the field effect transistor as the vital active component for GaAs
integrated circuits. Both monolithic microwave analogue and digital
circuits are covered involving both normally-on, depletion mode and
normally-off, enhancement mode MESFETs. Chapter 10 reviews the technologies being adopted for both linear and digital integrated

circuits with examples of state-of-the-art circuits (in 1981). Along-side the exciting progress being made in GaAs IC's is the parallel activity of the improvement of basic FET performance and the extension of its useful frequency range by the use of other III-V compounds. The final chapter (Chapter 11) deals with this subject introducing such materials as indium phosphide, ternary and quaternary alloys, MISFETs and IGFETs and introduces the reader to the concept of the permeable base transistor which is heralded as the successor to the FET.

This book has only been made possible by the extensive co-operation which the author has received from his colleagues worldwide, parti-cularly those who have given permission for the publication of hereto-fore unpublished material. The I.E.E.E. is, in particular, acknow-ledged for its co-operation.

I would like to thank, in particular, my colleagues at Plessey Research (Caswell) Ltd. who include R.S. Butlin, I.R. Sanders, A.Peake, J. Arnold, H.J. Finlay, M.G. Stubbs, J.R. Suffolk, C.W. Suckling, J. Singleton, J.R. Cockrill, S.G. Greenhalgh, K. Vanner, P. Cooper, J.A. Turner, D. Wilcox and A. Hughes. Special thanks are due to J. Joshi for his contribution to the writing of Chapter 7 and J.Arnold for reading the final manuscript. The following colleagues amongst many have supplied information for which I offer my thanks: W.R. Wisseman, R. Pucel, D. Ch'en, D. Maki, P. Harrop. J. Oakes, R. Eden, K. Weller, H. Phillips, C. Liechti, H. Huang, M. Kumar, J. Magarshak, J. Mun, W. Kellner, H.Q. Tserng, R. Zucca and R. Zuleeg.

The considerable task of typing the manuscript was undertaken by Mrs. H. Barbour to whom I offer my grateful thanks. The line drawings and photographs were prepared by Mr. K. Jenkins of Plessey Research (Caswell) Ltd.'s report section for whose meticulous work I am greatly indebted. I would also like to acknowledge the co-operation of Drs. G. Gibbons and J. Bass in making the preparation of this book possible and C. Aitchison for his detailed editing of the various chapters.

Table of Contents

1. INTRODUCTION...................   1
   1. Introduction................   1
   2. Semiconductor Theory........   1
   3. Intermetallic Compounds.....   5
   4. Metal-Semiconductor Contacts   6
   5. Semiconductor-Semiconductor
      Contacts....................   10
   6. Conclusions.................   12

2. GaAs FET THEORY - SMALL SIGNAL.   13
   1. Introduction................   13
   2. Materials for MESFETs.......   14
   3. Principles of Operation of
      the Schottky Barrier MESFET.   15
      A. Saturation Current, Small
         signal parameters and
         switching time...........   20
      B. Small signal Equivalent
         Circuit..................   25
      C. Noise Theory of GaAs
         MESFETs..................   28
      D. Minimum Noise Figure of
         the GaAs FET.............   30
      E. Practical Equations for
         Noise Parameters.........   37
      F. Example of Low Noise GaAs
         FET Design...............   39
      G. The GaAs FET Versus the
         Bipolar Transistor as a
         Low Noise Device.........   43
   4. The Dual-Gate FET...........   45
      A. Equivalent Circuit Para-
         meters...................   47
      B. Gain versus Second-Gate
         Terminations.............   50
      C. Gain Control with Second-
         Gate Bias................   51

    D. Noise performance......... 54
  5. Conclusions.................. 56
  6. Bibliography................. 56

3. GaAs FET THEORY - POWER......... 61
  1. Introduction.................. 61
  2. Principles of Operation...... 62
  3. Modelling of GaAs FET to Predict Large Signal Performance....................... 65
    A. Channel Capacitance (C_{gs}). 66
    B. Transconductance (g_m)..... 68
    C. Feedback Capacitance (C_{FB}) 68
    D. Intrinsic Channel Resistance, R_I and Output Resistance, R_O............ 69
    E. Gunn Domain Resistance, R_{GD}....................... 69
  4. Predictions of Non-Linear Performance of GaAs Power FETs - Fundamental and Harmonic performance........... 69
  5. Intermodulation Performance.. 71
  6. Power FET Device Performance. 77
    A. Structures used to Increase Gate Width............... 78
    B. The minimization of Parasitics.................... 83
    C. Thermal Impedance......... 87
    D. Source-to-Drain Burnout and Gate-to-Drain Avalanching..................... 88
  7. Power FET Results............ 91
  8. Conclusions.................. 94
  9. Bibliography................. 94

4. MATERIAL REQUIREMENTS AND FABRICATION OF GaAs FETs............. 99
  1. •Introduction................. 99
  2. Material Requirements........ 99
    A. Epitaxial Layers.......... 99
    B. Liquid Encapsulated Czochralski Growth........ 113
    C. Ion Implantation.......... 114
  3. FET Fabrication Techniques... 129
    A. Self-aligned Gate Technology............... 129
    B. Recessed Channel Technology.................... 130
    C. Ion Implanted FET Processing....................... 132
  4. Conclusions.................. 134
  5. Bibliography................. 134

5. THE DESIGN OF TRANSISTOR
   AMPLIFIERS...................... 141
   1. Introduction................ 141
   2. Low Noise/Small Signal
      Amplifiers................... 141
      A. S-parameters............. 141
      B. Stability of a 2 Port.... 143
      C. Transducer Power Gain.... 147
      D. Circles of Constant Uni-
         lateral Gain............. 149
      E. Unilateral Figure of
         Merit.................... 150
      F. Variation of Gain with
         Drain Current and Temper-
         ature.................... 151
      G. Optimum Load Conditions
         for Output power......... 152
      H. Equivalent Circuit of the
         GaAs FET................. 153
   3. Example of Narrow-Band Ampli-
      fier design................. 155
      A. Input Matching Circuit... 156
      B. Output Matching Network.. 157
   4. Example of Broadband Ampli-
      fier Design................. 158
      A. Lumped Element Designs –
         Design Where Stability is
         'not Considered.......... 158
      B. Design Where Stability is
         Considered............... 159
      C. Distributed Designs...... 161
   5. Designing an Amplifier for
      Optimum Noise Figure........ 161
      A. Introduction............. 161
      B. Constant Noise Figure
         Circles.................. 162
      C. Noise Modelling.......... 163
   6. Example of Broadband Ampli-
      fier Designed for Minimum
      Noise Figure................ 165
      A. Input Matching.......... 166
      B. Output Matching Network.. 170
   7. Computer-Aided Design
      Practice.................... 172
      A. General Format of Micro-
         wave CAD Programs........ 172
   8. Network Synthesis........... 174
   9. The Use of Single Ended and
      Balanced Amplifiers......... 180
   10. Variations in Amplifier
       Performance................ 182

A. Transistor Variations
and Circuit Sensitivity.. 182
B. Variations in Amplifier
Performance with Tempera-
ture..................... 182
11. Designing an Amplifier for a
Specified Linear Output
Power....................... 183
12. The use of Feedback, Common-
Gate and Source-Follower
GaAs FET Configurations in
Amplifier Design............ 185
A. Feedback Amplifiers...... 185
B. Common-Gate and Source-
Follower Configurations.. 187
13. Power Amplifiers............ 192
A. Introduction............. 192
B. D.C. Characteristics..... 192
C. R.F. Characteristics of
Power GaAs FETs.......... 194
D. Circuit Topologies for
Matching Power FETs...... 196
14. Narrow Band Power FET Ampli-
fier Design................. 197
15. Broadband Power FET Amplifier
Design...................... 197
16. Designing an Amplifier for
Maximum Spurious Free Dynamic Range................... 199
17. Power Combining Techniques.. 201
18. Thermal Considerations in
Power Amplifier Design...... 203
19. Pulsed Operation of Power
FETs........................ 203
20. Reflection Amplifiers....... 205
21. Conclusions................. 209
22. Bibliography................ 209

6. FET MIXERS..................... 213
1. Introduction................ 213
2. The GaAs FET as a Mixing
Element..................... 213
A. LO Applied Between Gate
and Source............... 213
B. LO Applied Between Drain
and Source............... 222
3. Experimental Results on
Gate Mixers................. 227
4. Noise Figure................ 228
5. Signal Handling of FET
Mixers...................... 230

6. Further Mixer Configurations Using Single Gate
   FETs......................... 231
7. The Dual-Gate Mixer......... 233
8. Image Rejection Mixers...... 236
9. Frequency Up-Conversion
   using Dual-Gate FETs........ 238
10. Frequency Multiplication
    using Dual-Gate FETs........ 239
11. Conclusions.................. 241
12. Bibliography................ 242

7. GaAs FET OSCILLATORS........... 245
1. Introduction................ 245
2. Induced Negative Resistance. 246
3. S-parameter Mapping......... 246
4. Oscillator Design........... 250
   A. Theoretical Analysis..... 250
   B. Small Signal and Large
      Signal Analysis.......... 255
5. Free-Running Oscillator -
   Performance Review.......... 256
   A. Output Power............. 261
   B. Noise.................... 262
6. Stabilized Oscillators...... 263
   A. Stabilisation Tech-
      niques................... 264
7. Dielectric Resonators....... 265
   A. Resonant Frequencies of
      Dielectric Resonators.... 266
8. Dielectric Resonator
   Stabilised FET Oscillators.. 271
9. Electronic Tuning of GaAs
   FET Oscillators............. 280
10. Varactor Tuned FET
    Oscillators................. 281
11. YIG Tuned GaAs FET
    Oscillators................. 284
    A. YIG Resonators........... 284
    B. Performance Review....... 286
12. Pulsed r.f. Oscillators..... 289
13. Conclusions.................. 293
14. Bibliography................ 293

8. MICROWAVE FET PACKAGING........ 299
1. Introduction................ 299
2. Packages and Sealing....... 300
3. Package Modelling........... 303
4. Prematched GaAs FETs........ 306
5. Packaging and Thermal
   Resistance.................. 309
6. Conclusions................. 312
7. Bibliography................ 312

9. NOVEL FET CIRCUITS............. 315
   1. Introduction................ 315
   2. Switches.................... 315
   3. Phase Shifters.............. 323
   4. Discriminators............. 329
   5. GaAs FET Osciplier.......... 333
   6. Pulsed Oscillators.......... 334
   7. Conclusions................. 336
   8. Bibliography................ 336

10. GALLIUM ARSENIDE INTEGRATED
    CIRCUITS....................... 339
    1. Introduction............... 339
    2. Monolithic Microwave Circuit
       Design...................... 339
       A. Lumped Components........ 340
       B. Distributed Components... 348
       C. GaAs Planar Diodes....... 350
       D. Low Frequency Circuit
          Techniques............... 353
       E. High Frequency Circuit
          Techniques............... 357
    3. Digital Circuits............ 366
       A. Introduction............. 366
       B. GaAs Digital Circuit
          Techniques............... 370
    4. Technology of GaAs Inte-
       grated Circuits............. 380
       A. Some IC Technologies -
          A Review................. 381
       B. Resistor Technology...... 389
       C. Capacitor Technology..... 391
       D. Plasma Etching........... 397
       E. Ion Milling.............. 397
       F. Inductors................ 398
       G. Interconnections......... 398
    5. Integrated Circuit
       Examples.................... 398
       A. Small Signal Amplifiers.. 398
       B. Power Amplifiers......... 407
       C. Oscillators.............. 415
       D. Switches................. 417
       E. Mixers................... 418
       F. Further Levels of
          Integration.............. 422
       G. Digital Circuits......... 422
    6. Conclusions................. 429
    7. Bibliography................ 431

11. NEW MATERIALS AND NEW STRUCTURES 439
    1. Introduction............... 439
    2. The InP MESFET............. 439
    3. The InP MISFET............. 444

4. Ternary and Quaternary
   Compounds for MESFETs....... 448
5. Permeable Base Transistor... 454
6. Ballistic Electron Transistors..................... 458
7. Conclusions................. 462
8. Bibliography................ 463

INDEX............................ 467

CHAPTER 1
Introduction

1. INTRODUCTION

A monograph on microwave field effect transistors would be incomplete
without an introductory chapter on basic semiconductor theory.   Thus,
this chapter gives a basic review of energy bands in solids and
introduces the reader to the concepts of intrinsic and impurity semiconductors and metal-to-semiconductor contacts.

2. SEMICONDUCTORY THEORY

For many years the energy levels of electrons in solids have been
treated in discrete bands separated by gaps in which ordinarily no
energy levels occur.

The band structure of occupied energy levels of electrons in solids
have been extensively investigated by using X-ray emission spectroscopy. When the crystal structure of the solid is bombarded by high-
energy electrons, electrons are ejected from the innermost part of
the atom thus allowing transitions to take place between the outer
and inner atomic shells. Such transitions give rise to narrow
discrete spectroscopic lines, except for certain lines that are`
broader due to the transition of certain outer electrons which have
a range of energies.

In ionic crystals, such as NaCl, there are two energy bands occupied by electrons where one energy band is due to the outer electrons
of the positive ion and the other to the outer electrons of the
negative ion. In valence-bond crystals, such as diamond, only one
energy band exists and this is attributed to the electrons in the
valence bond.

The band structure of unoccupied energy levels has been investigated
by the use of X-ray absorption spectroscopy. When the crystal is
irradiated by X-rays at certain frequencies there are more or less
abrupt jump s in the absorption spectra. This is because the X-ray
quanta bring electrons from one of the inner shells to unoccupied
levels of the crystal. The minimum energy needed is that needed to

transfer the electron from a given shell to the lowest unoccupied energy level of the crystal.

Ionic crystals and valence-bond crystals have their highest occupied energy levels several electron volts lower than the highest unoccupied energy level. In metals no such difference, or gap, exists. In the first materials if an electric field is applied to the crystal the electrons in an occupied energy band, separated from the lowest unoccupied energy band by a gap of several electron volts, cannot gain energy from the field because there are no unoccupied energy levels to which they can go. In metals, however, there are unoccupied energy levels in the immediate vicinity of the occupied levels and hence electrons gain energy from the applied electric field and move through the crystal quite freely.

If the energy gap, Eg, is less than 2 electron volts (eV), the material is a good insulator at low temperatures but becomes a conductor at elevated temperatures since some of the electrons can gain enough energy from the crystal lattice vibrations to reach the unoccupied energy levels. The number of electrons that can do this varies as exp(-eEg/kT) where k is Boltzmann's constant and T is the absolute temperature. Accordingly the conductivity increases rapidly with increasing temperature. Such materials are called SEMICONDUC-TORS.

The band of unoccupied energy levels is called the CONDUCTION BAND. The band of occupied energy levels in valence-bond crystals is called the VALENCE BAND. The bands in ionic crystals are named after the ions responsible for them. In insulators the conduction band is empty at temperatures below several hundred degrees centigrade, in semiconductors the conduction band is empty at low temperatures, and in metals the conduction band is partly occupied continuously (Figs. 1.1(a),(b)).

If the energy gap, Eg, between the filled band and the conduction band of an insulator is relatively small the material will be a good insulator at low temperatures but at high temperatures the crystal will start to conduct. This happens because the electrons in the filled band and the lattice vibrations interact causing the former to gain enough energy to transfer from the valence band to the conduction band. Conductors having this property are referred to as 'intrinsic semiconductors' where intrinsic means that the semiconducting characteristic is a property of the pure material. However, many materials become semiconducting due to the introduction of impurities or deviations from stoichiometry. Such materials are called 'extrinsic semiconductors'.

Table 1.1 lists, for example, Group IV elements where the gradual transition from insulating properties through semiconductor properties to metallic properties are shown.

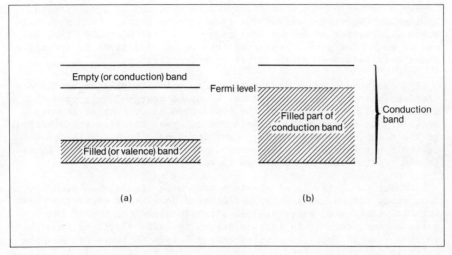

FIG. 1.1. (a) Energy Level Diagram of an Insulator at T = 0
(b) Energy Level Diagram of a Metal at T = 0

TABLE 1.1. Energy Gap and Properties of Group IV Materials

Element	Eg (V)	Property
C (diamond)	7	Insulator
Si	1.1	Semiconductor
Ge	0.7	Semiconductor
Sn (grey tin)	very small	Semiconductor
Pb	–	Conductor

The current in semiconductors is carried by two types of carriers.
One type is the electrons in the conduction band. For every electron
in the conduction band there is an electron missing in the valence
band. These vacancies are called 'holes'. A hole acts in many
respects like a positive charge – it may move and gives rise to a
current.

Semiconductors in which the current is carried predominantly by
holes are called p-type semiconductors, whilst those in which the

current is carried predominantly by electrons are called n-type.

n-type or p-type semiconductors can be produced in ionic crystals by introducing slight deviations from stoichiometry. Such materials are called excess or defect semiconductors. Valence-band crystals can be made n or p type semiconductors by the additions of impurities. These materials are called impurity semiconductors.

If a semiconductor is at absolute zero with its valence band completely empty together with its occupied energy levels only slightly below the bottom of the conduction band, then it is only necessary to increase the temperature in order to raise the electrons in the occupied levels to the conduction band. Such levels are called donor levels. Because such electrons occupy fixed positions in the crystal they are referred to as localized.

If a semiconductor is at absolute zero with its valence band completely filled and its conduction band completely empty together with its unoccupied energy levels slightly above the top of the valence band, then it is only necessary to raise electrons from the valence band to these unoccupied energy levels to leave behind free holes in the valence band. These unoccupied energy levels are called acceptor levels and they are localized.

In impurity semiconductors the donor or acceptor levels are provided by impurities. The most notable semiconductors such as silicon and germanium exist in Group IV of the periodic table and, therefore, have a crystal structure like diamond where each atom has four outer electrons which are shared by its four nearest neighbours to form a covalent type of bond.

If impurity atoms from group V of the periodic table that have five outer electrons are added (such as P, As or Sb) they take the place of some of the regular atoms of the lattice. Four of their outer electrons are used to form the four covalent bonds with their neigh-bours leaving one loosely bound electron behind. Such impurity atoms give rise to occupied donor levels close to the bottom of the conduction band - thus the semiconductor becomes n-type.

If impurity atoms from Group III of the periodic table (B, Al, In) are added they also will take the place of some of the regular atoms of the lattice. Their three outer electrons are used to make three covalent bonds where the electron for the fourth covalent bond is missing - thus a loosely bound hole is left forming unoccupied accep-tor levels close to the top of the filled band; the semiconductor therefore becomes p-type.

The most useful property of such impurity semiconductors is that their properties can be very well controlled. The type of conduction depends on the kind of impurity added where the magnitude of the conductivity depends on the concentration of the impurity atoms.

3. INTERMETALLIC COMPOUNDS

Semiconductors such as silicon crystallize in the diamond structure where each atom is surrounded by four like neighbours and the four outer electrons of each atom form electron-pair (covalent) bonds with each neighbour.

A similar structure exists in compounds of the AB type, where A is an element from Group III and B is an element from Group V of the periodic table. Each atom of one type is surrounded by four neighbours of the other type - the structure showing close similarity to the diamond structure being known as the zinc blende (ZnS) structure. These compounds resemble the crystals of the group IV elements in that they are semiconductors. Since the constituent elements are metallic the compounds are called intermetallic compounds.

Having four electron-pair bonds around each A atom and around each B atom requires the transfer of one electron from a B atom to an A atom. The A-B bond is not truly covalent but is partly ionic. This results in a larger forbidden gap width than would be expected otherwise. For example, if we compare Si with AlP formed by its neighbours in the periodic table, we find that the gap width of AlP is 3 eV whereas Si is only 1.1 eV. Table 1.2 lists some of the intermetallic compounds. The crystals can be made p-type by replacing some of the Group III atoms by Group II atoms or made n-type by replacing some of the Group V atoms by group VI atoms.

TABLE 1.2. Properties of Some Intermetallic Compounds

Compound	Melting Point ^{o}C	Energy Gap at 300°K eV	Electron Mobility, 10^{-4} m^2/Vsec	Hole Mobility 10^{-4} m^2/Vsec
InSb	523	0.17	70000	500
InAs	936	0.4	23000	100
InP	1070	1.25	34000	650
GaSb	720	0.75	4000	700
GaAs	1240	1.35	4000	200
GaP	–	2.2	–	–
AlSb	1080	1.6	100	200

Much of this monograph concentrates on the application of one of these III-V compounds, gallium arsenide (GaAs) to an active three terminal microwave device.

4. METAL-SEMICONDUCTOR CONTACTS

When two materials are brought into contact, a redistribution of charge occurs, and finally a new equilibrium condition is reached in which the Fermi-levels of the two materials are at equal heights, where the Fermi level of an n-type semiconductor is given by

$$E_f = - \frac{1}{2} E_o + \frac{1}{2} kT \ln \frac{N_d}{N_c} \qquad 1.1$$

where k is Boltzmann's constant, T is the absolute temperature, N_d is the donor concentration, N_c is the carrier concentration and E_o is the binding energy of the electrons to the donors.

Owing to the redistribution of charge, a dipole layer will form at the contact between the two materials. In a metal to metal contact the ohmic contact is formed by surface charges on both sides of the contact and electrons can move freely from one metal to the other. In a metal-semiconductor contact, the contact may be either ohmic or rectifying; in the latter case the current flowing more easily in one direction than the other.

Consider, for example, a contact between a metal and an n-type semiconductor where the donor concentration in the semiconductor is relatively large with most of the donors ionized at room temperature. Let Φ_m be the work function of the metal and Φ_s the work function of the semiconductor, where in general

$$\Phi = \chi - E_f \qquad 1.2$$

χ being the electron affinity, i.e. the energy difference between the vacuum level and the bottom of the conduction band.

Consider the case where $\Phi_m > \Phi_s$. The situation prior to contact is shown in Fig. 1.2(a) where the Fermi level of the semiconductor is above the Fermi level of the metal by $(\Phi_m - \Phi_s)$. On contact electrons from the surface layer of the semiconductor enter the metal, leaving ionized donors behind in the surface layer. The Fermi levels of both materials are at the same height after the exchange of charge which results in the energy levels in the bulk semiconductor being lowered by $(\Phi_m - \Phi_s)$. A potential barrier is formed at the surface where, on the semiconductor side the height of the barrier is $(\Phi_m - \Phi_s)$ and on the metal side the barrier height is $(\Phi_m - \Phi_s) + (\Phi_s - \chi_s) = (\Phi_m - \chi_s)$ (Fig. 1.2(b)). The height of the barrier $\Phi_m - \Phi_s$ is often called the diffusion potential. The potential difference is maintained by the electric dipole layer at the contact.

The positive charge at the semiconductor side of the contact is caused by ionized donors having a lower density than the ionized atoms in a metal. The donors being bound to fixed positions results in the positive charge being distributed in a so-called space-charge layer. Because of the potential barrier at the contact this surface layer is also known as the barrier layer.

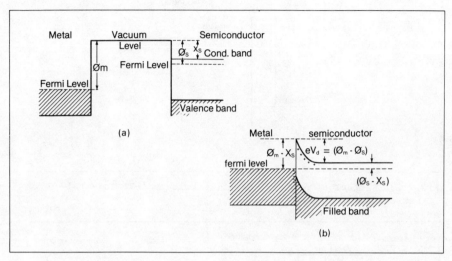

FIG. 1.2. Energy-Level Diagrams of a metal n-type Conductor Contact (with $\Phi_m > \Phi_s$)

Owing to the thermal agitation some electrons of the metal will have enough energy to cross the potential barrier into the semiconductor and vice versa. In equilibrium this gives rise to equal and opposite currents I_o crossing the barrier.

Applying a voltage -V to the semiconductor (Fig. 1.3) results in the energy levels in the conduction band being raised by an amount eV, such that the barrier for electrons going from right to left is lowered by an amount eV. Consequently, the corresponding current from left to right changes by a factor exp (eV/kT). Since the barrier for electrons going from left to right remains the current from right to left is unchanged. Thus, a current characteristic

$$I = I_o \left(\exp \left(\frac{eV}{kT} \right) - 1 \right) \qquad\qquad 1.3$$

results which is that of a rectifying or Schottky contact since for V>>kT/e I is large and positive but for V<<-kT/e the current is small and virtually equal to $-I_o$. The first condition is called forward bias and the second back or reverse bias.

If $\Phi_m < \Phi_s$ no rectifying barrier is formed. Before contact the energy levels are as shown in Fig. 1.4(a) where the Fermi level of the semiconductor is below the Fermi level of the metal by the amount $(\Phi_s - \Phi_m)$.

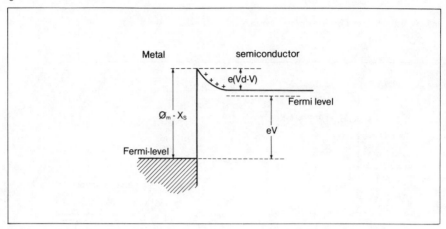

FIG. 1.3. Effect of Applying Negative Voltage (-V) to Metal-
Semiconductor Contact

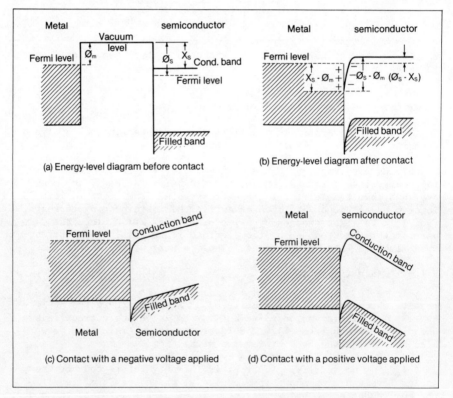

FIG. 1.4. Energy-Level Diagrams of a Metal n-type Semiconductor
Contact ($\phi_m < \phi_s$)

On contact electrons flow from the metal into the surface layer of the semiconductor leaving a positive surface charge behind on the metal side of the contact and, thus, causing a negative surface charge at the semiconductor side of the contact. The Fermi level in the semiconductor bulk material is raised by an amount $(\Phi_s - \Phi_m)$ (Fig. 1.4(b)). Applying a voltage V results in the potential difference being distributed across the semiconductor as shown in Fig. 1.4(c),(d). If $(\Phi_s - \chi_s)$ is small the electrons can move across the barrier relatively easily and for this reason the contact can be considered as ohmic (i.e. linearly resistive). Thus, a contact is ohmic if $\Phi_m < \Phi_s$ and Schottky (rectifying) if $\Phi_m > \Phi_s$.

One would expect that if two metals are compared with different values of Φ_m that the one with the largest Φ_m would give the smallest value of I_o. This does not necessarily happen in practice since the semiconductor can have a natural surface barrier. The way in which this can happen is that a large number of electron energy levels, the so-called surface states, are located at the surface of the semiconductor. Many of these surface states are occupied resulting in a distributed positive charge due to the ionized donors in the surface layer. Making contact with metals of different work functions, Φ_m results in a different portion of the occupied surface states being emptied into the metal – thus leaving the space-charge barrier at the surface unchanged (Fig. 1.5).

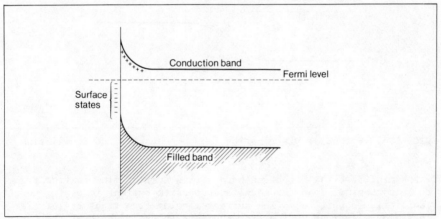

FIG. 1.5.   Energy-Level Diagram of an n-type Semiconductor with Surface States

An ohmic contact can also be formed by alloying the metal and semiconductor together resulting in a gradual transition from the one material to the other. This occurs, for example, if Ge-Au is alloyed to n-type GaAs. It can also be shown that if the donor concentration in the semiconductor surface is above a certain critical value then an ohmic contact can be formed.

5. SEMICONDUCTOR-SEMICONDUCTOR CONTACTS

If n and p-type semiconductors are brought into contact, electrons
will flow from the n-type material into the p-type material until the
Fermi levels are equal in height.

In practice such rectifying contacts are made by diffusing n-type
impurities into p-type material or vice versa. The structure that
results is partly p-type and partly n-type; such a structure is
called a p-n junction. Fig. 1.6 shows the energy level diagram of
such a junction in which there is a sudden change from p-type to
n-type material at x = 0. There is a region of negative space charge
for $-x_1 < x < 0$ due to ionized acceptors and a region of positive charge
for $0 < x < x_2$ due to ionized donors. The n region has a positive potential with respect to the p region due to the space charge region when
there is no external bias. This potential is called the diffusion
potential, V_d.

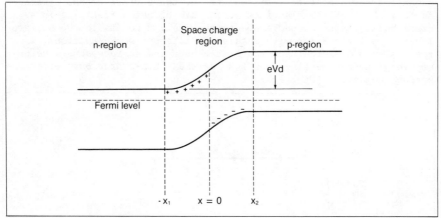

FIG. 1.6.   Potential Distribution in a p-n Junction at Equilibrium

Assuming that the direction of positive current flow is from right
to left there will be two equal and opposite current flows I_{no} under
equilibrium conditions. The current flowing from right to left is
caused by electrons moving from the left to the right over the
potential energy barrier of height eV_d. The current flowing from
left to right is caused by electrons generated in the p-region going
downhill into the n-region. This latter current does not change if
an external potential V is applied to the p-region. However, the
first current obeys an exp (eV/kT) law since the potential energy
barrier is lowered by an amount eV.

Consequently the characteristic is

$$I_n = I_{no} \ (\exp \ (\frac{eV}{kT}) - 1)$$

1.4

A similar expression occurs for the hole currents.

If different semiconducting materials are grown one on top of the other another form of junction is formed. An n-type or p-type compound when grown on another p-type or n-type compound of approximately the same lattice spacing forms a so-called hetero-junction. There are four possibilities, n-n, n-p, p-n and p-p junctions.

For example, Fig. 1.7 shows two n-type semiconductors with work functions ϕ_1 and ϕ_2 (where $\phi_2 < \phi_1$ and $\chi_2 < \chi_1$). On contact, electrons flow from semiconductor 2 to semiconductor 1 until the Fermi level heights are the same (Fig. 1.7(b)). There is now a potential barrier of height $\chi_1-\chi_2$ on the side of the semiconductor 1 and a barrier of height $\phi_1-\phi_2$ on the side of semiconductor 2. Such a contact can be shown to be rectifying.

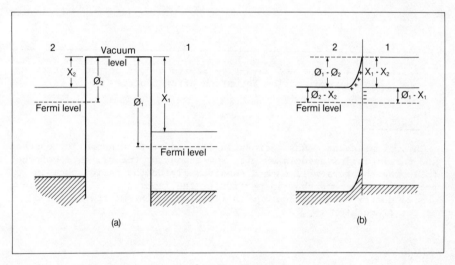

FIG. 1.7. (a) Two n-type Semiconductors Before Contact
              (b) Heterojunction Formation After Contact

Fig. 1.8 shows a p-n junction where the work functions of the n and p-type semiconductors are ϕ_1 and ϕ_2 respectively ($\phi_2 > \phi_1$). On contact, electrons flow from the n-type to the p-type semiconductor, leaving ionized donors behind on the n side of the contact and ionized acceptors on the p side. As the gap widths are different there is no match at the contact between the bottoms of the conduction band or the tops of the valence band. Even so this contact is also rectifying.

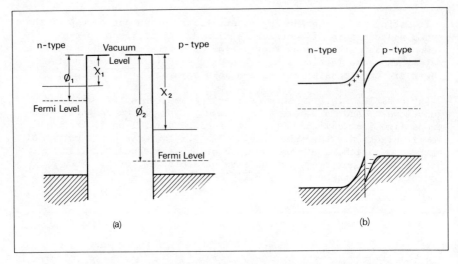

FIG. 1.8. (a) A p-type and an n-type Semiconductor Before Contact
(b) Heterojunction Formation After Contact

6. CONCLUSIONS

This first chapter has introduced the reader to a number of concepts and terminologies associated with semiconductor theory. Some of the following chapters will expand considerably on the basic concepts presented above and show in particular the importance of the metal-semiconductor Schottky contact in relation to the microwave field effect transistor .

CHAPTER 2.
GaAs FET Theory—
Small Signal

1. INTRODUCTION

Over ten years ago the first gallium arsenide transistors which used
a diffused-gate structure were reported by Turner et al (1967) giving
useful gains in the lower megahertz frequency bands. In 1969
Middelhoek realized a silicon Metal Semiconductor Field Effect Transistor (MESFET) with a 1μm gate length by projection masking which
had a maximum frequency of oscillation of 12 GHz (Middelhoek, 1970a,
b). This was comparable to the maximum frequency of oscillation,
f_{max}, of the best bipolar transistors at that time.

In 1971 a significant step was made by Turner et al, when 1μm gate
length FETs on GaAs were made having f_{max} equal to 50 GHz and useful
gains up to 18 GHz.

The substantial improvement in FET performance over the silicon
bipolar transistor is due mainly to two reasons:-

1. In gallium arsenide the conduction electrons have a six times
   larger mobility and twice the peak drift velocity of those in
   silicon (Ruch et al, 1970).

2. The active layer is grown on a semi-insulating GaAs substrate with
   resistivity larger than 10^7 Ω cm. This compares with a typical
   value of 30 Ω cm for intrinsic silicon.

The first property results in lower parasitic resistances, larger
transconductances and shorter electron transit times.

The second property results in lower gate-bonding pad parasitic
capacitance when the gate pad is on the semi-insulating (SI) substrate.

By the early 1970's it was clear that the gallium arsenide MESFET
could be used in low noise amplifiers up to X-band and Leichti et al
(1972) reported a noise figure of 3.5 dB with 6.6 dB associated gain
at 10 GHz.

A scanning–electron micrograph of a typical low noise FET is shown in Fig. 2.1.

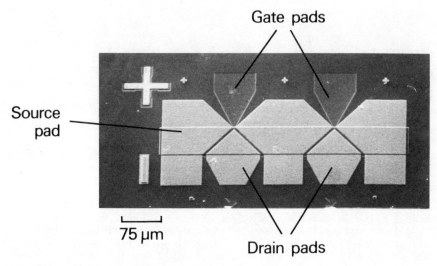

FIG. 2.1.   Plessey GAT6 GaAs FET

2. MATERIALS FOR MESFETs

Besides Si and GaAs, InP has been investigated as a substrate for MESFET fabrication. InP has a 50 percent higher maximum drift velocity than GaAs (Lam et al, 1971; Fawcett et al, 1974). The f_T for InP FETs is therefore expected to be higher and indeed Barrera et al (1975) has measured f_T's which are 1.6 times larger than in GaAs. However, the maximum frequency of oscillation, f_{max}, is 20 percent lower. Also degenerate feedback resulting from a larger gate-to-drain capacitance and a smaller output resistance degrade the overall microwave performance. Further details of InP FETs are contained in Chapter 11.

A field effect transistor requires a semiconductor having large mobility, large maximum drift velocity and a large avalanche breakdown field. This means that the electrons must have a small effective mass, and the material must have a large intervalley separation as well as a large energy gap. However a large energy gap implies a large electron effective mass. Materials which are closer to the ideal in this respect than either GaAs or InP are the so-called quaternary crystals such as InAs-InP and InAs-GaAs (Fawcett et al, 1969; Glicksman et al, 1974). Further details of such materials are examined in Chapter 11.

3. PRINCIPLES OF OPERATION OF THE SCHOTTKY BARRIER MESFET

The first part of the analysis of the operation of the Schottky barrier FET deals with the characteristics of the silicon device such that the differences between Si and GaAs can be outlined especially in relation to short gate lengths.

Fig. 2.2(a) shows the current-voltage relationship of a thin n-type silicon layer in which the electrons are carrying the current and where the layer is grown on an insulating substrate. Ohmic contacts, the source and drain, are fabricated on the surface of the conducting layer. A cross section of the device is shown in Fig. 2.2(a) where band bending at the free surface of the n-type layer and the depleted region at the substrate-layer interface are neglected.

Applying a positive voltage V_{DS} between drain and source causes electrons to flow. For small values of V_{DS} the layer appears to be a linear resistor, but as larger voltages are applied the electron drift velocity does not increase at the same rate as the electric field, E (Ruch, 1972). The current-voltage relationship therefore falls below the initial 'resistor' line. As V_{DS} is increased further E reaches a critical value, E_C at which the electrons reach a saturation velocity, v_s as shown in Fig. 2.3. - correspondingly the drain to source current saturates.

By adding a metal-to-semiconductor contact between the source and drain, the so-called gate, a layer is created in the semiconductor which is depleted completely of free-carrier electrons depending on the barrier height between the gate stripe and the n-type layer (Fig. 2.2(b)). This so-called depletion layer has the same action as an insulating layer restricting the current flow in the n-layer. The width of this depletion region is related to the voltage applied between the gate and the source electrodes. By connecting the gate to the source as in Fig. 2.2(b) the depletion layer will have a finite width and the conducting channel beneath has a smaller cross section d than in Fig. 2.2(a) - the source to drain distance being larger.

The drain to source current, I_{DS}, can be determined by considering a 'sheet' of charge (of dimension W by $d(x)$) moving at the drift velocity $v(x)$, thus

$$I_{DS} = Wqn(x)v(x)d(x) \qquad 2.1$$

where W is the gate width (Fig. 2.6), q is the electronic charge, n is the density of conduction electrons, v is the electron drift velocity, d is the conducting layer thickness and x is the distance from the source to drain. The electron density n is equal to N_D, the doping density, for values of field E less than the critical value, E_c.

The metal-to-semiconductor junction becomes increasingly reverse

16

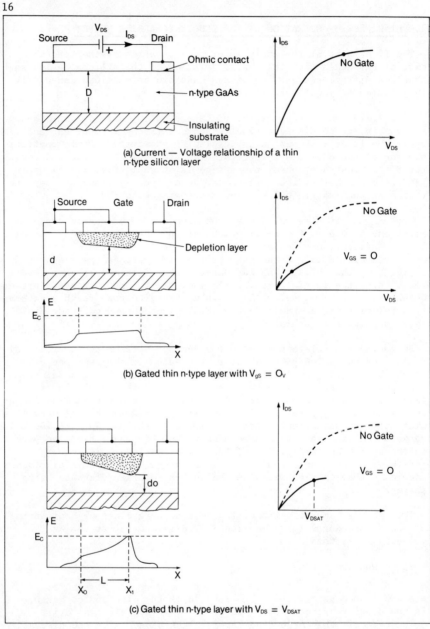

(a) Current — Voltage relationship of a thin n-type silicon layer

(b) Gated thin n-type layer with $V_{gS} = O_v$

(c) Gated thin n-type layer with $V_{DS} = V_{DSAT}$

FIG. 2.2(a). Current-voltage relationship of a thin n-type silicon layer

(b). Gated thin n-type layer with $V_{GS} = OV$

(c). Gated thin n-type layer with $V_{DS} = V_{DSAT}$

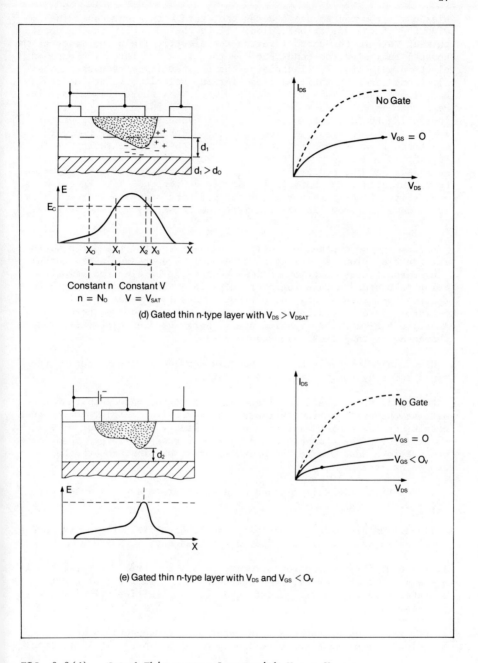

(d) Gated thin n-type layer with $V_{DS} > V_{DSAT}$

(e) Gated thin n-type layer with V_{DS} and $V_{GS} < O_v$

FIG. 2.2(d). Gated Thin n-type Layer with $V_{DS} > V_{DSAT}$
        (e). Gated Thin n-type Layer with V_{DS} and $V_{GS} < OV$

biased   from the source towards the drain with a corresponding widening of the depletion region. Since the device is taking constant current through the channel region the electric field increases as the channel region narrows and there is therefore a related increase in electron velocity, v. Increasing the drain voltage results in the electrons reaching their maximum limiting velocity, v_s at the drain end of the gate. This is shown in Fig. 2.2(c) where the channel is restricted to a cross-section d_o at the drain end of the gate. By increasing the drain to source voltage beyond V_{DSAT} the depletion region widens towards the drain contact and the point x_1 where the electrons reach saturation velocity moves towards the source as in Fig. 2.2(d). As x_1 moves towards the source the voltage at x_1 decreases and the conducting channel widens resulting in more current being injected into the velocity-saturated region. The I_{DS} curve has a positive slope in this regime with a finite drain-to-source resistance beyond current saturation (Grebene et al, 1969; Lehovec et al, 1975).

Beyond x_1 the channel potential increases towards the drain widening the depletion layer with a corresponding decrease in the conducting channel cross section to less than d_1. The electron velocity being saturated in this region results in a change in carrier concentration to maintain constant current. Thus from equation 2.1 an electron space charge layer must form between x_1 and x_2 where $d < d_1$. At x_2, $d = d_1$ and the negative space charge becomes a positive space charge again to preserve constant current.

Electron velocity is still saturated between x_2 and x_3 due to the field added by the negative space charge.

Now consider a negative voltage applied to the gate such that the gate to channel junction is reverse biased (Fig. 2.2(e)). Under such conditions the depletion region becomes wider. Again under small values of V_{DS} the channel acts as a linear resistor-larger than in the previous case due to the narrower cross section beneath the depletion region.

The critical field is reached at a lower drain current than when $V_{gs} = 0$.

In gallium arsenide the analysis in the high-field region is more complicated than in silicon due to two reasons:-

1. The equilibrium electron velocity versus electric field reaches a peak value at 3 kV/cm, then decreases and levels off at a saturated velocity that is about equal to the limiting velocity in Si (Fig. 2.3).

2. For gate lengths shorter than about 3μm, a non-equilibrium velocity-field characteristic has to be considered (Maloney et al, (1975).

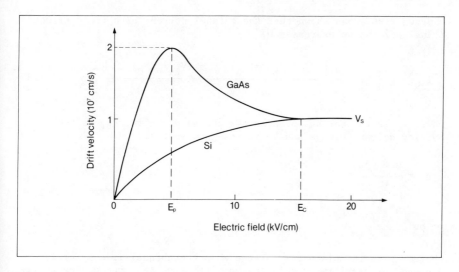

FIG. 2.3. The Equilibrium Electron Drift Velocity in Si and GaAs

Various analytical solutions for the voltage-current characteristics of short gate length MESFETs have been developed. Many use a one-dimensional analysis based on the gradual-channel approximation of Shockley (1952).

Considerable work has taken place in the development of accurate two-dimensional numerical solutions for GaAs and other materials (Mo et al, 1970; Alley et al, 1974; Shur et al, 1978).

Recently more emphasis has been placed on very short gate length GaAs FETs where the electrons do not reach equilibrium transport in the high field region of the channel.

Experimental data and two dimensional calculations (Englemann et al, 1976; Yamaguchi et al, 1976) show that the formation of a stationary Gunn domain at the drain side of the gate (rather than a channel pinch-off) is responsible for the current saturation in GaAs MESFETs.

Current saturation is assumed to occur when the average electric field under the gate reaches the Gunn domain sustaining field, E_s:

$$E_s = \frac{v_s}{\mu} \qquad\qquad 2.2$$

where μ is the low field mobility.

20

A. SATURATION CURRENT, SMALL-SIGNAL PARAMETERS AND SWITCHING TIME

Grove (1967) has shown that the channel current of a field effect transistor can be expressed by

$$I_{ch} = g_o \left[V_i - \frac{2}{3} \frac{\left[(V_i + V_{Bi} - V_G)^{3/2} - (V_{Bi} - V_G)^{3/2} \right]}{V_{po}^{\frac{1}{2}}} \right] \qquad 2.3$$

where V_i is the voltage drop across the gate region, V_{Bi} is the barrier height voltage and V_G is the gate voltage.

$$g_o = \frac{q\mu N_D WA}{L}$$

$$\qquad \qquad 2.3a$$

$$V_{po} = \frac{qN_D A^2}{2\varepsilon_o \varepsilon}$$

where V_{po} is the pinch off voltage when the channel is completely depleted of carriers, $\varepsilon_o \varepsilon$ is the permittivity, A is the device thickness, L is the gate length and W is the gate width (see Fig. 2.4).

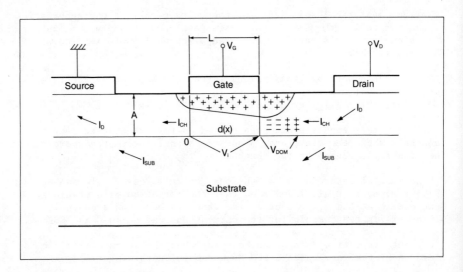

FIG. 2.4.   MESFET Geometry

Equation 2.3 is valid when

$$V_i \leq V_s \quad \text{where } V_s = E_s L \qquad \qquad 2.4$$

For a typical GaAs FET with $L \simeq 1\mu m$

$$V_s \ll (V_{Bi}-V_G) \qquad 2.5$$

and the channel current is almost linear up to the saturation point as explained earlier. Thus, from Equns. 2 and 2.3a we can write

$$I_{ch} \simeq g_o (1-\frac{A_o}{A})V_i = g_d V_i \qquad 2.6$$

where

$$A_o = \left[\frac{2\varepsilon_o \varepsilon (V_{Bi}-V_G)}{qN_D}\right]^{\frac{1}{2}}$$

$$= A \left[\frac{V_{Bi}-V_G}{V_{po}}\right]^{\frac{1}{2}}$$

and $g_d \simeq g_o (1-\frac{A_o}{A})$ is the drain conductance.

The saturation current I_{SAT} is equal to

$$I_{SAT} = g_d V_s \qquad 2.9$$

The small signal transconductance in the saturation region is given by

$$g_m \equiv \frac{\partial I_{ch}}{\partial V_G}\bigg|_{V_i=V_s} = g_o \left[\frac{(V_s + V_{Bi}-V_G)^{\frac{1}{2}} - (V_{Bi}-V_G)^{\frac{1}{2}}}{V_{po}^{\frac{1}{2}}}\right] \qquad 2.10$$

or

$$g_m \simeq g_o \frac{V_s}{2\left[V_{po}(V_{Bi}-V_G)\right]^{\frac{1}{2}}} = \left[\frac{qN_D\varepsilon_o\varepsilon R}{2(V_{Bi}-V_G)}\right]^{\frac{1}{2}}.V_s.W \qquad 2.11$$

by taking into account equation 2.5.

Table 2.1 indicates the satisfactory agreement between experimental data and this theory for the MESFET parameters given. The smaller value of g_m measured by Tsironis (1977) is probably due to series gate-to-source resistance R_S and source contact resistance R_{SC} since the observed value $g_m(obs)$ is given by (Grove, 1967)

$$g_m(obs) = \frac{g_m}{1 + (R_S + R_C) g_m} \qquad 2.12$$

TABLE 2.1. Measured and Calculated Parameters of GaAs MESFETs

I_{SAT} mA		g_m mS		C_{gs} pF	
Measured	Model	Measured	Model	Measured	Model
(a) 42	42.3	53	46	0.28	0.28
(b) 15	12.6	11.8	24.2	-	0.15

(a) $N_D = 1.1 \times 10^{17}/cc$ (b) $N_D = 2 \times 10^{17}/cc$
    $A = 0.16\mu m$ $A = 0.1\mu m$
    $L = 1\mu m$ $L = 1\mu m$
    $W = 500\mu m$ $W = 200\mu m$

$V_G = 0V$, $V_{Bi} = 0.8V$, $V_s = 0.8 \times 10^7$ cm/sec, $V_i = V_s = 0.15V$.

The depletion layer width $d(x)$ (Fig. 2.4) can be expressed as

$$d(x) = A\left[\frac{V(x) + V_{Bi} - V_G}{V_{po}}\right] \qquad 2.13$$

where $V(x)$ is the potential drop between points 0 and x $(V(L)=V_i)$.
The electric field at x is given by

$$E(x) \simeq V_i/L \;(\simeq E_i)$$

$$V(x) \simeq E_i \cdot x \qquad 2.14$$

Thus

$$d(x) = A\left[\frac{E_i x + V_{Bi} - V_G}{V_{po}}\right]^{\frac{1}{2}} \qquad 2.15$$

Since the current is continuous under the gate

$$d(x)\, E(x) = constant.$$

The total charge Q under the gate in the linear region when $V_i \lessgtr V_s$ is given by

$$Q = qN_D W \int_0^{W_G} d(x)dx$$
$$= \frac{2\sqrt{2}}{3} WL\, (\varepsilon_o \varepsilon q N_D)^{\frac{1}{2}}\left[(V_i + V_{Bi} - V_G)^{\frac{3}{2}} - (V_{Bi} - V_G)^{\frac{3}{2}}\right] \qquad 2.16$$

from Eq. 2.3.

Since $V_S \leqslant (V_{Bi} - V_G)$

$$Q \simeq qN_D A_o WL \qquad 2.17$$

We are now in a position to derive expressions for the drain-to-gate and gate-to-source capacitances C_{dg} and C_{gs} using the simple equivalent circuit of the GaAs FET shown in Fig. 2.5.

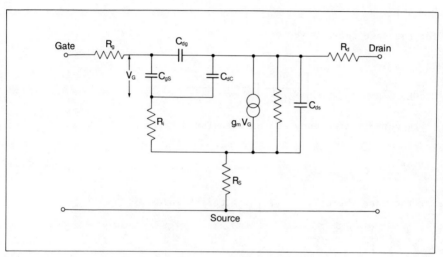

FIG. 2.5.   Simplified Equivalent Circuit of GaAs FET

$$C_{dg} = \left(\frac{\partial Q}{\partial V_i}\right)_{V_G = const.} = \frac{2\sqrt{2}}{3} \frac{L(\mathcal{E}_o \mathcal{E} q N_D)^{\frac{1}{2}}}{V_i^2}$$

$$\cdot \left[\frac{3}{2} V_i (V_i + V_{Bi} - V_G)^{\frac{1}{2}} - (V_i + V_{Bi} - V_G)^{\frac{3}{2}} + (V_{Bi} - V_G)^{\frac{3}{2}}\right]$$

$$2.18$$

and

$$C_{gs} = \left(\frac{\partial Q}{\partial V_G}\right)_{V_i - V_G = const.} = \frac{2\sqrt{2}}{3} \frac{wL(\varepsilon_0\varepsilon q N_D)^{\frac{1}{2}}}{V_i^2}$$

$$\cdot \left[(V_i + V_{Bi} - V_G)^{\frac{3}{2}} - (V_{Bi} - V_G)^{\frac{3}{2}} \right.$$

$$\left. - \frac{3}{2}(V_{Bi} - V_G)^{\frac{1}{2}} V_i \right] \qquad 2.19$$

for the case where $V_i \ll V_{Bi} - V_B$

$$C_{dg} = C_{gs} = \frac{1}{2\sqrt{2}} WL \left[\frac{\varepsilon_0\varepsilon q N_D}{V_{Bi} - V_G}\right]^{\frac{1}{2}} = \frac{1}{2} \frac{\varepsilon_0\varepsilon WL}{A_0} \qquad 2.20$$

The cut-off frequency can be calculated to be

$$f_T \simeq \frac{1}{2\pi} \frac{g_m}{C_{gs}} \qquad 2.21$$

For the case where $V_i \ll (V_{Bi} - V_a)$ this becomes

$$f_T \simeq \frac{1}{\pi} \frac{V_s}{L} \qquad 2.22$$

so that $f_T \sim 25.5$ GHz for a 1μm device. This agrees well with estimates given by Englemann (1976) where

$$f_T \simeq \frac{1}{2\pi} \frac{V_p}{L} \qquad 2.23$$

where V_p is a peak electron velocity.

The characteristic switching time of the GaAs FET is given by

$$\tau = \frac{Q(V_s)}{I_{SAT}} \qquad 2.24$$

Using 2.17 and 2.6

$$\tau \simeq \frac{L}{V_s} \cdot \frac{A_0}{A - A_0} \qquad 2.25$$

Thus the switching time is proportional to the transit time under the gate and the saturation velocity determines the switching time.

We can therefore decrease the switching time and hence increase the cut-off frequency by making the gate shorter.

B. SMALL SIGNAL EQUIVALENT CIRCUIT

A simple lumped element equivalent circuit is capable of modelling the FET's S-parameters up to 20 GHz (Wolf, 1970; Dawson, 1975; Vendelin, 1976; Kuvas, 1980; Vendelin, 1975). The equivalent circuit of the GaAs MESFET when operating under saturated current conditions is shown in Fig. 2.5 for the case where the FET is connected in common source. Fig. 2.6 shows a diagram indicating the physical origin of the circuit elements.

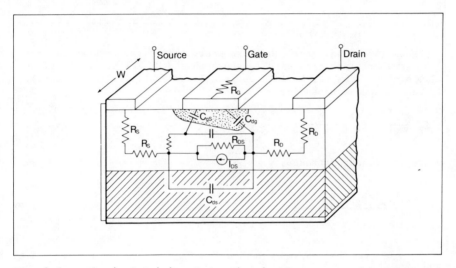

FIG. 2.6. Physical Origin of the Circuit Elements of the MESFET

In the intrinsic FET model, the elements ($C_{dg} + C_{gs}$) represent the total gate-to-channel capacitance; R_i and R_{DS} show the effects of the channel resistance and i_{ds} defines the voltage controlled current source. The transconductance g_m relates i_{ds} to the voltage across C_{gs}. The extrinsic (parasitic) elements are: R_s, the source resistance, R_D the drain resistance, R_g the gate metal resistance and C_{ds} the substrate capacitance. Typical element values for a 1μm gate length and 300μm gate width device are listed in Table 2.2. The extrinsic elements are those caused by the fabrication of the device not being ideal, for example R_S and R_D are due to the contact resistances of the 'ohmic' source and drain electrodes.

TABLE 2.2. Equivalent Circuit Parameters of a Low Noise GaAs FET
with a 1μm x 300μm Gate

Intrinsic Elements	Extrinsic Elements
g_m = 30 mS	C_{ds} = 0.07 pF
τ_o = 3 pS	R_g = 2Ω
C_{gs} = 0.4 pF	R_d = 5Ω
C_{dg} = 0.01 pF	R_s = 5Ω
C_{dc} = 0.015 pF	d.c. bias
R_i = 3Ω	V_{ds} = 5V
R_{ds} = 500Ω	V_{gs} = 0V
	I_{ds} = 50 mA

Ohkawa (1975) has shown that the equivalent circuit shown in Fig.
2.5 has a critical frequency f_K above which the FET is unconditionally stable (Chapter 5 includes details of stability factor etc.).
f_K can be approximated by:

$$f_K \simeq \frac{1}{2\pi(\tau_o + \tau_1 + \tau_2)} \qquad 2.26$$

where the transadmittance, $y_m = g_m\, e^{-j\omega\tau_o}$ \qquad\qquad 2.27

$$\tau_1 = \frac{C_{dg}(2R_g + R_i + R_s)}{\left[C_{dg}/C_{gs} + R_s/R_{ds}\right]} \qquad 2.28$$

and

$$\tau_2 = \frac{2}{\frac{g_m}{C_{gs}}\left[C_{dg}/C_{gs} + R_s/R_{ds}\right]} \cdot \frac{R_g + R_i + R_s}{R_{ds}} \qquad 2.29$$

The GaAs FET detailed in Table 2.2 has an f_K of 5.7 GHz. Below this
frequency the device becomes unstable since a larger fraction of the
output voltage is fed back to the input over the C_{dg}-R_{in} voltage
divider where

$$R_{in} = \frac{1}{2\omega^2 C_{gs}^2 (R_g + R_i + R_s)} \qquad 2.30$$

i.e. $R_{in} \propto 1/\omega^2$

and the feedback capacitance,

$C_{dg} \propto 1/\omega$

The device unilateral power gain can be defined as

$$G_u \simeq (f_u/f)^2 \qquad\qquad 2.31$$

where f_u is the maximum frequency of oscillation. Now the frequency for unity current gain, f_T, is given by equn. 2.21 and

$$f_u \simeq \frac{f_T}{2\sqrt{r_1 + f_T \tau_3}} \qquad\qquad 2.32$$

where

$$r_1 = \frac{R_g + R_i + R_s}{R_{DS}} \qquad\qquad 2.33$$

and

$$\tau_3 = 2\pi R_g C_{dg} \qquad\qquad 2.34$$

For example for the device of Table 2.2, f_u is 69 GHz. To maximise f_u, the frequency f_T and the resistance ratio R_{ds}/R_i must be optimised. In addition the extrinsic resistances R_g and R_s and the feedback capacitance C_{dg} must be minimised.

The maximum available gain at a frequency f of the FET is given by

$$MAG \simeq \left(\frac{f_T}{f}\right)^2 \left[\frac{4}{R_{DS}} (R_i + R_s + R_G + R_G + \frac{\omega_T L_S}{2}) \right.$$

$$\left. + 2\omega_T C_{dg}(R_i + R_s + 2R_g + \omega_T L_S) \right]^{-1} \qquad\qquad 2.35$$

where L_S is the inductance in the common source lead of the device.

Thus it may be appreciated that in order to increase MAG the value of f_T must be optimised whilst R_g, R_s and C_{dg} are minimised.

It has been found empirically for GaAs FETs having gate lengths of about 1μm, that

$$f_u \simeq \frac{40}{L} \quad (GHz) \qquad \text{where L is in microns.} \qquad\qquad 2.36$$

f_{max} is a figure of merit for the FET and is comparable to the figure of merit for the microwave bipolar transistor which is given by $40/S$ where S is the emitter strip width and S is in microns. In the FET, f is directly related to the saturated drift velocity, V_S, by equation 2.22 and it is for this reason that GaAs is preferred to Si, since V_S is 1.4×10^7 cm.sec^{-1} for GaAs whereas V_S is 8×10^6 cm. sec^{-1} for Si.

C. NOISE THEORY OF GaAs MESFETs

The noise properties of a linear two port can be represented by a noiseless two port with noise current generators connected across the input and output ports. This is particularly meaningful in the case of the MESFET, corresponding to noise sources at gate, source and drain as shown in Fig. 2.7.

The noise-current generator at the output of the FET represents the short-circuit channel noise generated in the drain-source path. The mean square of i_{ND} can be expressed (Van der Ziel, 1962) as:-

$$\overline{i_{ND}^2} = 4KT_o \Delta f g_m P \qquad \qquad 2.37$$

where K is Boltzmann's constant

T_o is the lattice temperature

Δf is the bandwidth

g_m is the transconductance

and   P is a dimensionless factor depending on the device geometry and the d.c. bias conditions.

For zero drain voltage, i_{ND}^2 represents the thermal noise generated by the drain conductance, G_{ds}. It can be shown that $P = G_{ds}/g_m$. For positive drain voltages the noise generated in the channel is larger than the thermal noise generated by G_{ds} since:

1. A thermal noise voltage generated locally in the channel modulates the conductive cross section of the channel resulting in an amplified noise voltage at the drain.

2. The electrons are accelerated in the electric field and are scattered due to interactions with lattice phonons. Their random drift-velocities and the free-carrier temperature increase with the applied field to values considerably higher than the lattice temperature (so called hot electron noise (Baechtold, 1971)).

3. In GaAs the carriers undergo field dependent transitions from the central valley in the conduction band to satellite valleys and vice versa. Such a transferred electron suffers a dramatic velocity change. These transitions generate so-called 'inter-valley - scattering noise' (Baechtold, 1972).

4. For large drain voltages, the electrons reach their limiting velocity on the drain side of the channel. Thus this channel section cannot be treated as ohmic. In this region the noise is analysed as high-field diffusion noise (Van der Ziel, 1971; Statz et al, 1974). The mean square value of the noise current is proportional to the high-field diffusion coefficient in the semiconductor.

Noise voltages generated in the channel cause fluctuations in the depletion-layer width. The resulting charge fluctuation in the depletion layer induces a compensating charge variation on the gate electrode. The total induced-gate charge fluctuation is shown in Fig. 2.7 by a noise generator i_{NG} at the gate terminal where

$$\overline{i_{NG}^2} = 4\, KT_o\, \Delta f\ \frac{\omega^2\, C_{gs}^{\,2}\, R}{g_m} \qquad\qquad 2.38$$

where R is a factor depending on the FET geometry and the bias conditions. For zero drain voltage, $R \sim g_m R_i$.

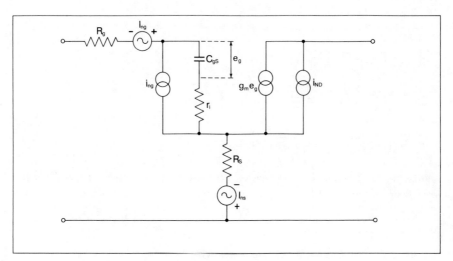

FIG. 2.7. Noise Equivalent Circuit of GaAs MESFET

Since the two noise currents i_{NG} and i_{ND} are caused by the same noise voltages in the channel partial correlation can be expected. A correlation factor C is defined as (Van der Ziel, 1963)

$$jC = \frac{\overline{i_{NG}^*\cdot i_{ND}}}{\sqrt{\overline{i_{NG}^2}\cdot\overline{i_{ND}^2}}} \qquad\qquad 2.39$$

where j is the imaginary unit and the asterisk denotes the complex conjugate.

C is purely imaginary since i_{NG} is caused by the capacitive coupling of the gate circuit to the noise source in the drain circuit.

The minimum noise figure for the intrinsic MESFET can be expressed by

$$F_{min} = 1 + 2\sqrt{PR(1-C^2)}\ \frac{f}{f_T} + 2g_m R_i P(1-C\sqrt{\frac{P}{R}})\ (\frac{f}{f_T})^2 \qquad 2.40$$

For actual GaAs MESFETs, however, this expression is no longer effective. With an exhaustive treatment using the equivalent circuit of Fig. 2.7, Pucel et al (1975) obtained an expression for the minimum noise figure as:

$$F_{min} = 1 + 2(2\pi f C_{gs}/(g_m))\sqrt{K_g[K_r + g_m(R_g+R_s)]}$$
$$+ 2(2\pi f C_{gs}/(g_m))[K_g g_m(R_g+R_s+K_c R_i)] \qquad 2.41$$
$$+ \ldots$$

where

$$K_g = P[(1-C\sqrt{R/P})^2 + (1-C^2)R/P]$$

$$K_r = R(1-C^2)/[(1-C\sqrt{R/P})^2 + (1-C^2)R/P]$$

and

$$K_c = (1-C\sqrt{R/P})/[(1-C\sqrt{R/P})^2 + (1-C^2)R/P]$$

where R_g is the gate resistance, R_s is the source series resistance and g_m is in Siemens.

In 2.41 R_g and R_s are parasitic and remain unchanged under normal operating conditions.

D. MINIMUM NOISE FIGURE OF THE GaAs FET

Consider the case where the FET is operating at a frequency below its cut-off frequency at room temperature.

A simple expression for the minimum noise figure, F_{min}, can be found from the equivalent circuit elements of the device from Equn. 2.41.

TABLE 2.3. Design Parameters of GaAs MESFETs used for Noise Calculations

Parameter		Device				
Symbol	Units	1	2	3	4	5
L	μm	0.9	0.9	0.5	0.5	0.25
L_g	μm	0.9	1.2	0.8	0.8	0.4
L_2	μm	1.0	0.75	0.75	0.75	0.4
L_3	μm	0	0.4	0.3	0.3	0.2
h	μm	0.5	1.0	0.65	0.65	0.4
N_D	$10^{16}/cc$	7	4	8	8	18
N_1	$10^{16}/cc$	7	200	200	200	200
N_2	$10^{16}/cc$	7	200	200	200	200
N_3	$10^{16}/cc$	–	4	8	8	18
a	μm	0.3	0.27	0.15	0.15	0.1
a_1	μm	0.3	0.15	0.15	0.15	0.15
a_2	μm	0.17	0.12	0.12	0.12	0.12
a_3	μm	–	0.27	0.15	0.15	0.1
W	mm	0.25	0.25	0.25	0.1	0.065
W_1	nm	0.24	0.23	0.14	0.14	0.065

TABLE 2.4. Comparison of the Predicted Value of Minimum Noise Figure for the Devices having the Parameters of Table 2.3 with the Measured Values

Parameters			Device				
Symbol		Units	1	2	3	4	5
PREDICTED	F_{min}	dB*	1.72	1.8	2.12	1.56	1.7
	F_{min}	dB**	1.77	1.85	2.18	1.62	1.76
MEASURED	F_{min}	dB	1.75	1.76	2.22	1.51	1.74

* using Eq. 2.54

** using Eq. 2.59

$$F_{min} = 1 + 2\pi K_f fC_{gs} \sqrt{\frac{R_g + R_s}{g_m}} \times 10^{-3}$$ 2.42

where K_f is a fitting factor approximately equal to 2.5 representing the quality of the channel material. Essentially 2.42 is a special case of 2.41 where R = 0 and or C = 1 neglecting higher order terms. Since

$$f_T = \frac{g_m}{2\pi C_{gs}}$$

equn. 2.42 can be rewritten as

$$F_{min} = 1 + K_f \frac{f}{f_T} \sqrt{g_m(R_g + R_s)}$$ 2.43

Also since f_T is related to the gate length L (see equn. 2.49) we have

$$F_{min} = 1 + K_1 Lf \sqrt{g_m(R_g + R_s)}$$ 2.44

where $K_1 \sim 0.27$ when L is in microns.

Consider Table 2.3 where the parameters of five GaAs FETs are shown for the purposes of noise figure calculations. These devices had optimised gate recess structures (see Chapter 4) and as a result the effective gate length reduces to the physical length of the gate metal when the gate is rectangular in cross section on a planar channel. Parameters such as L, g_m, R_g and R_s can be measured using the method of Fukui (1979a).

The predicted and the measured values of F_{min} for these FETs shown in Table 2.4 are in excellent agreement when measured at a frequency of 6 GHz (i.e. approximately $f_T/3$). Equation 2.44 shows that a short gate length is essential for low noise as well as minimising the parasitic resistances. This can be most conveniently achieved by narrowing the unit gate width through a gate paralleling scheme even though there is an increased fabrication complexity.

From the circuit design viewpoint the GaAs FET can be treated as a black box of noisy two ports. The noise properties of this black box can be characterised by the use of four noise parameters (Rothe et al, 1956). A variant of this expression gives:-

$$F = F_{min} + \frac{R_n}{R_{ss}} \left[\frac{(R_{ss} - R_{OP})^2 + (X_{ss} - X_{OP})^2}{R_{OP}^2 + X_{OP}^2} \right]$$ 2.45

where F_{min} is as previously defined

R_n is the equivalent noise resistance

R_{ss} is the signal source resistance

R_{OP} is the optimum signal source resistance

X_{ss} is the signal-source reactance and

X_{OP} is the optimum signal source reactance.

In this expression F_{min}, R_n, R_{OP} and X_{OP} are the characteristic noise parameters of the device.

As explained in Chapter 5 Equn. 2.45 can be represented on the source impedance Smith Chart as a family of circles each of which corresponds to a constant value of F.

A small R_n is essential for a device to be used in a broadband amplifier where a large tolerance is required in the input match. Furthermore, the smaller the value of R_n the higher the gain in a given gate structure.

The four noise parameters can be expressed as

$$F_{min} = 1 + K_1 f C_{gs} \sqrt{(R_g + R_s)/g_m}$$

$$R_n = K_2/g_m^2$$

2.46

$$R_{OP} = K_3 \left[\frac{1}{(4g_m)} + R_g + R_s \right]$$

2.47

$$X_{OP} = \frac{K_4}{f C_{gs}}$$

2.48

where K_1, K_2, K_3 and K_4 are fitting factors.

Figs. 2.8 to 2.11 show the fits obtained for six different FET structures having different channel carrier concentrations, gate lengths and channel thicknesses.

Good fits are obtained for values of:-

$K_1 = 0.016$ $K_3 = 2.2$

$K_2 = 0.03$ and $K_4 = 160$

where R_n, R_{OP}, X_{OP}, R_g and R_s are in ohms, g_m in mhos and C_{gs} in pico-farads with f in gigahertz.

In order to design a GaAs FET conveniently it is advantageous to have expressions for g_m, C_{gs} and f_T.

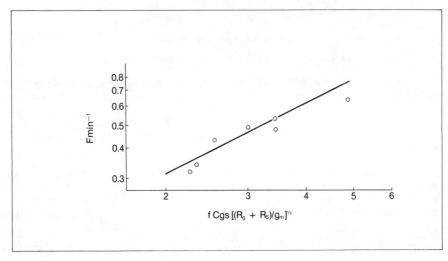

FIG. 2.8. Correlation Between the Minimum Noise Figure F_{min} and Equivalent Circuit Elements C_{gs}, g_m, R_S and R_g (after Fukui)

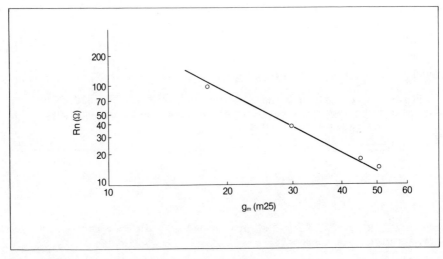

FIG. 2.9. Correlation Between the Equivalent Noise Resistance R_n and the Transconductance g_m (after Fukui)

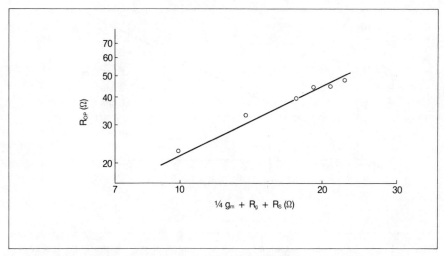

FIG. 2.10. Correlation Between the Optimum Source Resistance R_{OP} and Equivalent Circuit Elements R_g, R_S and g_m

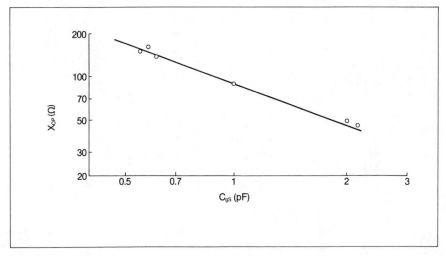

FIG. 2.11. Correlation Between the Optimum Source Reactance X_{OP} and the Gate-to-Source Capacitance C_{gs}.

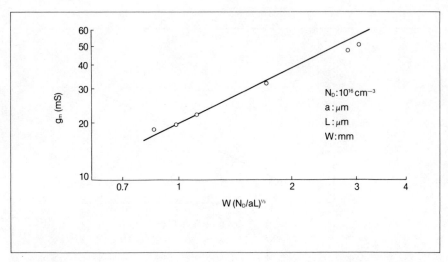

FIG. 2.12. Transconductance g_m as a Function of Channel Parameters W, L, a and N_D

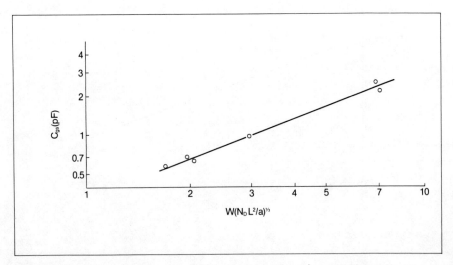

FIG. 2.13. Gate-Source Capacitance as a Function of Channel Parameters W, L, a and N_D

We have already seen that the maximum frequency of oscillation is approximately related to the gate length L by equn. 2.36. Similarly

$$f_T = 10^3 \; g_m/2\pi C_{gs} = \frac{9.4}{L} \quad \text{GHz} \qquad\qquad 2.49$$

$$g_m = k_5 W \left[N_D/aL \right]^{\frac{1}{3}} \quad \text{mhos} \qquad\qquad 2.50$$

$$\text{and } C_{gs} = k_6 W \left[N_d L^2/a \right]^{\frac{1}{3}} \quad \text{pF} \qquad\qquad 2.51$$

where k_5 and k_6 are found to be 0.02 and 0.34 respectively.

Fig. 2.12 and 2.13 show the agreement between the measured values of g_m and C_{gs} and the empirical results of Equns. 2.50 and 2.51.

Simplified expressions for the gate metallisation resistance R_g and the source resistance R_s can be found as

$$R_g = 17 \; W_1^{\;2}/hLW \quad \Omega \qquad\qquad 2.52$$

and

$$R_s = \frac{1}{W} \left[\frac{2.1}{a^{\frac{1}{2}} N_D^{\;2}/3} + \frac{1.1 \; L_{sg}}{(a-a_s) \; N_D^{\;0.82}} \right] \Omega \qquad\qquad 2.53$$

where h is the gate metallisation thickness in microns,

L_{sg} is the distance between source and gate in microns,

a_s is the depletion layer thickness in microns at the surface in the source-gate gap and

W_1 is the unit gate width for a multi-parallel gated FET.

E. PRACTICAL EQUATIONS FOR NOISE PARAMETERS

We can now substitute Eq. 2.49 to 2.53 together with the fitting factors $k_1, k_2, \ldots\ldots k_6$ into Equns. 2.44, 2.46 – 2.48 giving

$$F_{min} = 1 + 0.038f \left[N_D L^5/a \right]^{\frac{1}{6}}$$

$$x \left[\frac{17 W_1^{\;2}}{hL} + \frac{2.1}{a^{0.5} N_D^{\;0.66}} + \frac{1.1 \; L_{sg}}{(a-a_s) N_D^{\;0.82}} \right]^{\frac{1}{2}} \qquad\qquad 2.54$$

$$R_n = 75 W^{-2} \left[\frac{aL}{N_D} \right]^{\frac{2}{3}} \quad \Omega \qquad\qquad 2.55$$

$$R_{OP} = 2.2W^{-1}\left[12.5\ \left(\frac{aL}{N_D}\right)^{\frac{1}{3}} + \frac{17W_1^2}{hL} + \frac{2.1}{a^{0.5}N_D^{0.66}} + \frac{1.1}{(a-a_s)N_D^{0.82}}\right]\quad 2.56$$

and

$$X_{OP} = \frac{450}{fW}\left|\left[\frac{a}{N_D L^2}\right]^{\frac{1}{3}}\right|\ \Omega \qquad\qquad 2.57$$

where f is in gigahertz; W_1 and W are in millimetres; a, a_s, h, L and L_{sg} are in microns and N_D is in 10^{16} cm^{-3}. (Fukui, 1979b).

F_{min} is invariant to the total device width but varies with the unit gate width.

As the operating frequency increases, the skin effect on the gate metallisation can no longer be ignored and Equn. 2.54 becomes:

$$F_{min} = 1 + 0.038f\left[N_D L^5/a\right]^{\frac{1}{6}} \cdot \left[\frac{17W_1^2}{hL} + 1.3W_1^2\ \left(\frac{f}{hL}\right)^{\frac{1}{2}}\right.$$
$$\left. + \frac{2.1}{a^{0.5}N_D^{0.66}} + \frac{1.1\ L_{sg}}{(a-a_s)N_D^{0.82}}\right]^{\frac{1}{2}} \qquad 2.58$$

Consider now a practical device as shown in Fig. 2.14 which has a recessed gate structure with the geometrical parameters as shown.

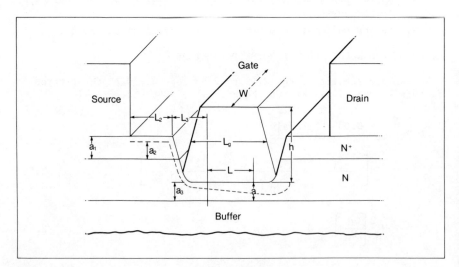

FIG. 2.14. Optimised Low-Noise MESFET Structure

Equn. 2.58 can be extended to cover this practical case by the expression (Fukui, 1979c)

$$F_{min} = 1 + 0.038f\left[N_D L^5/a\right]^{\frac{1}{6}}$$

$$x\left\{\frac{17W_1^2}{hL_g} + \frac{2.1}{a_1^{0.5}N_1^{0.66}} + \frac{1.1\ L_2}{a_2 N_2^{0.82}} + \frac{1.1\ L_3}{a_3 N_3^{0.82}}\right.$$

$$\left. + 1.3W_1^2(f/hL_g)^{\frac{1}{2}}\right\}^{\frac{1}{2}} \qquad\qquad 2.59$$

where L_g is the average gate metallisation length in microns, L_2 and L_3 are the effective lengths of each sectional channel between the source and gate electrodes in microns; a, is the effective channel thickness under the source electrode, a_2 and a_3 are the effective thicknesses of the sectional channel in microns; N_1 is the effective free carrier concentration in the channel under the source electrode and N_2 and N_3 are the effective free carrier concentrations of the sectional channel, all in 10^{16} cm^{-3}.

For the case where an n$^+$ GaAs layer exists between the ohmic metal and n-GaAs, L_2 can be approximated to the distance of the n layer between the edge of the n$^+$ layer and the effective edge of the gate electrode.

F. EXAMPLE OF LOW NOISE GaAs FET DESIGN

Let us firstly examine the variation of F_{min} and noise resistance, R_n as a function of the carrier concentration N_D and the active channel thickness a.

Consider a 1μm GaAs FET where the source-to-gate separation L_{sg} is also 1μm. The source resistance is assumed to be 4 ohms and f = 3.8 GHz. a_s is assumed to be approximately equal to the gate depletion layer thickness, a_o. Fukui (1979a) has shown that this parameter is given by

$$a_o = \left[\frac{0.706 + 0.06\ \log N_D}{7.23\ N_D}\right]^{\frac{1}{2}} \text{μm} \qquad\qquad 2.60$$

for aluminium devices.

Thus equn. 2.54 reduces to

$$F_{min} = 1 + 0.15\left[N_D/a\right]^{\frac{1}{6}} \cdot \left[1.82 + \frac{1.9}{a^{0.5}N_d^{0.66}} + \frac{1}{(a-a_s)N_D^{0.82}}\right]^{\frac{1}{2}}$$

40

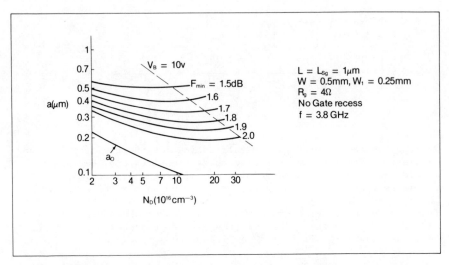

FIG. 2.15. Contours of F_{min} as a Function of N_D and a for a 1μm
Gate Length FET

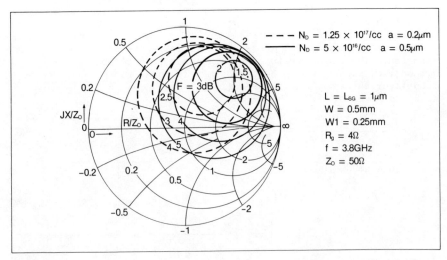

FIG. 2.16. Constant Noise Figure Circles for two 1μm Gate Length
MESFETs

Fig. 2.15 shows the values of F_{min} as a function of N_D and a. It may be seen that F_{min} is lowest for the highest values of a/N consistent with the avoidance of drain-source breakdown at around 10 volts. This condition implies a non-optimum value for R_n by Equn. 2.55 so there is clearly a compromise value of a/N_D.

Let us examine this a little more closely. Consider two devices with the first device having $N_D = 1.25 \times 10^{17}$ cm^{-3} with a = 0.2μm and the second device having $N_D = 5 \times 10^{16}$ cm^{-3} and a = 0.5μm.

The noise performance of these devices can be plotted on the Smith Chart (Fig. 2.16) following equations 2.54 to 2.57.

As may be seen the second device, even though having a smaller F_{min}, has a larger R_n leading to the fact that the 50 ohm noise figure is over 1 dB higher than the first device.

Referring to the 50 ohm noise figure as

$$F_{50} = F_{min} + \frac{R_n}{50} \left[\frac{(50 - R_{OP})^2 + X_{OP}^2}{R_{OP}^2 + X_{OP}^2} \right]$$

it is possible to calculate the variation of both F_{min} and F_{50} with the total and unit gate widths W and W_1. Again for the two devices used as examples previously Fig. 2.17 results where it has been assumed that the gate metallisation resistance is related to the total gate width by the simple expression:

$R_S = 8W$, W being in millimetres

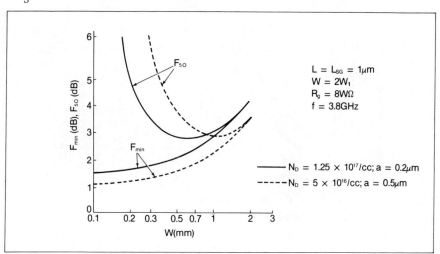

FIG. 2.17. F_{min} and F_{50} as a Function of Total Device Width for two 1μm Gate Length MESFETs

42

As may be seen F_{50} is a strong function of W and indicates that the optimum value for the total gate width is approximately 450μm for the first device and 900μm for the second FET.

The maximum value of unit gate width W_{1max} can be defined as the limit above which R_g becomes greater than R_s. Thus equating 2.52 and 2.53 and accounting for equn. 2.59 we obtain

$$W_{1max} = 0.25\sqrt{hL}_g \left[\frac{1.9}{a_1^{0.5}N_1^{0.66}} + \frac{L_2}{a_2N_2^{0.82}} + \frac{L_3}{a_3N_3^{0.82}} \right] \qquad 2.61$$

when $W_1 = W_{1max}$ Eq. 2.59 reduces to

$$F_{min} = 1 + 0.057f\left[N_D L^5/a\right]^{1/6}$$

$$\times \left[\frac{1.9}{a_1^{0.5}N_1^{0.66}} + \frac{L_2}{a_2N_2^{0.82}} + \frac{L_3}{a_3N_3^{0.82}} \right]^{\frac{1}{2}} \sqrt{1 + S} \qquad 2.62$$

where $S = 0.08\sqrt{fhL_g}$, the perturbation due to the skin effect.

Fig. 2.18 shows the calculated minimum noise figures versus frequency for the devices of Table 2.3 (Fukui, 1979).

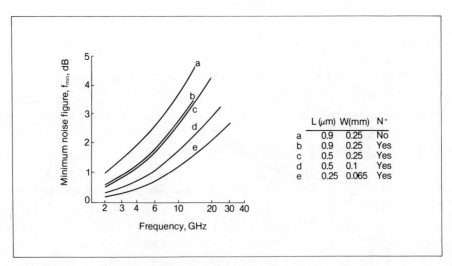

FIG. 2.18.   Calculated Minimum Noise Figure as a Function of Frequency for the Devices of Table 2.3

Four of the devices have n[+] layers under the ohmic contacts. Device a has a Cr-Au gate metallisation with a resistivity of 2.5 x 10^{-6} Ω cm whilst the remaining devices have aluminium gate metal of resistivity 5 x 10^{-6} Ω cm. Device b is an optimized design for a gate length of 0.9μm. Device c has a W_1 value greater than W_{1max} negating the effect of the 0.5μm gate length. Device d is an improved version of device c with a W_1 of 100μms. Device e is a 0.25μm gate length device with a unit gate width of 65μms. This is similar to a device of 0.3μm gate length with unit gate width of 50μms described by Butlin et al (1978) which promises to give noise figures close to 2 dB at 20 GHz (Fig. 2.19).

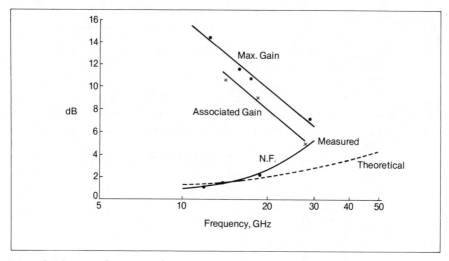

FIG. 2.19.   Gain and Noise Performance of 0.3μm GaAs FET

G. THE GaAs FET VERSUS THE BIPOLAR TRANSISTOR AS A LOW NOISE DEVICE

It was mentioned earlier in this chapter that the improved maximum cut off frequency of a GaAs FET when compared to a silicon BJT for similar geometries was due to the larger saturated drift velocity of GaAs.

The minimum noise figure for the bipolar junction transistor at room temperature is given approximately by

$$F_{min}\big|_{BJT} \simeq 1 + bf^2 \left[1 + \sqrt{1 + 2/bf^2} \right]$$

2.63

where $b = 40I_C r_b/f_T^2$.

r_b being the parasitic base resistance in ohms and I_C being the collector current in amperes. An equivalent figure for the GaAs FET is

44

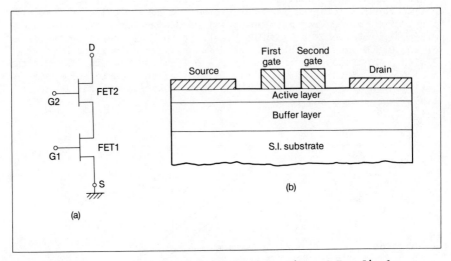

FIG. 2.20(a). Equivalent Circuit Configuration of Two Single
    Gated FETs
    (b). Cross Section of Dual-Gate MESFET

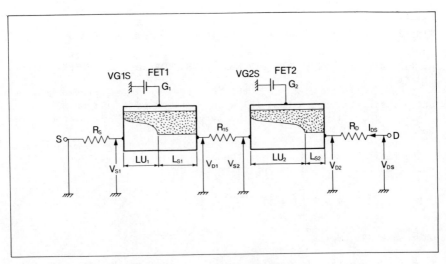

FIG. 2.21. Analytic Model of the Dual-Gate FET

$$F_{min}\big|_{FET} \simeq 1 + mf$$

$$\text{where } m = \frac{2.5}{f_T} \sqrt{g_m(R_g + R_s)} \qquad\qquad 2.64$$

It may be appreciated therefore that the minimum noise figure for the GaAs FET increases linearly with frequency compared with a quadratic function for the BJT. Thus, the GaAs FET not only has a better intrinsic noise performance but also degrades much more slowly with frequency than the bipolar device.

4. THE DUAL GATE FET

Although the single gate structure has become the most widely manufactured and used device the dual gate FET has the advantages of an increased capability due to the ability of the two independent gates to perform such functions as gain control and mixing as well as reduced feedback and improvement in signal gain. The dual gate FET, however, is in general much less well understood than the single gate FET mainly due to the r.f. interaction which can occur between its three accessible ports when it is used in a particular configuration, usually common source. The applications of the dual gate FET are numerous. It has been used in up and down converters, modulators, pulse regenerators, phase shifters, and discriminators as well as in oscillator applications. Its major use in mixers and multipliers is covered in Chapter 6, whilst it figures considerably in Chapters 8 and 10.

The first GaAs dual gate MESFETs were made in 1971 by Turner et al but it was some time before this pioneering work was followed up by the first serious attempt at analysis of the device by Liechti in 1975.

The dual gate device can be considered as two separate FETs connected in cascade as shown in Fig. 2.20(a) where the current characteristics of the bottom FET are determined by the top device. Fig.2.20(b) shows a representation of a dual-gate device where the first and second Schottky gates G_1 and G_2 are formed between the source and drain ohmic contacts.

The operation and characteristics of the dual-gate FET can be analysed by combining the analyses of the two single gate FETs. As we have already seen in this chapter the performance of such single gate devices has been extensively examined.

Consider the analytic model of the dual gate MESFET shown in Fig. 2.21. It is composed of 2 single gate FETs and three parasitic resistors, i.e. FET1, FET2, a source series resistor R_S, the intergate resistance R_{IS} and the drain resistor, R_D.

Using the model of Statz et al (1974) it is possible to calculate

1. The drain currents I_{D1} and I_{D2} of FETs 1 and 2.

2. The channel lengths L_{U1} and L_{U2} where the carriers move at the unsaturated velocity.

3. The regions L_{S1} and L_{S2} where the carriers have reached saturation velocity.

These 3 quantities are calculated as functions of the gate-source biases V_{G1S} and V_{G2S} and the applied voltages across each single gate FET, i.e. $V_{D1} - V_{S1}$ and $V_{D2} - V_{S2}$.

Thus by setting $I_{D1} = I_{D2} = I_{DS}$ for given bias conditions V_{G1S}, V_{G2S} and V_{DS}. The voltage drops across the 2 FETs and the 3 resistors will equal V_{DS}.

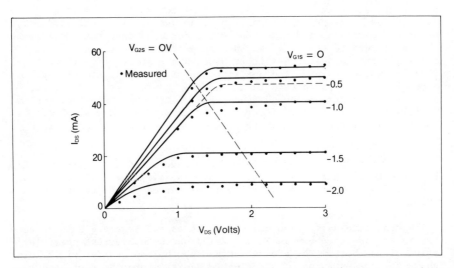

FIG. 2.22. Drain Current–Drain Voltage Characteristic as a Function of First Gate Bias for a Dual–Gate MESFET

Fig. 2.22 shows calculated and measured I–V characteristics for a dual gate FET having the parameters:

$$L_{G1} = L_{G2} = 1\mu m$$
$$L_{IG} = 1\mu m$$
$$W_1 = W_2 = 300\mu m$$
$$V_{p1} = V_{p2} = -2.5V$$
$$N_D = 2.5 \times 10^{17} \, cm^{-3} \quad \text{and} \quad a = 0.14\mu m.$$

where L_{Gi}, L_{IG}, W_i, V_{pi}, N_D and a are the gate lengths, intergate spacing, gate widths, pinch-off voltages, and the carrier concentration in the active region of layer thickness, a. (i = 1 or 2).

Various other parameters were as used by Statz (1974) or Pucel (1975). It may be seen from Fig. 2.22 that the agreement between measured and theoretical characteristics is satisfactory.

A. EQUIVALENT CIRCUIT PARAMETERS

The equivalent circuit of the GaAs dual gate MESFET is somewhat more complicated than the single gate device. Fig. 2.23 shows the equivalent circuit where parasitic circuit elements and noise sources have also been included.

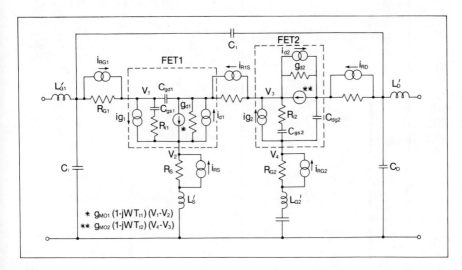

FIG. 2.23. Equivalent Circuit of GaAs Dual-Gate FET

The intrinsic circuit elements of the two FETs, FET1 and FET2, enclosed by dotted lines in Fig. 2.23 have been analyzed using Statz's model (Furutsuka et al, 1978).

The gate resistance R_{Gi} (i = 1 or 2) can be estimated using the equation

$$R_{Gi} = \frac{1}{4} \frac{\rho W_i}{3hL_{Gi}} \qquad 2.65$$

where two unit gate widths, each $W_i/2$, are assumed, and where ρ is the resistivity of the gate metal (2.75×10^{-6} Ω cm for aluminium).

TABLE 2.5. Comparison of S-Parameters of Single and Dual-Gate FETs at 12 GHz

Device*	S_{11}	S_{21}	S_{31}	S_{12}	S_{22}	S_{32}	S_{13}	S_{23}	S_{33}
Single gated GAT4	.55 −180°	1.39 22°	—	.08 42°	.67 −86°	—	—	—	—
Dual-Gate DUGAT 10−000	.6 170°	1.49 −72°	1.12 30°	.06 −16°	.69 −81°	.24 −21°	.06 −13°	.61 12°	.85 −83°

* Both devices have L_g = 1μm and W = 300μm

Frequency = 12 GHz

(Courtesy Plessey Co. Ltd.)

Gate 2 is port 3 for the dual gate FET

The source series resistance R_S is given by

$$R_S = \frac{1}{W_1} \left[\left(\frac{\rho_c}{q\mu N_D a} \right)^{\frac{1}{2}} + \frac{L_{SG1}}{q\mu N_D a} \right] \qquad 2.66$$

where L_{SG1} is the spacing between the first gate and source and ρ_c is the specific ohmic contact resistance. The drain series resistance is given by a similar expression. (Furutsuka et al, 1978).

The electrical characteristics of the dual gate FET can thus be expressed in terms of all the circuit elements of FET1 and FET2. For example, the transconductance associated with the first gate is given by:

$$g_m = \frac{g_{mo1}}{\left[1 + g_{mo1} R_s + g_{d1} (R_s + R_{1G}) + \frac{g_{d1}(1 + g_{d2} R_D)}{(g_{mo2} + g_{d2})} \right]} \qquad 2.67$$

$$\simeq g_{mo1}, \quad g_{mo1} \text{ being the d.c. transconductance of FET1}$$

where both FETs are operating under current saturation. Also, for example, the drain conductance g_d of the dual-gate FET is given by:

$$g_d = \frac{g_{d2}}{\left[1 + g_{d2} R_D + (g_{mo2} + g_{d2})(R_s + R_{IG}) + \frac{(g_{mo2} + g_{d2})(1 + g_{mo1} R_s)}{g_{d1}} \right]}$$

$$\qquad 2.68$$

$$\simeq \frac{g_{d1}}{(g_{mo2}/g_{d2})}$$

where both FETs are in current saturation.

The variation of g_m with V_{G1S} is shown in Fig. 2.24 as a parameter of V_{G2S} for the same MESFET structure used in Fig. 2.22. As both devices in the cascade enter the current saturation region the g_m rises to a maximum value.

The dual gate FET is usually characterised as a 3-port device with the source connected to ground. Table 2.5 shows the S-parameters of a single gate FET and a dual gate FET where the dual gate FET is similar in geometry to the single gate device with $L_{G1} = L_{G2} = 1\mu m$ and $L_{IG} = 1.5\mu m$.

It may be seen that, as with a single gate FET, there is a high transmission coefficient between gate 1 and drain. However, there is also considerable transmission from gate 1 to gate 2. The reverse scattering parameters are small, that between drain and gate 1 being lowered with respect to the single gate FET due to the shielding effect of gate 2.

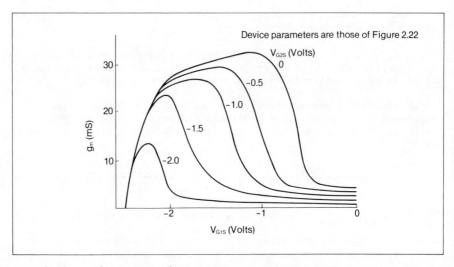

FIG. 2.24.   First Gate Bias Dependence of Transconductance on
   Second Gate Bias, V_{G2S}

The forward transmission coefficient between gate 2 and drain is
lower than gate 1 to drain because the output impedance of FET1,
$1/g_{d1} + R_S + \omega L_{S1}$ is in series with the second gate capacitance C_{gs2}
and the intergate resistance R_{IG}, thus lowering the transconductance
of the second gate. The output impedance of the first FET also acts
as a series feedback impedance increasing the reverse transmission
coefficient from drain to gate 2.

B. GAIN VERSUS SECOND-GATE TERMINATIONS

When the dual gate FET is used as an amplifier the device is converted into a 2 port by terminating the second gate with an impedance
Z_{G2}. As with a single gate FET gate 1 is the input and the drain is
the output. The S-parameters of this 2 port $S_{iK}{}'$ are related to the
original 3 port S-parameters S_{iK} by the relationship (Bodway, 1968)

$$S_{iK'} = S_{iK} + \frac{(S_{i3} \cdot S_{3K})}{((1/\Gamma_{G2}) - S_{33})} \qquad\qquad 2.69$$

where Γ_{G2} is the reflection coefficient of the load Z_{G2} with respect
to the characteristic impedance Z_o, i.e.

$$\Gamma_{G2} = \frac{(Z_{G2} - Z_o)}{(Z_{G2} + Z_o)} \qquad\qquad 2.70$$

The maximum available gain, MAG, and associated reverse isolation, RI, between the matched input and output ports can be calculated as:-

$$MAG = |\frac{S_{21}'}{S_{12}'}| \left[K \pm (K^2-1)^{\frac{1}{2}} \right]$$
2.71

and
$$RI = |\frac{S_{12}'}{S_{21}'}| \left[K \pm (K^2-1)^{\frac{1}{2}} \right]$$
2.72

where
$$K = \frac{1 + |S_{11}' \cdot S_{22}' - S_{12}' \cdot S_{21}'|^2 - |S_{11}'|^2 - |S_{22}'|^2}{2 \cdot |S_{12}'| \cdot |S_{21}'|}$$

For an RF shorted second gate the dual gate MESFET is essentially a cascode connection with a common source input and a common gate output. The properties of such a connection are:

1. S_{11}' and $|S_{21}'|$ are similar in value to those of an equivalent single gate device.

2. The phase angle of S_{21}' is smaller than that of the single gate FET due to the increased length between gate 1 and drain.

3. $|S_{12}'|$ is smaller than in the single-gate FET because of second gate shielding.

4. $|S_{22}'|$ is somewhat higher than in the single gate device.

A map of forward gains (Fig. 2.25) has been computed for the device in Table 2.5 (Plessey DUGAT10-000) at a frequency of 12 GHz for various second gate terminations. It may be seen that with the second gate grounded ($\Gamma_{G2} = -1$) the maximum available gain is 13.4 dB with 20 dB reverse isolation between drain and gate 1. Up to 20.5 dB forward gain can be obtained from the dual gate FET with an inductive termination. The reverse S-parameters decrease however and the limits are set by the boundary k = 1 beyond which the transistor is potentially unstable. Also, there is a region where the transistor will yield forward loss. Fig. 2.25 shows this area where the signal flowing via gate 2 to the drain interferes destructively with the signal flowing from gate 1 to drain.

C. GAIN CONTROL WITH SECOND-GATE BIAS

At 12 GHz, for the device above, the second gate termination effect on the forward transmission coefficient can be mapped when the second gate bias is increased negatively to the point where the channel under the second gate is completely depleted and the drain current is cut-off.

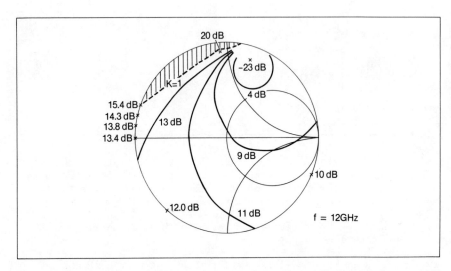

FIG. 2.25.   Forward Gain Mapping for (DUGAT10-000) 1μm Dual Gate FET
when V_{G2S} = 0 volts (Courtesy - Plessey) as a Function of Second
Gate Impedance

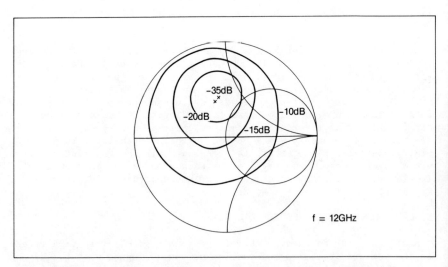

FIG. 2.26.   Forward Insertion Loss Mapping for (DUGAT10-000) 1μm
Dual Gate FET when V_{G2S} = V_p as a Function of Second Gate
Impedance

Essentially the dual gate FET then behaves like a passive, recipro-
cal device, i.e. the forward loss equals the reverse loss in which
the drain to first-gate capacitance C_{gd}, dominates the coupling
between the 3 ports. However, there is still a signal component
present in the second gate circuit which interferes with the gate 1
to drain signal causing the off-isolation of the device to vary accor-
ding to Γ_{G2}. This is shown in Fig. 2.26.

Fig. 2.26 is a mapping of forward insertion loss for the same dual
gate FET when $V_{G2S} = V_p$ and can be compared with Fig. 2.25 showing
that there is a termination which will produce an acceptable forward
gain, when gate 2 bias is zero and also an acceptable insertion loss
at a value such that the channel under gate 2 is pinched-off. The
input impedance versus V_{G2S} stays within close limits because the
reverse coupling coefficients S_{12} and S_{13} remain small in the bias
range. Fig. 2.27(a) demonstrates this whilst Fig. 2.27(b) shows that
the forward transmission phase of the dual gate FET can vary considerably as the second gate voltage is changed. Again a second gate
termination exists which minimises the phase and input impedance
variations. For the device of Table 2.5 a variation in gain from a
maximum of 10 dB to below −20 dB can be produced whilst keeping the
input VSWR below 1.5:1.

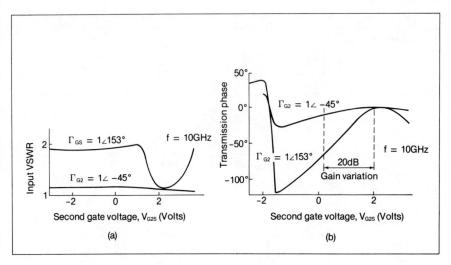

FIG. 2.27(a). Input VSWR Versus Second Gate Voltage for Two Second
    Gate Terminations (after Liechti (1975)).
        (b). Transmission Phase Versus Second Gate Voltage for Two
Second Gate Terminations

54

D. NOISE PERFORMANCE

As with the noise analysis of the single gate FET, the dual gate FET noise figure can be predicted by evaluating the equivalent noise conductances and correlation factors within the intrinsic FETs 1 and 2 and includes the effects of the parasitic resistances shown in the equivalent circuit of Fig. 2.23. Denoting each noise current which appears at the drain $\bar{I}_x$, the noise figure of the dual gate FET is given by:

$$F = 1 + \frac{|(\bar{I}_{RS} + \bar{I}_{RG1} + \bar{I}_{R1G} + \bar{I}_{G1} + \bar{I}_{d1} + \bar{I}_{RG2} + \bar{I}_{RD} + \bar{I}_{G2} + \bar{I}_{d2}|^2}{|\bar{I}_{in}|^2}$$

2.74

where $\bar{I}_{in}$ is the noise current at the drain generated by the input signal source at a standard temperature. The correlation coefficients between the various noise sources must also be determined.

Fig. 2.28 shows the calculated and measured dependence of the minimum noise figure on drain current (i.e. on V_{G1S}) for a device having the same parameters as those used to compute the I-V characteristics of Fig. 2.22. Again it is seen that there is good agreement between theoretical and measured results. The large increase in F_{min} for $I_{DS}/I_{DSS} > 0.9$ is due to FET1 working in the unsaturated regime and therefore having a low g_m.

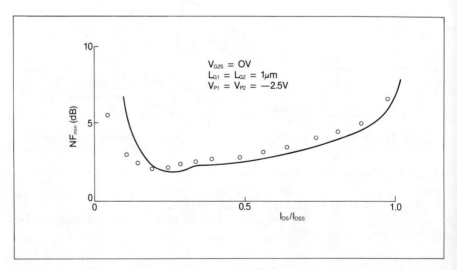

FIG. 2.28. Computed Drain Current, I_{DS}/I_{DSS} Dependence of Minimum Noise Figure, F_{min}

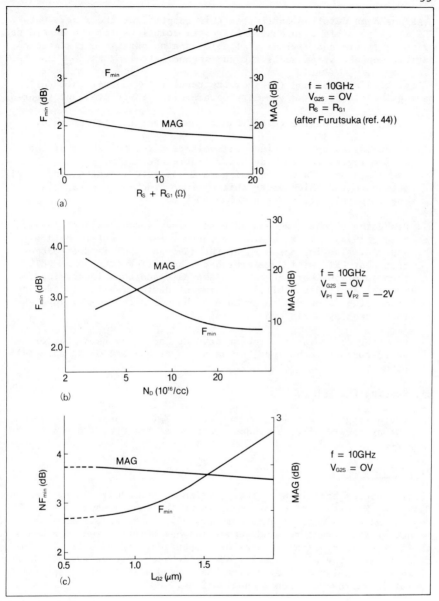

FIG. 2.29(a). Computed Parasitic Resistance Dependence of F_{min} and MAG at $V_{G2S} = OV$

(b). Computed Carrier Concentration, N_d Dependences of F_{min} and MAG at $V_{G2S} = OV$

(c). Computed Second Gate Length Dependences of F_{min} and MAG at $V_{G2S} = OV$

As has been detailed earlier in this chapter for the single gate
FET it is possible to produce analytical solutions for the dependence
of noise figure and maximum available gain on parasitic resistances,
carrier concentration and first and second gate lengths, L_{G1} and L_{G2}.

Fig. 2.29(a), (b) and (c) show the results for a dual gate FET
having the material and geometry parameters used in previous figures.

In summary the dual gate FET can be optimised in its design by:-

a). Producing an optimum pinch-off voltage (i.e. thickness of the
active layer) to minimise noise figure and maximise gain. For
example, for FETs with equal gate 1 and gate 2 lengths and equal
active region thicknesses this optimum pinch-off voltage is
approximately -1.5V for n = 8 x 10^{16}/cc.

b). Producing a longer second gate and wider second channel (Asai
et al, 1975). This leads to a further improvement in both F_{min}
and MAG. The increased g_m of the dual-gate FET is due to the
fact that FET2 will have a higher pinch-off voltage than FET1
thus supplying more current to the pair and thus increasing the
voltage applied to FET1. In terms of device fabrication further
processing is needed to produce a channel with two thicknesses
thus lowering yields and raising costs.

c). Lowering the parasitic resistances R_S and R_G by using better
ohmic contact technology, or an n^+ layer and employing parallel
gates.

d). Keeping $L_{G2} < L_{G1}$.

e). Increasing the carrier concentration, N_D, the upper limit being
set by the required breakdown voltage of the drain to the Schottky
gate.

5. CONCLUSIONS

This chapter has given an introduction to the theory of small signal
GaAs MESFETs with particular emphasis on the models used to predict
the FET gain and noise figure performance. Such calculations have led
to better FET structures and geometries being introduced as a means to
exploit the intrinsically superior properties of GaAs over Si. The
properties of the dual-gate FET have also been covered in some depth
as this particular type of device has proved itself to be especially
useful in microwave circuit applications.

6. BIBLIOGRAPHY

Alley, G. and Talley, H. A theoretical study of the high frequency
performance of a Schottky barrier field effect transistor fabricated on a high-resistivity substrate. IEEE Trans. Microwave Theory
and Techniques, Vol. MTT-22, pp.183-189, March 1974.

Asai, S., Murai, F. and Kodera, H. GaAs dual-gate Schottky barrier
FETs for microwave frequencies. IEEE Trans. Electron Devices,
Vol. ED-22, pp.897-904, October 1975.

Baechtold, W. Noise behaviour of Schottky barrier gate field-effect
transistors at microwave frequencies. IEEE Trans. Electron Devices
Vol. ED-18, pp.97-104, February 1971.

Baechtold, W. Noise behaviour of GaAs field effect transistors with
short gate lengths. IEEE Trans. Electron Devices, Vol. ED-19,
pp.674-680.

Barrera, J. and Archer, R. InP Schottky gate field effect transistors. IEEE Trans. Electron Devices, Vol. ED-22, pp.1023-1030,
November 1975.

Bodway, G. Circuit design and characterisation of transistors by
means of three port scattering parameters. Microwave Journal,
Vol. 11, pp.55-63, May 1968.

Butlin, R.S., Hughes, A.J., Bennett, R.H., Parker, D. and Turner, J.A.
J-band performance of 300 nm gate length GaAs FETs. 1978, Int.
Electron. Devices Meeting, Dig. Tech. Papers, pp.136-139.

Dawson, R. Equivalent circuit of the Schottky-barrier field effect
transistor at microwave frequencies. IEEE Trans. Microwave Theory
and Techniques, Vol. MTT-23, pp.499-501, June 1975.

Englemann, R.W.H. and Liechti, C.A. Gunn domain formation in the
saturated current region of GaAs MESFETs. IEDM Tech. Digest,
pp. 351-354, Dec. 1976.

Fawcett, W., Hilsum, C. and Rees, H. Optimum semiconductors for
microwave devices. Electronics Letters, Vol. 5, pp.313-314,
July 1969.

Fawcett, W. and Herbert, D. High-field transport in gallium arsenide
and indium phosphide. J. Phys. C: Solid State Phys., Vol. 7,
pp.1641-1654, May 1974.

Fukui, H. Determination of the basic device parameters of a GaAs
MESFET. Bell Syst. Tech. J. Vol. 58, pp.771-797, March 1979.

Fukui, H. Design of microwave GaAs MESFETs for broadband low-noise
amplifiers. IEEE Trans. Microwave Theory and Techniques, Vol.
MTT-27, No. 7, July 1979, pp.643-650.

Fukui, H. Optimal noise figure of microwave GaAs MESFETs. IEEE Trans.
Electron Devices, Vol. ED-26, No. 7, July 1979, pp.1032-1037.

Furutsuka, T., Ogawa, M. and Kawamua, N. GaAs Dual-gate MESFETs.
IEEE Trans. Electron Devices, Vol. ED-25, No. 6, pp.580-586,
June 1978.

Glicksman, M., Enstrom, R., Mittleman, S. and Appert, J. Electron. mobility in $In_xGa_{1-x}As$ alloys. Phys. Rev. B., Vol. 9, pp.1621-1626, Feb. 1974.

Grebene, A. and Ghandi, S. General Theory for pinched operation of the junction-gate FET. Solid State Electron. Vol. 21, pp.573-589, July 1969.

Grove, A.S. Physics and technology of semiconductor devices. New York, Wiley, 1967.

Kuvas, R.L. Equivalent circuit model of the GaAs FET including distributed gate effects. IEEE Trans. Electron Devices, Vol. ED-27 No. 6, June 1980.

Lam, H. and Acket, G. Comparison of the microwave velocity field characteristics of n-type InP and n-type GaAs. Electronics Letts, Vol. 7, pp.722-723, Dec. 1971.

Lehovec, K. and Miller, R. Field distribution in junction field-effect transistors at large drain voltages. IEEE Trans. Electron Devices, Vol. ED-22, pp.273-281, May 1975.

Liechti, C., Gowen, E. and Cohen, J. GaAs microwave Schottky-gate FET 1972 Int. Solid State Circuits Conf. Digest of Tech. Papers, pp.158-159.

Liechti, C.A. Performance of dual gate GaAs MESFETs as gain control-led amplifiers and high-speed modulations. IEEE Trans. Microwave theory and techniques, Vol. MTT-23, pp.461-469, June 1975.

Maloney, T. and Frey, J. Frequency limits of GaAs and InP field effect transistors. IEEE Trans. Electron Devices, Vol. ED-22, pp.357-358, July 1975, and Vol. ED-22, p.620, August 1975.

Middelhoek, S. Projection masking, thin photoresist layers and interface effects. IBM J. Res. Develop. Vol. 14, pp.117-124, March 1970.

Middelhoek, S. Metallization processes in fabrication of Schottky barrier FETs. IBM J. Res. Develop. Vol. 14, pp.148-151, March 1970.

Mo, D. and Yanai, H. Current-voltage characteristics of the junction-gate field effect transistor with field dependent mobility. IEEE Trans. Electron Devices, Vol. ED-17, pp.577-586, August 1970.

Ohkawa, S., Suyama, K. and Ishikawa, H., Low noise GaAs field effect transistors. Fujitsu Sci. Tech. J., Vol. 11, pp.151-173, March 1975.

Pucel, R., Haus, H. and Statz, H. Signal and noise properties of gallium arsenide microwave field effect transistors. Advances in Electronics and Electron Physics, Vol. 38, New York: Academic Press, 1975, pp.195-265.

Rothe, H. and Dahlke, W. Theory of noisy fourpoles. Proc. IRE Vol. 44, pp.811-818, June 1956.

Ruch, J. and Fawcett, W. Temperature dependence of the transport properties of gallium arsenide determined by a Monte Carlo method. J. Appl. Phys. Vol. 41, pp.3843-3849, August 1970.

Ruch, J. Electron dynamics in short channel field effect transistors. IEEE Trans. Electron Devices, Vol. ED-19, pp.652-654, May 1972.

Shockley, W. A unipolar field effect transistor. Proc. IRE, Vol. 40, p.1365-1367, November 1952.

Shur, M.S. and Eastman, L.F. Current-voltage characteristics, small-signal parameters and switching times of GaAs FETs. 1978 IEEE MTT-5 Int. Microwave Symp. Digest, June 1978, pp.150-152.

Statz, H., Haus, H. and Pucel, R. Noise characteristics of gallium arsenide field effect transistors. IEEE Trans. Electron Devices, Vol. ED-21, pp.549-562, September 1974.

Tsironis, C. and Beneking, H. Avalanche noise in GaAs MESFETs. Electron Lett. Vol. 13, No. 15, pp.438-439, 1977.

Turner, J.A., 1967 Gallium Arsenide (Inst. Phys. Conf. Ser. 3).

Turner, J., Waller, A., Bennett, R. and Parker, D. An electron beam fabricated GaAs microwave field effect transistor. 1970 Symp. GaAs and Related Compounds (Inst. Phys. Conf. Serial No. 9, London 1971) pp.234-239.

Turner, J.A., Waller, A.J., Kelly, E. and Parker, D. Dual-gate gallium arsenide microwave field effect transistor. Electronics Letters, Vol. 7, pp.661-662, November 1971.

Van der Ziel, A. Thermal noise in field effect transistors. Proc. IRE Vol. 50, pp.1808-1812, August, 1962.

Van der Ziel, A. Gate noise in field effect transistors at moderately high frequencies. Proc. IEEE Vol. 51, pp.461-467, March 1963.

Van der Ziel, A. Thermal noise in the hot electron regime in FETs. IEEE Trans. Electron Devices, Vol. ED-18, p.977, October 1971.

Vendelin, G. and Omore, M. Circuit model for the GaAs MESFET valid to 12 GHz. Electronics Letts., Vol. 11, pp.60-61, February 1975.

Vendelin, G.D. Feedback effects in the GaAs MESFET model. IEEE Trans. Microwave Theory and Techniques, Vol. MTT-24, No. 6, June 1976, pp.383-385.

Wolf, P. Microwave properties of Schottky-barrier field effect transistors. IBM J. Res. Develop. Vol. 14, pp.125-141, March 1970.

Yamaguchi, K., Asai, S. and Kodera, H.   Two-dimensional numerical analysis of stability criteria of GaAs FETs.   IEEE Trans. Electron Devices, Vol. ED-23, pp.1283-1290, December 1976.

CHAPTER 3
GaAs FET Theory—
Power

1. <u>INTRODUCTION</u>

The low noise and small signal properties of GaAs MESFET's have been
the subject of the previous chapter. In contrast to the low noise
device the power FET has only recently started to receive attention
from theoreticians. The outcome of such studies is now beginning to
show in the results of power, efficiency and reliability of power
FETs at frequencies up to 20 GHz.

For example the progress of power obtained from GaAs FETs as a
function of time is shown in Fig. 3.1 where the results of several
companies have been collected together.

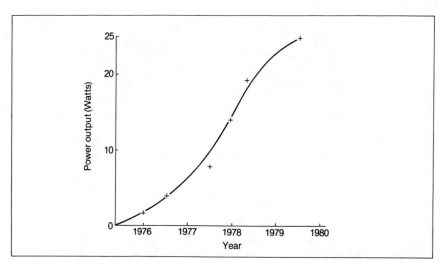

FIG. 3.1.   Progress of Power Obtained from GaAs FETs as a Function
of Time

Power FETs with output powers in excess of 2 watts at 12 GHz are available commercially and over 5 watts has been achieved in the laboratory.

This chapter will review the operating principles of the power GaAs FET and discuss the various structures that are being used to increase device performance. The large signal performance of the device will be analysed together with its third order intermodulation distortion and gain compression characteristics.

2. PRINCIPLES OF OPERATION

As has been seen in Chapter 2, Shur (1978) and Shur and Eastman (1978) have modelled the GaAs FET as a device where a Gunn domain forms at the drain contact thus effecting the current saturation and breakdown voltage of the FET. Recently, Willing et al (1978) have considered the power performance of the FET by calculating the large signal performance from the experimentally observed bias dependences of equivalent circuit model components.

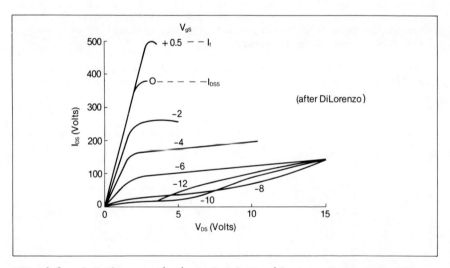

FIG. 3.2. I-V Characteristics of a 1 mm wide Gate Power GaAs FET

Fig. 3.2 shows the I-V characteristics of a power FET having a 1 mm wide gate. From this figure two important static characteristics can be seen:-

1. The values of the saturated drain current, I_{DSS}, with V_{GS} = 0V and the maximum forward drain current I_F with V_{GS} > 0V, and

2. The behaviour of the drain current I_{DS} with applied gate voltage V_{GS} at a fixed drain voltage V_{DS}.

When V_{DS} is approximately 9V and V_{GS} is approximately -8 volts, i.e. $V_{DG} \simeq 17$ volts there is a significant change in gradient of I_{DS} indicating that there is avalanche breakdown between the gate and drain.

Fig. 3.3 shows the I-V characteristics of Fig. 3.2 displayed in a somewhat simpler form where a representation of the maximum drain current I_F is shown for a forward gate bias of V_F.

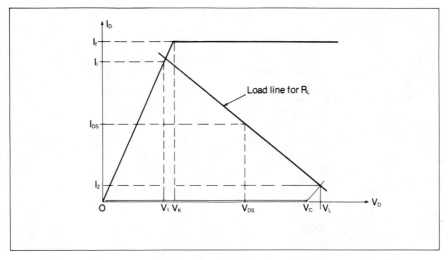

FIG. 3.3. Simple Representation of the I_D-V_{DS} Characteristic of a Power FET

The load line for a load resistance R_L is also shown as well as V_K - a knee voltage at which current saturation is deemed to have occurred.

V_L is the limiting source-drain voltage given by

$$V_L = V_{GD} - V_P \qquad\qquad 3.1$$

where V_{GD} is the gate-drain avalanche breakdown voltage for a FET biased to its pinch-off voltage V_P. Beyond V_L excess current flows between drain and source which cannot be modulated by the r.f. voltage on the gate and thus is a limiting value of V_{DS} for further output power.

It is possible to calculate the output power of the device from this I-V characteristic.

Assuming that the d.c. bias points are V_{DS} and I_{DS} and that the total output power is made up of harmonics as well as the fundamental frequency, the output power available to the load is

$$P_o = \frac{1}{8} (V_L - V_1)(I_1 - I_2) = \frac{(V_L - V_1)^2}{8R_L}$$

$$= \frac{R_L}{8} (I_1 - I_2)^2 \qquad\qquad 3.2$$

where

$$R_L = \frac{V_L - V_1}{I_1 - I_2} \qquad\qquad 3.3$$

and $I_1 - I_2$ is the current swing associated with the voltage swing $V_L - V_1$ (Di Lorenzo et al, 1979).

The maximum output power P_m is given by

$$P_m = \frac{I_F}{8} (V_{GD} - V_P - V_K) \qquad\qquad 3.4$$

where

$$R_L = \frac{V_{GD} - V_P - V_K}{I_F} \qquad\qquad 3.5$$

and

$$V_{DS} = \frac{V_{GD} - V_P - V_K}{2} \qquad\qquad 3.6$$

For example, consider a device with an I_{DS} = 350 mA, V_{GD} = 30V, V_P = 5V, V_K = 2V, where the channel of the FET is not thermally limited.

Thus

$$P_m = \frac{350}{8} (30-5-2) = 1 \text{ watt}$$

This result corresponds closely to the best output powers found experimentally for 1 mm gate width FETs having power gains of approximately 4 dB up to frequencies approaching 18 GHz.

Thus the important device parameters for power FETs are:

1. I_F, the maximum drain current available with a forward biased gate, V_F;

2. V_K, the knee voltage at which I_F saturates;

3. V_{GD}, the gate to drain avalanche voltage;

4. V_P, the pinch-off voltage; and

5. V_L, the limiting drain-source voltage at which the gate can no longer modulate the channel current effectively.

Unfortunately, some of the above variables are inter-related. For example increasing I_F by increasing the carrier concentration N_D generally decreases the breakdown voltage, V_{GD}. However we shall see later in this chapter that the geometrical design of the device plays a major role in its power handling and efficiency.

3. MODELLING OF GaAs FET TO PREDICT LARGE SIGNAL PERFORMANCE

For large signal performance evaluation it is necessary to derive a nonlinear circuit-type device model. In order to achieve such results the bias and frequency dependences of the device S-parameters have to be measured. The model thus derived is then used in a time-domain analysis computer program to produce large-signal waveforms. These results can be transformed into the frequency domain by the usual Fourier analysis.

Consider, for example, the measured I-V characteristics of a Texas Instruments power MESFET shown in Fig. 3.4(a) together with the FET model of Fig. 3.4(b).

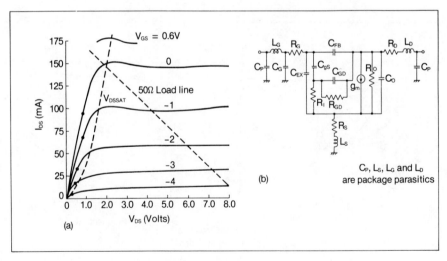

FIG. 3.4(a).   Static I-V Characteristic
     (b).   Equivalent Circuit of Power GaAs FET

In this model the package parasitics are L_S, L_G, L_D and C_P representing the source common lead inductance, gate wire bonds, drain wire bonds and shunt parasitic capacitances due to the package respectively. The bias dependent elements in this equivalent circuit are:

C_{FB} the gate to drain feedback capacitance

C_{GS} the gate-to-source capacitance

g_m the transconductance

R_I the intrinsic channel resistance

R_{GD} the Gunn domain resistance

R_0 the output resistance.

The equivalent circuit is the same as that used for small signal FET analysis which included the formation of a Gunn domain at the drain contact (Willing et al, 1977).

Fig. 3.5 shows the dependence of the small signal S-parameters of the device as a function of frequency for both the model and the measured device at a V_{DS} = 6V and V_{GS} = -2V. It is seen that there is a good agreement over the 1 to 18 GHz frequency range. The components in the model having the most significant bias dependences are those mentioned above and the relationship of their values to drain to source voltage is shown in the plots of Fig. 3.6(a) to (f). Each curve shows both the measured results and the computer-fitted small-signal model element values as a function of both drain to source voltages and gate to source voltages.

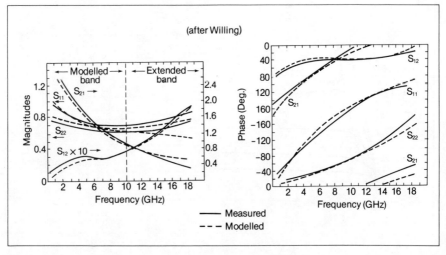

FIG. 3.5.   S-Parameters for a Power GaAs MESFET

A. CHANNEL CAPACITANCE (C_{GS})

The influence of bias on the active layer or channel capacitance is shown in Fig. 3.6(a). C_{GS} initially decreases with increasing V_{DS} at a constant V_{GS} due to the depletion width increasing. C_{GS} approaches

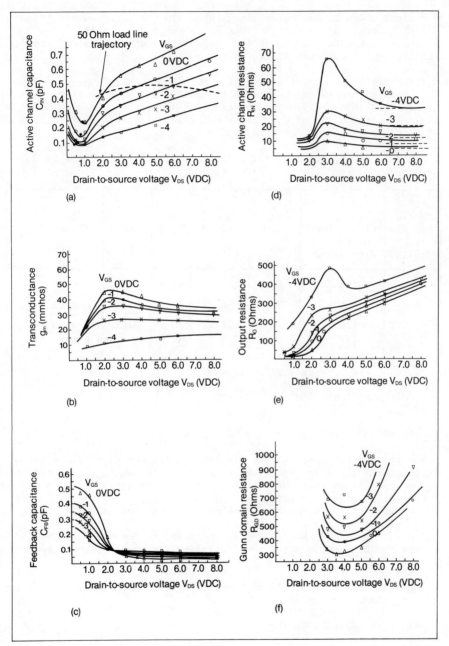

FIG. 3.6.   Bias Dependences of C_{GS}, g_m, C_{FB}, R_I, R_O and R_{GD}
(after Willing, 1978)

a minimum close to the value of V_{DS} at which the device drain to source current saturates. Further increase in V_{DS} results in an increase in C_{GS} due to the charge accumulation effects as described in Chapter 2 (Engelmann et al, 1976).

The total input capacitance of the FET is made up of C_{GS}, C_{EX} and C_G. This can be calculated using the analytical techniques of Lehovec and Zuleeg (1970) at the onset of saturation. Table 3.1 shows the calculated value of C_{TOT} and the fitted value of C_{TOT} for the set of bias conditions given where $C_{TOT} = C_{GS} + C_{EX} + C_G$.

TABLE 3.1. Dependence of FET Equivalent Circuit Parameters on DC Bias at Onset of Current Saturation

Gate to Source Voltage V_{gs} (volts)	Drain to Source Saturation Current V_{ds} (volts)	$g_m{}^*$ (calc) mS	$g_m{}^{**}$ (fitted) mS	g_m (DC) mS	$C_{gs}{}^*$ (calc) pF	$C_{gs}{}^{**}$ (fitted) pF
0	2.3	50	47	45	0.76	0.76
-1	2.0	37	40	41	0.65	0.63
-2	1.65	31	32	35	0.49	0.53
-3	1.2	25	22	25	0.35	0.4

* After Lehovec and Zuleeg (1970)

** After Willing et al (1978).

B. TRANSCONDUCTANCE (g_m)

The transconductance can be calculated using the expressions given in Chapter 2 and compared to the computer fitted g_m ($g_{m(FIT)}$) and the $g_{m(DC)}$ calculated from the static I-V characteristics. Table 3.1 compares the values of $g_{m(CALC)}$, $g_{m(FIT)}$ and $g_{m(DC)}$ at the indicated bias conditions. It may be seen that there is good agreement.

Fig. 3.6(b) shows that as V_{DS} is increased g_m reaches a maximum around the knee voltage and then decreases for values of V_{DS} greater than V_K.

Engelmann and Liechti (1976) have attributed this effect to a reduction in charge accumulation.

C. FEEDBACK CAPACITANCE (C_{FB})

Charge accumulation also accounts for the decrease in the feedback capacitance seen in Fig. 3.6(c) as V_{DS} is increased. Charge accumulation at the drain edge of the gate is seen in the sharp reduction in the value of C_{FB} as V_{DS} approaches V_K. Above V_K the feedback capacitance increases slightly with increasing V_{GS}.

D. INTRINSIC CHANNEL RESISTANCE R_I AND OUTPUT RESISTANCE R_O

Fig. 3.6(d) shows the variation of R_I with V_{DS} and V_{GS}. For small values of V_{DS}, R_I initially increases at a greater rate for values of V_{GS} close to the pinch-off voltage. At higher values of V_{DS} the value of R_I decreases and approaches a constant value which depends on the initial open channel conductance as well as the ratio of the channel width at the drain edge of the gate when biased at V_K to the undepleted channel width.

Fig. 3.6(e) indicates that the R_O of the device increases from the d.c. resistance of the channel to a final value dependent on $|V_{GS}|$.

E. GUNN DOMAIN RESISTANCE R_{GD}

Willing and De Santis (1978) have reported the bias dependence of the Gunn domain resistance.

As $|-R_{GD}|$ becomes the same order as R_O the magnitude of the output reflection coefficient S_{22} may approach or even become greater than unity. Normally $|S_{22}|$ increases monotonically with increasing V_{DS} but the influence of the Gunn domain resistance results in a non-monotonic variation of $|S_{22}|$, the effect being most noticeable in the region where $V_{DS} \simeq V_K$ and when $|V_{GS}| \ll V_p$, i.e. when the drain current is high. The relative magnitudes of $|-R_{GD}|$ and R_O can be seen by comparing Fig. 3.6(e) and (f).

4. PREDICTIONS OF NON-LINEAR PERFORMANCE OF GaAs POWER FETs FUNDAMENTAL AND HARMONIC PERFORMANCE

Large signal circuit models simulating nonlinear device performance depend on the use of expressions which relate the current-voltage relationships for each of the nonlinear elements.

Instantaneous currents through each of the nonlinear elements can be expressed as the product of an instantaneous element value and an instantaneous voltage.

Consider, for example, the simplified equivalent circuit of a tuned GaAs power FET amplifier (Higgins et al, 1980). As has already been shown the dominant contributions to nonlinear response come from the variations in transconductance, g_m, with gate voltage, in the drain conductance, G_D, with drain voltage, and the voltage dependence of the gate-source capacitance, C_{GS}.

The variation in the equivalent circuit elements is most conveniently represented by a Taylor series. Thus, for example the transconductance can be expressed as

$$g_m(V) = g_{m1} + g_{m2}V_{GS} + g_{m3}V_{GS}^2 + g_{m4}V_{GS}^3 + \cdots \tag{3.7}$$

Thus the FET r.f. drain current is given by:

$$I_{DS(t)} = \int_o^{V_{GS}(t)} g_m(V)\,dV$$

$$= g_{m1}V_{GS} + g_{m2}\frac{V_{GS}^2}{2} + g_{m3}\frac{V_{GS}^3}{3} + g_{m4}\frac{V_{GS}^4}{4} + \cdots \tag{3.8}$$

where $V_{G(t)} = V_{GS(t)} - V_{GO}$ \hfill 3.9

specifies the instantaneous deviation of the gate-source voltage from the gate bias V_{GO}.

The expansion coefficients g_{m1} are derived from the polynomial fit to the characteristic of Fig. 3.6(b), for example. The variation in g_m with drain bias is neglected in the simplest calculations by assuming an average value for the typical drain voltage range.

An expression for the drain conductance G_0 ($= 1/R_0$) as a function of the drain voltage is given by:

$$G_0(V) = G_{01} + G_{02}V_{DS} + G_{03}V_{DS}^2 + G_{04}V_{DS}^3 + \cdots \tag{3.10}$$

where $V_{D(t)} = V_{DS(t)} - V_{DO}$ \hfill 3.11

represents the instantaneous deviation in the drain voltage from the quiescent value of V_{DO}. The dependence of G_0 on gate bias has been neglected by using an average value for the gate voltage range.

The active channel capacitance, C_{GS}, as a function of the instantaneous gate voltage $V_{G(t)}$ can be expressed as

$$C_G(V) = C_{GS1} + C_{GS2}V_{GS} + C_{GS3}V_{GS}^2 + C_{GS4}V_{GS}^3 + \cdots \tag{3.12}$$

The resulting impedance associated with $C_{G(V)}$ is derived from

$$I_{G(t)} = \frac{d}{dt}\left[\int_o^{V_{GS}(t)} C_{GS}(V)\,dV \right] \tag{3.13}$$

g_{mn}, G_{0n} and C_{Gn} are dependent on the doping profile of the active layer. All the expansion coefficients such as g_{mn}, G_{0n} and C_{Gn} are used to calculate the IMD products using the methods of Tucker and Rauscher (1977).

Let us consider the typical results of such a nonlinear model following time-domain analysis. A 600μm gate width device with a 1.7μm gate length and an epitaxial thickness of 0.32μm was used where the doping density N_D = 7.5 x 10^{16}/cc. Pinch-off voltage was calculated to be −5.1V with a built-in potential V_{Bi} of 0.7V resulting in a channel pinch-off voltage of −5.8V.

Now consider the device to be biased at V_{GS} = −2V and V_{DS} = +6V with r.f. inputs at a frequency of 2 GHz with drive levels up to the point where 6 dB gain compression occurs. The sequence of steps leading to the fundamental and harmonic output powers with input power P_{in} is shown in Fig. 3.7.

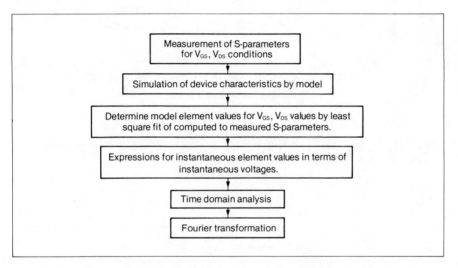

FIG. 3.7. Simulation of Non-Linear Power GaAs FET Performance

Fig. 3.8 shows the excellent agreement between the measured fundamental output power and that calculated.

Fig. 3.9(a)-(d) shows the excellent agreement between the modelled and measured results for harmonic powers up to the 5th order.

5. INTERMODULATION PERFORMANCE

In Equation 3.10 the drain coefficients are greatly reduced if the drain voltage level is increased so that the instantaneous drain voltage never approaches the saturation drain voltage. This is a commonly observed effect of power GaAs FETs where intermodulation levels can be reduced by increasing the drain to source voltage. The coefficients of both g_m and G_0 are dependent on the carrier profiles of the GaAs active layer.

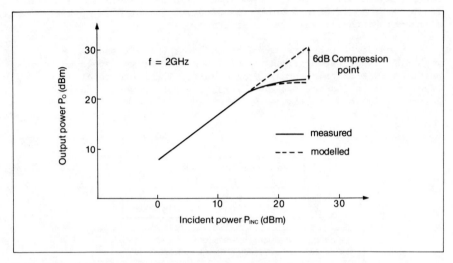

FIG. 3.8.   Measured and Modelled Fundamental Output Power Versus
Input Power of GaAs FET

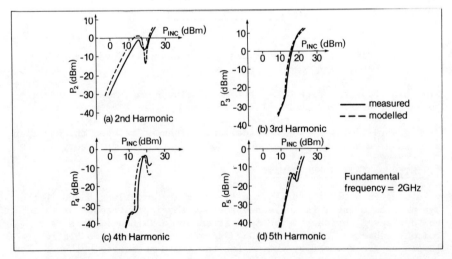

FIG. 3.9.   Measured and Modelled Harmonic Output Power Levels Versus
Input Power for a GaAs Power FET

Most of the modelling of GaAs FETs use expressions derived from Shockley's work (1952) but recently Pucel et al (1975) have included the effects of velocity saturation. Most of the analytical work assumes that the carrier profile from the surface immediately beneath the gate to the semi-insulating GaAs is flat.

Fig. 3.10 represents the model used to calculate the effects of a non flat profile (Higgins, 1978). The model divides up the active layer into 150 layers. The electrons' motion under the gate region is assumed to be such that they reach saturation velocity after an initial short channel section where the velocity is proportional to electric field. As explained in Chapter 2 the necessary boundary conditions in the directions along and normal to the charge flow establish the current at a given bias condition.

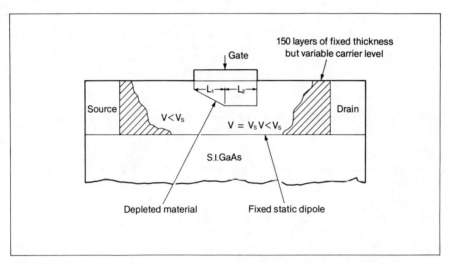

FIG. 3.10.   Layer Model used in the Model of GaAs FETs with Variable Carrier Concentration Profile

The numerical integration adaptation of Pucel's model is extremely useful in predicting the way in which the intermodulation distortion (IMD) of a GaAs FET depends on the carrier profile.

Fig. 3.11 shows the profiles for three different layers:

1. An idealized epitaxial profile described by Williams and Shaw (1978);

2. A flat epitaxial profile with an abrupt doping transition at the substrate; and

3. An ion implanted layer where a 500 keV Se implant has been

purposely compensated at the surface with a shallow 40 keV Be implant
(see Chapter 4).

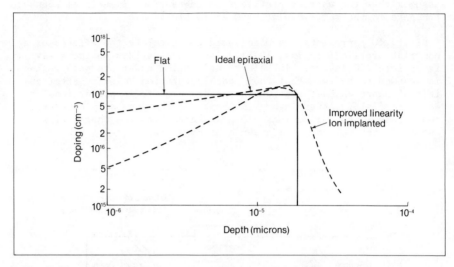

FIG. 3.11.   Three Different Profiles used in Calculating Inter-
    modulation Characteristics of Power GaAs FETs

The devices modelled were 500μm gate width, 1μm gate length FETs
with specific ohmic contact resistances of 10^{-6} Ω cm^2.

In calculating the third order intermodulation products it is normally assumed that the third order coefficient (g_{m3}, for example) is
much larger than the fifth or seventh order coefficients (g_{m5} and
g_{m7}). From Table 3.2 it may be appreciated that g_{m5} and g_{m7} can, in
fact, be comparable in magnitude or larger than g_{m3}. The g_{m5} and g_{m7}
coefficients are least for the ion-implanted profile, promising a
better IMD performance.

In fact, for moderate to high signal levels the transconductance
contributes mainly from its fifth order (g_{m5}) term. The analysis
therefore takes account of the contributions to third order IMD products due to the higher order terms in the nonlinear device model.

The two tone intermodulation products are given at frequencies
$2f_1-f_2$, $3f_1-2f_2$ and $4f_1-3f_2$ where f_1 and f_2 are the two input tones
to the FET. g_{m3} contributes to the $2f_1-f_2$ product whilst g_{m5} and g_{m7}
contribute to the $2f_1-f_2$ and $3f_1-2f_2$ products and the $2f_1-f_2$, $3f_1-2f_2$
and $4f_1-3f_2$ products respectively. For example, it may be shown that
if we neglect the g_{mn} terms for n greater than 5, the intermodulation
current is given by

$$I_{2f_1-f_2} = 0.25\ g_{m3}A^2B + 0.25\ g_{m5}A^4B$$

$$+ 0.375\ g_{m5}A^2B^3$$

where $V = A\cos\omega_1 t + B\cos\omega_2 t$.

TABLE 3.2. Polynomial Coefficients of Transconductance* and Drain
Output Conductance**

Coefficient	Flat EPI	Profile Improved EPI	Ion Implant Se + Be
g_{m1}	3.5×10^{-2}	3.55×10^{-2}	3.1×10^{-2}
g_{m2}	5.8×10^{-3}	4×10^{-3}	3.3×10^{-3}
g_{m3}	-4.5×10^{-4}	-7×10^{-4}	7.5×10^{-4}
g_{m4}	3.3×10^{-4}	5.8×10^{-4}	-5.4×10^{-5}
g_{m5}	1.46×10^{-3}	9×10^{-4}	-2×10^{-4}
g_{m6}	5×10^{-5}	-1×10^{-4}	4.2×10^{-5}
g_{m7}	-2×10^{-4}	-1×10^{-4}	4.48×10^{-5}
g_{m8}	5×10^{-6}	1.5×10^{-5}	-4.7×10^{-6}
G_{01}	1.64×10^{-4}	1.99×10^{-4}	4.28×10^{-4}
G_{02}	-3.07×10^{-5}	-4.19×10^{-5}	-4.9×10^{-5}
G_{03}	1.13×10^{-5}	5.18×10^{-6}	-2.23×10^{-6}
G_{04}	-1.37×10^{-6}	7.62×10^{-7}	1.23×10^{-7}
G_{05}	-3.92×10^{-7}	-2.25×10^{-7}	1.26×10^{-7}
G_{06}	5.8×10^{-8}	-8.47×10^{-9}	-2.02×10^{-8}
G_{07}	6.2×10^{-9}	5.4×10^{-9}	2×10^{-9}
G_{08}	-8.21×10^{-10}	-3.4×10^{-10}	-1.08×10^{-10}

$$*g_m(V) = g_{m1} + g_{m2}V + g_{m3}V^2 + g_{m4}V^3 + \ldots$$
$$**G_0(V) = G_{01} + G_{02}V + G_{03}V^2 + G_{04}V^3 + \ldots$$

Calculated IMD products are based on the usual two tone method.
Third order products have been calculated by Higgins and Kuvas (1980).
Several conclusions have been drawn:

1. Optimum tuning and loading conditions change with signal power
   level because of the corresponding changes in the admittance

matrix of the device.

2. The IMD products are sensitive to tuning and loading since the peak RF voltage levels will change with these two parameters.

3. At low r.f. input powers the IMD contributions from the $G_0(V)$ polynomial dominate.

4. At higher r.f. input powers the IMD products are due to both the $g_m(V)$ and $G_0(V)$ terms. A correlation exists between the gate and drain voltages which are effective in the respective polynomials and this results in a decrease in the IMD versus input level relationship accompanied by a steep rise in the IMD power output.

5. The sign of the fifth order coefficient relative to the third order coefficient contributes to the determination of the low signal level IMD products. Cancellation effects can occur in third order IMD products from the drain (or gate) alone due to sign differences of the different coefficients.

Fig. 3.12 shows the calculated gain and third order intermodulation products versus input power using the active layer profiles of Fig. 3.11.

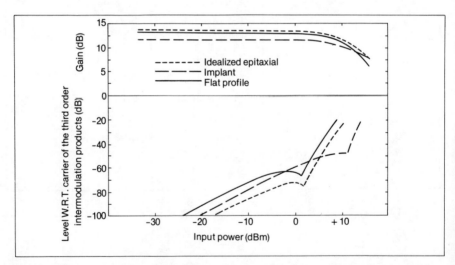

FIG. 3.12. Calculated gain and Third Order Intermodulation Products Versus Input Power for the Three Profiles of Fig. 3.11.

These IMD products have three regions. At very low signal levels the third order products rise with a gradient of 3. In the second region the cancellation effects of 4. above are seen. In the region where the signal levels are becoming large the contribution from the $g_m(V)$

polynomial dominates and the rate of rise of the IMD product is greater than 3. This is because $g_{m5} > g_{m3}$ as shown by Table 3.2.

The dependence of IMD on drain bias level has also been calculated for the implanted profile of Fig. 3.11. The result of a rise in the drain bias level is a reduction in the G_0 coefficients with the resulting gain and third order intermodulation products of Fig. 3.13.

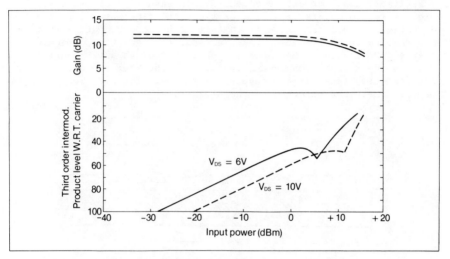

FIG. 3.13. Dependence of Gain and Intermodulation Distortion on Drain Voltage

6. POWER FET DEVICE PERFORMANCE

The performance of GaAs power FETs is dominated by four items:-

(a) Gate width.

(b) Minimization of electrical parasitics, particularly the common lead inductance.

(c) Reduction of the device thermal impedance.

(d) A device structure capable of high drain to source and gate to drain potentials.

Devices with excessive common lead inductance, for example, exhibit narrow bandwidth capability, low gain and spurious oscillations at low frequencies.

A. STRUCTURES USED TO INCREASE GATE WIDTH

There are basically three techniques available for increasing total gate width. The first approach involves the parallelling of a number of gate fingers by the use of crossover structures which can utilise
(a) Dielectric crossovers
(b) Air bridge overlays
(c) Wrap around or edge plating.

Fujitsu (Fukata et al, 1976) and NEC (Higashisaka et al, 1980) use dielectric crossovers with SiO_2 as the insulating material. The advantage of this technique is that it leads to a compact structure but results in additional parasitic capacitances at the crossover points. Devices with up to 26 mm total gate width have been fabricated using such a technique (Fig. 3.14). The source electrodes are all connected to a large grounding electrode. The SiO_2 insulates the overlaid gate conductors from the sources.

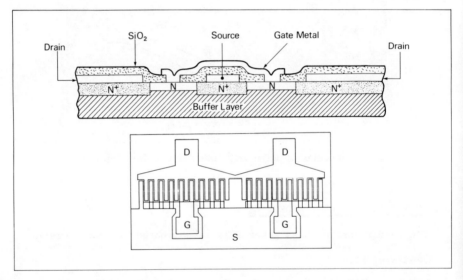

FIG. 3.14. Power GaAs FET Incorporating Crossovers (after Fukuta)

The air bridge overlay shown in Fig. 3.15 is superior to the dielectric crossover in that it has a lower capacitance/unit area. The air bridge is formed by first depositing a thick resist layer and then evaporating a thin gold film. A second resist layer follows to define the bridge area which is then electroplated for strength. Finally the resist is removed leaving the bridge suspended in air as shown in Fig. 3.15. Hence for the same parasitic capacitance the air bridge overlay can have a larger area leading to a reduced inductance per unit length as well as an increased current handling capability. The processing requirement of such a connection is also less than the dielectric crossover since both dielectric deposition and vias (holes) through the SiO$_2$ to make source to source contact are no longer needed. Wrap around or edge plating of the chip is a low inductance approach because of the large periphery connection occurring between the source area and the ground plane. However, a crossover technique is still required to interconnect the various gate and drain fingers as shown in Fig. 3.16.

FIG. 3.15.　Air-Bridge Source Interconnections (Courtesy Plessey Co)

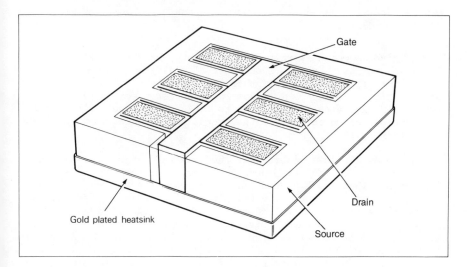

FIG. 3.16.　Wrap Around or Edge Plated Power FET

The second approach simply involves wire bonding a number of FET cells in parallel. The device in Fig. 3.17 has multiple source bonding pads and drain and gate pads. Devices as large as 9.6 mm for operation at 9 GHz (Macksey et al, 1977) and 24 mm for operation at 4 GHz (Wemple et al, 1978) have been constructed in this way. Although the multiple bond approach uses a simpler processing technology, with plated-up drain contacts to remove the possibility of wire bonds touching the gate fingers, the device assembly is difficult.

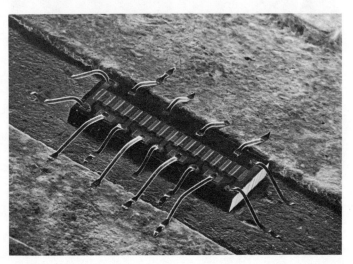

FIG. 3.17. Wire Bonded 4 Cell Q3FET in BMH60 Package (Courtesy Plessey Co.)

The third approach is that first attempted by RCA (Drukier et al, 1975) and now used by MSC (Drukier et al, 1979) and Mitsubishi (Mitsui et al, 1979) amongst others. This is the 'flip-chip' mounting of the chip to produce a very low inductance connection since the source pads are attached directly to the ground plane. The source pads are plated-up to prevent the ground plane contacting the channel areas (Fig. 3.18).

D'Asaro et al (1977) introduced the 'via-FET' approach where the source connections to the ground plane are made through the semi-insulating substrate. Several laboratories are now using this technique, the general conclusion being that the 'via-FET' gives significant increases in output power and gain although the saturated output power is unchanged. Fig. 3.19 shows an array of 'vias' approximately 50μm in diameter, etched through a GaAs substrate which is of the order of 50μm thick. Fig. 3.20(a) shows the cross section of a typical via-FET whilst Fig. 3.20(b) shows a top view of an experimental device made by Plessey.

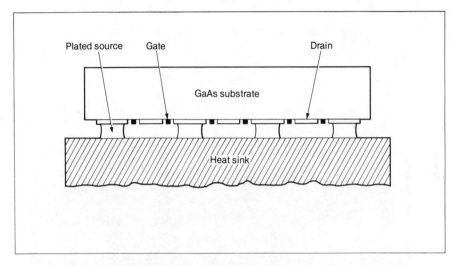

FIG. 3.18.   Flip Chip Mounted Power GaAs FET

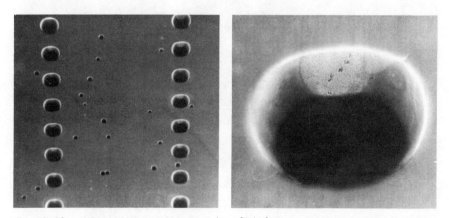

FIG. 3.19.   SEM Photographs showing 'Via' Holes before Metallisation
   (Courtesy Plessey Co. Ltd.)

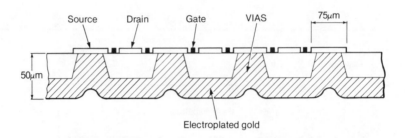

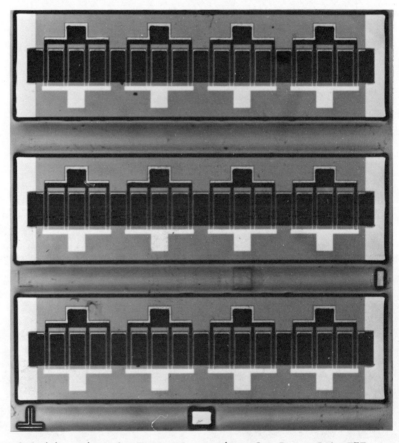

FIG. 3.20(a). Via Hole Source Connections for Power GaAs FET
(b). Device Structure (Via Plating Technology)
(Courtesy Plessey Co. Ltd.)

B. THE MINIMISATION OF PARASITICS

The parasitic inductances in the gate, L_G and drain L_D of Fig. 3.4(b) do not degrade the gain of the FET as much as the common-source lead inductance L_S. Hence the maximum available gain under small signal drive conditions can be approximated by

$$MAG = \left(\frac{f_T}{f}\right)^2 \frac{1}{\left[4(R_G+R_I+R_S + 2\pi f_T L_S)/R_0 + 4\pi f_T C_{FB}(2R_G+R_I+R_S+2\pi f_T L_S)\right]}$$

3.14

where f_T is the cut-off frequency,

     f is the operating frequency,

and all the other parameters are defined in Fig. 3.4(b).

For given device parameters, MAG falls off at a theoretical value of 6 dB per octave increase in frequency. Consider the effect of several short bonding wires approximately 0.4 mm long which are typical of the lengths associated with the source bonds on power FET chips. Fig. 3.21 shows the result of calculating the effect of common lead inductance on gain for a power FET of 3000 μm total gate width made up of twelve gates of 250 μm unit width (D'Asaro et al, 1977) at 4 GHz. It may be seen that an inductance of only 0.1 nH degrades the MAG by over 4 dB.

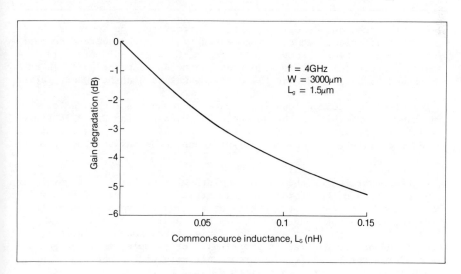

FIG. 3.21. Effect of Common Lead Source Inductance on the Maximum Available Gain of a Power GaAs FET.

The cell combining efficiencies (Macksey et al, 1978) of power FETs depends significantly on the common lead inductance, L_S. This can be seen by inspection of Table 3.3 (Di Lorenzo et al, 1979) where 4 cell FETs have been constructed either as one complete chip or as four bonded separately, the latter giving reduced source lead inductance due to the effectively shorter electrical path between the source electrode and ground.

TABLE 3.3.   Effect of Source Lead Inductance on Combining Efficiencies of Power FETs (Di Lorenzo et al, 1979)

	Gain (dB)	No. of Cells	P_{OUT} (Watts)	Power Added Efficiency η (%)	V_{DS} (volts)	Combining Efficiency (%)
	4	1	0.91	25.6	8	79.1
One		4	2.88	24.1	8	
chip	6	1	0.83	31.8	8	64.5
		4	2.14	19.9	8	
	4	1	0.96	39.3	8	96.9
Four		4	3.72	36.8	8	
chips	6	1	0.87	45.3	8	97.4
		4	3.39	40.0	8	

The via technique described in the previous section achieves this in a most efficient manner demonstrated by Fig. 3.22 where a comparison is made between a conventional power FET and a FET using via source connections. It is seen that the measured power gain can be increased by over 2 dB when the output power is 3 dB below saturation.

The effect of common lead inductance increases as the gate length is descreased. This can be appreciated by inspecting Eq. 3.14 since as gate length is reduced f_T becomes greater ($f_T \propto 1/L_G$) and thus the product of $2\pi f_T L_S$ becomes more significant in comparison to the other terms.

A gate much greater than $\lambda/10$ in width will suffer from transmission line effects and therefore it seems intuitive that such effects will reduce device performance. Fukata et al (1976) have suggested, for example, that a 1µm long gate should have a width no greater than 50µm at 10 GHz and 100µm at 5 GHz. This assumption has, in fact, been found not to be true in practice. For example, measured performance for power FETs with different unit gate widths where unit gate width is equal to total gate width divided by the number of gate fingers is shown in Fig. 3.23 at 6.4 and 8.3 GHz (Higashisaka et al, 1980) for devices with a total gate width of 3 mm. It may be seen that no significant degradation occurs in power gain until unit gate widths approach 200µm at 8 GHz, i.e. a factor of 5 wider than the intuitive approach. Such a factor is extremely useful in keeping the overall FET chip width low.

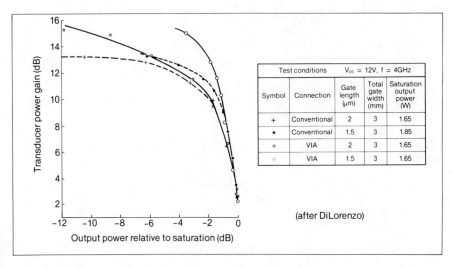

FIG. 3.22. Power Gain for Conventional Wire Bonded and Via-Source Power GaAs FETs

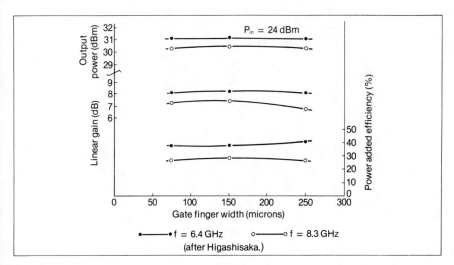

FIG. 3.23. Microwave Performance of Power GaAs FETs Versus Gate-Finger Width

The disagreement between the simple theoretical approach and experimental results is due to the difference between the assumed gate capacitance and the real device capacitance, as discussed below.

Consider the propagation constants α and β along the gate electrode to be given by the following equations (Aono et al, 1979):

$$\alpha = \omega\sqrt{C}_{GS}\left\{(-1_g + \frac{\sqrt{1_g^2 + r_g^2/\omega^2})}{2}\right\}^{\frac{1}{2}} \qquad 3.15$$

and

$$\beta = \omega\sqrt{C}_{GS}\left\{(1_g + \frac{\sqrt{1_g^2 + r_g^2/\omega^2})}{2}\right\}^{\frac{1}{2}} \qquad 3.16$$

where 1_g and r_g are the inductance and resistance per unit gate width.

C_{GS} dominates both equations 3.15 and 3.16. Fukata et al (1976) and Aono et al (1979) have employed a C_{GS} of 0.5×10^{-3} pF/μm, which was obtained theoretically from Wolf's model (1970). However measured values for 1μm gate length devices puts $C_{GS} = 1.5 \times 10^{-3}$ pF/μm. Fig. 3.24 shows the way in which the gain of a device is degraded with unit gate width as a function of the C_{GS}/unit gate width parameters.

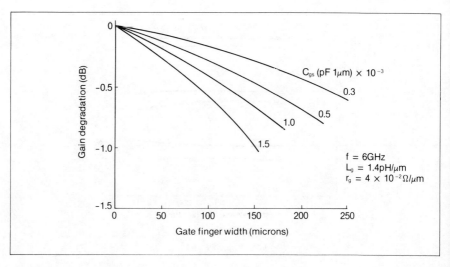

FIG. 3.24. Gain Degradation Versus Unit Gate Width as a Function of Gate Capacitance

It may be seen that there is a considerable difference between the
two values quoted. This knowledge has, therefore, led to the design
of FETs which produce the same power and gain as previous devices but
in a chip area which is one half the area of the best previous device
giving the same performance.

Power gain of a power FET does not appear to be particularly sensitive to gate length. For example, devices designed to operate at
X band with total gate widths of 2.4 mm have virtually indistingui-
shable maximum output powers and gains with gate lengths of 1.8 or
2.6μm. This may well be due to the fact that there are many other
effects masking the advantages to be gained in reducing gate length -
some of these have been discussed in this chapter, including common
lead inductance and unit gate widths.

C. THERMAL IMPEDANCE

Many of the power FETs discussed previously are mounted on a heat
sink such that the heat is removed through the GaAs surface. Devices
which are flip-chip mounted have their generated heat dissipated
through the source pads. Thus the heat must spread through the GaAs
and this spreading resistance dominates the thermal impedance. The
thermal impedance of power FET structures with 1 mm wide gates versus
GaAs substrate thickness have been calculated and the result is shown
in Fig. 3.25.

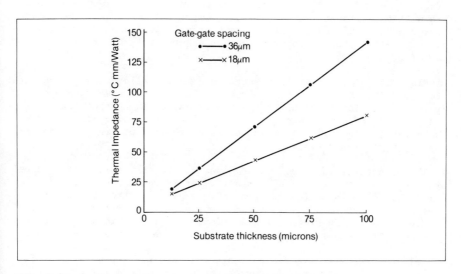

FIG. 3.25.   Thermal Impedance of Power GaAs FETs

It may be seen that the thermal impedance is a strong function of substrate thickness when thicknesses are less than 100μm. Above 100μm thickness the thermal impedance variation with substrate thickness is less pronounced. However, this results in substrates needing to be so thin that plated heat sink techniques have to be used to obtain the maximum benefit from this effect. In fact, for the thicker substrates substantial thermal impedance benefits are produced by decreasing the gate finger to gate finger spacing as demonstrated in Fig. 3.25.

Besides affecting the channel temperatures of the FET and hence the reliability of the device, the maximum output power of FET devices with gate lengths of 1.5μm reduces by approximately 0.02 dBm/$^{\circ}$C at an ambient temperature of 25°C. At least part of this performance degradation is due to decreasing mobility of GaAs as the temperature is increased.

D. SOURCE-TO-DRAIN BURNOUT AND GATE-TO-DRAIN AVALANCHING

Equation 3.4 has shown the importance of gate to drain voltage on the power handling capability of a power GaAs FET. Thus the performance of the device in this respect will be determined by the gate-to-drain and source-to-drain breakdown voltages. Much effort has been concentrated, therefore, on the improvement of the breakdown characteristics of power FETs by optimizing the structures and materials used. Breakdown voltages of over 70 volts have been obtained (Tiwari et al, 1979) to date with the theoretical promise of further improvements being possible.

Fig. 3.26 shows five kinds of GaAs MESFET structures that have been fabricated. The role of ohmic contacts in the drain-source burnout is well established (Wemple, 1976). Carrier densities in the active layer and the n^{+} contact layer are typically 10^{17} and 10^{18}/cc respectively, with a buffer layer which has a carrier level of less than 10^{14}/cc.

Fig. 3.26(a) shows a planar structure having alloyed ohmic contacts. Fig. 3.26(b) shows a similar structure but with n^{+} areas produced using ion implantation under the source and drain contacts to decrease the problem of avalanching and the resultant thermal runaway in the substrate. Fig. 3.26(c) is an epitaxial version of Fig. 3.26(b) with an n^{+}, n, n^{-} layer system.

The structures in Figs. 3.26(d) and (e) are of the recessed channel type but do not have the n^{+} contact layer (Furutsuka et al, 1978). Fig. 3.26(e) allows closer source to gate spacing than the fully recessed structure of Fig. 3.26(d) thus decreasing the source resistance and allowing simpler fabrication. Table 3.4 lists a summary of the contact technologies being used to improve source-drain burnout.

Bell Laboratories use the n^{+} contact technology achieving 52 volt source-drain breakdown at a current density of 100 mA/mm gate width whilst NEC who use the recessed channel technique of Fig. 3.26(d) and

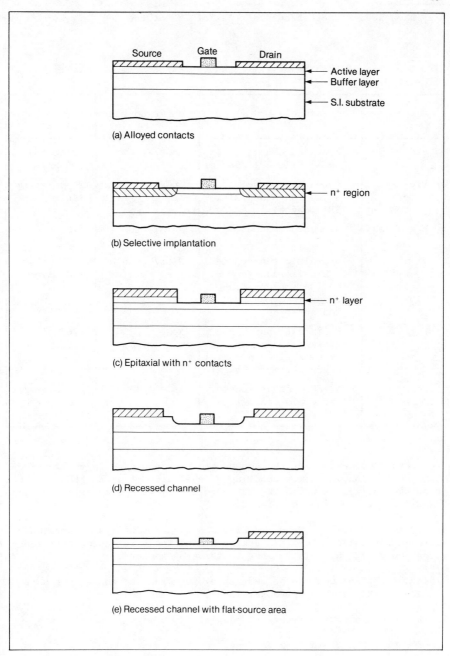

(a) Alloyed contacts

(b) Selective implantation

(c) Epitaxial with n⁺ contacts

(d) Recessed channel

(e) Recessed channel with flat-source area

FIG. 3.26.   Various GaAs MESFET Structures

(e) have achieved breakdown voltages of 25V at the same current rating.

TABLE 3.4.   Various Power FET Technologies Used to Produce High Burnout Voltages

Company	Technology	Ohmic Contact	Burn out Voltage
NEC	Recessed channel, No N^+ EPI, EPI Channel	AuGe/Ni	22 to 25V at 100 mA/mm
Plessey	Recessed channel, N^+ source and drain, EPI channel	AuGe	23 volts at 176 mA/mm
Mitsubishi	Recessed channel, graded profile	Au/Ni/AuGe	>20V
HP	N^+ inlaid source and drain islands formed by ion implantation, channel formed by implantation.	NiCr/Ge/Au	>30V at 100 mA/mm
Fujitsu	N^+ inlaid source and drain islands formed by selective epitaxial regrowth, EPI channel	AuGe	>26V at 100 mA/mm
Bell	N^+ EPI planar source and drain contacts, EPI channel	AuGe/Ag/Au	Up to 52V at 100 mA/mm

However, not only does the source-drain burn-out voltage have direct consequences on the power handling capabilities of a FET structure but also the voltage at which the gate-drain avalanche process takes place as seen by inspecting equation 3.4.

Wemple et al (1980) have recently shown that there is a correlation between the pulse gate-drain avalanche and the power added efficiency of a power FET which is related to three parameters:

1. Avalanche current can be reduced by reducing the charge per unit area in the active epitaxial layer;

2. By maintaining the smoothest possible gate edge structure; and

3. Reducing the drain bias voltage.

The first requirement implies that the gate metallisation should be placed in a recess etched to the zero-bias depletion depth. This also improves the source-drain breakdown voltage provided the material parameters are optimum.

Figure 3.27 shows the drain characteristics of a shallow recess power FET measured with 80µsec pulses.

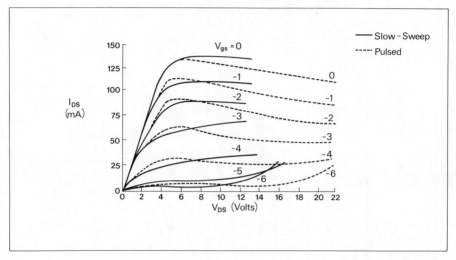

FIG. 3.27.   Slow Sweep and Pulsed Characteristics of Shallow
   Recess Power GaAs FET

Under such pulsed conditions these devices (White, 1980, private communication) typically withstand drain-source voltages of 35 volt before failure. Using a slow bias sweep, however, the drain current starts to increase in the region where gate voltages would ordinarily pinch-off the channel. This drain current increase is usually accompanied by light emission. This breakdown phenomenon is not only due to gate-to-drain avalanching but is also due to current multi-plication in the channel in the presence of deep electron traps in the buffer layer and also at the n-layer to buffer interface. Ladbrooke et al (1980) have shown that the electron trap densities (of the order of 10^{16} cm^{-3}) found in the buffer layers of VPE material based power FETs gave theoretical limiting output powers of 1 watt mm^{-1} at gate-to-drain voltages of 40 volts which is in reasonable agreement with the performance of good X-band power FETs (Hughes, 1981, private communication).

7. POWER FET RESULTS

Table 3.5 gives a summary of some of the power FET results which were obtained in laboratories worldwide in 1980. It should be noted that the best output power per millimetre gate width remains close to 1 watt up to 16 GHz or so for gains of 3 to 4 dB.

TABLE 3.5.   Performance of GaAs Power FETs (1980)

Frequency (GHz)	P_{OUT} (Watts)	Power Gain (dB)	Gate Width (mm)	P_{OUT}/unit gate width (W/mm)	Company
4	18.5	3.5	24	0.77	Bell
	15.0	5.0	26	0.58	Fujitsu
	14.4	3.5	16	0.9	Bell
	10.7	8.1	16	0.67	Bell
	9.6	5.0	13	0.74	Fujitsu
	3.6	3.0	3	1.2	RCA
	3.0	4.7	9.6	0.31	RCA
8	5.1	5.0	6.4	0.8	TI
	3.9	7.0	4.8	0.8	TI
	2.2	4.2	5.2	0.42	Fujitsu
	1.75	5.0	2.4	0.73	Plessey
	1.7	4.0	1.2	1.42	TI
	1.3	6.0	2.4	0.54	Plessey
	1.0	6.4	2.4	0.42	Plessey
10	5.5	3.0	12	0.46	Mitsubishi
	4.2	4.3	6.4	0.65	TI
	3.9	6.0	4.8	0.8	TI
	1.2	5.3	2.4	0.5	RCA
	1.2	4.0	2.8	0.43	Westinghouse
12	2.1	4.0	3.6	0.58	TI
14	0.6	4.0	3.0	0.2	NEC
15	1.1	4.0	1.2	0.92	TI
	1.0	3.0	2.4	0.42	Plessey
	1.9	3.8	4.8	0.4	Mitsubishi
	2.5	3.0	4.8	0.52	MSC
18	1.0	3.0	2.4	0.42	MSC
	0.85	4.0	1.2	0.71	TI

Obviously greater output powers can be achieved by combining FET chips using internally matched structures as discussed in Chapter 5 or by power combining individual power amplifiers. However the frequency performance of power FETs tends to be limited by the ability to either match or uniformly feed large transistors as the frequency increases.

Fig. 3.28 indicates that the maximum output power shows a dependence on frequency of the form

$$P_{OUT} \propto 1/f^2$$

where f is frequency.

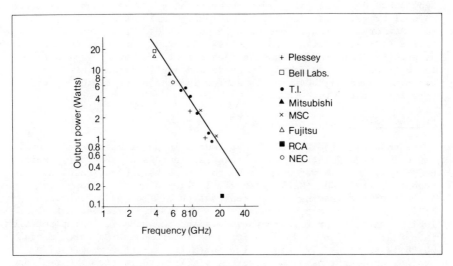

FIG. 3.28. Output Power of GaAs FETs as a Function of
Frequency (1980)

TABLE 3.6.   Power Output Versus Total Gate Width for Power FETs
Operated at 24V Source-Drain Bias (Di Lorenzo et al, 1979)

Gate Width (mm)	Output Power at 3 dB gain (watts)	Power/mm at 3 dB gain (watts/mm)	Gain at 106 mW/mm Input (dB)	Output power at 106 mW/mm Input (watts)
4	4.2	1.05	9.4	3.6
6	5.4	0.9	8.7	4.7
8	7.2	0.9	8.6	6.2
16	13.5	0.84	8.1	10.7

In Table 3.6 the results of a detailed study of the effect of over-all gate width done by Wemple et al (1978) are shown. This table indicates the power/mm gate width, the gain at approximately 100 mW/mm input power and the corresponding output power at that gain. It may be seen that the degradation in gain is 1.3 dB when comparing the 16 mm and 4 mm total gate width devices whilst the output power/mm gate width for 3 dB gain is reduced by approximately 1 dB. This dependence on total gate width observed at 4 GHz has also been noted on smaller gate width devices at 10 GHz (Wemple et al, 1977). The reasons for the gradual decrease in power added efficiency with device size at a given frequency is undoubtedly due to a combination of effects including the common lead effect, the inability to feed large transistors uniformly in phase and amplitude and the effect of circuit losses on the very low device impedances presented by the very large gate width FETs.

8. CONCLUSIONS

The understanding of voltage breakdown in power FETs is continually being improved resulting in substantial increases in output powers being achieved. Power outputs up to 250 mW have been achieved recently, at 22 GHz (Macksey et al, 1981) and new cell combining techniques are leading to the realisation of good power combining efficiencies at frequencies well into J-band (Schellenberg et al, 1981). Material profile improvements and the use of ion implantation into GaAs are also leading to improved intermodulation performance which is particularly encouraging for multicarrier communication systems applications. The power GaAs FET is now a firmly established member of the microwave solid-state device world and is now being integrated with other devices into GaAs monolithic microwave circuits (see Chapter 10).

9. BIBLIOGRAPHY

Aono, Y., Higashisaka, A., Ogawa, T. and Masegawa, F. X and Ku-band performance of submicron gate GaAs power FETs. Japan J. Appl. Phys., Vol. 17, Supplement 17-1, pp.147-152, 1979.

D'Asaro, L.A., Di Lorenzo, J.V. and Fukui, H. Improved performance of GaAs microwave field effect transistors with via-connections through the substrate. Int. Electron. Devices Meeting Tech. Digest, pp.370-371, December 1977.

Di Lorenzo, J.V. and Wisseman, W.R. GaAs Power MESFETs: design, fabrication and performance. IEEE Trans. on Microwave Theory and Techniques, Vol. MTT-27, No. 5, May 1979, pp.367-378.

Drukier, I., Camisa, R., Jolly, S., Huang, H. and Narayan, W. Medium power GaAs field effect transistors. Electronics Letters, Vol. 11, pp.104-104, March 1975.

Drukier, I., Wade, P.C. and Thompson, J.W. A high power 15 GHz GaAs

FET. Proceedings of the 1979 European Microwave Conference, Brighton, England, pp.282-286.

Engelmann, R.W.H. and Liechti, C.A. Gunn domain formation in the saturated region of the GaAs MESFET. IEEE Int. Electron Devices Conf. Digest, 1976, pp.351-354.

Fukata, M., Suyama, K., Suzuki, H. and Ishikawa, H. GaAs microwave power FET. IEEE Trans. Electron Devices, Vol. ED-23, pp.388-394, April 1976.

Fukata, M., Suyama, K., Suzuki, H., Nakayama, Y. and Ishikawa, H. Power GaAs MESFET with high drain-source breakdown voltage. IEEE Trans. Microwave Theory and Techniques, Vol. MTT-24, pp.312-317, June 1976.

Furutsuka, T., Tsuji, T. and Hasegawa, F. Improvement of the drain breakdown voltage of GaAs power MESFETs by a simple recess structure. IEEE Trans. Electron Devices, Vol. ED-25, No. 6, June 1978, pp.563-567.

Higashisaka, A., Takayama, Y. and Hasegawa, F. A high-power GaAs MESFET with an experimentally optimized pattern. IEEE Trans. on Electron Devices, Vol. ED-27, No. 6, June 1980, pp.1023-1029.

Higgins, J.A. Intermodulation distortion in GaAs FETs. IEEE MTT-S International Microwave Symposium Digest, 1978, Ottawa, Canada, pp.138-141.

Higgins, J.A. and Kuvas, R.L. Analysis and Improvement of Intermodulation distortion in GaAs power FETs. IEEE Trans. on Microwave Theory and Techniques, Vol. MTT-28, No. 1, January 1980, pp.9-17.

Hughes, A.J. private communication, 1981.

Ladbrooke, P.H. and Martin, A.L. Material and structure factors affecting the large-signal operation of GaAs MESFETs. Semi-insulating III-V Materials, Nottingham 1980, Shiva Publishing 1980, pp.313-320.

Lehovec, K. and Zuleeg, R. Voltage-current characteristics of GaAs J-FETs in the hot electron region. IEE Solid State Electron. Vol. 13, pp.1415-1429, 1970.

Macksey, H.M., Adams, R.L., McQuiddy, D.N., Shaw, D.W. and Wisseman, W.R. Dependence of GaAs power MESFET Microwave performance on device and material parameters. IEEE Trans. Electron Devices, Vol. ED-24, pp.113-122, Feb. 1977.

Macksey, H.M., Blocker, T.G. and Doerbeck, F.H. GaAs power FETs with electron beam defined gates. Electronics Letters, Vol. 13, p.312, May 1977.

Macksey, H.M., Blocker, T.G., Doerbeck, R.H. and Wisseman, W.R. Optimisation of GaAs power FET performance. 1978 Workshop on Compound Semiconductor Microwave Materials and Devices (WOCESEMMAD) Feb. 1978.

Macksey, H.M., Tserng, H.Q. and Nelson, S.R. GaAs power FET for K-band operation. ISSCC 1981 Digest of Technical Papers, New York, Feb. 1981, pp.70-71.

Mitsui, Y., Otsubo, M., Ishii, T., Mitsui, S. and Shirahata, K. Flip-chip mounted GaAs power FET with improved performance in X to Ku band. Proceedings of the 1979 European Microwave Conference, Brighton, England, pp.272-276.

Pucel, R.A., Hans, H.A. and Statz, H. Signal and noise properties of gallium arsenide microwave field effect transistors. Advances Electron and Electron. Physics, Vol. 38, New York: Academic Press, 1975, pp.195-265.

Shockley, W. A unipolar field effect transistor. Proc. IRE, Vol. 40, pp.1365-1376, 1952.

Shur, M.S. and Eastman, L.F. Current voltage characteristics, small signal parameters and switching times of GaAs FETs. IEEE Trans. Electron Devices, Vol. ED-25, pp.606-611, June 1978.

Shur, M.S. Analytical model of GaAs MESFETs. IEEE Trans. Electron Devices, Vol. ED-25, pp.612-618, June 1978.

Schellenberg, J.M. and Yamasaki, H. An FET chip-level cell combiner. ISSCC 1981 Digest of Technical Papers, New York, Feb. 1981, pp.76-77.

Tucker, R.S. and Rauscher, C. Intermodulation distortion properties of GaAs FETs. Electronics Letters, Vol. 13, p.509, 1977.

Tiwari, S., Woodard, D.W. and Eastman, L.F. Domain formation in MESFETs - Effect of device structure and materials parameters. Proceedings of the 7th Biennial Cornell Electrical Engineering Conference, Cornell University, New York, 1979, pp.237-248.

Wemple, S.H. and Nichaus, W.C. Source-drain burn-out in GaAs MESFETs Gallium Arsenide and Related Compounds (St. Louis), 1976, Institute of Phys. Conf. Series No. 33b, pp.262-270.

Wemple, S.H. Nierhaus, W.C., Schlosser, W.O., Di Lorenzo, J.V. and Cox, H.M. Performance of GaAs power MESFETs. Electron Lett. Vol. 14, pp.104-105, March 1975.

Wemple, S.H., Steinberger, M.L. and Schlosser, W.O. Relationship between power added efficiency and gate-drain avalanche in GaAs MESFETs. Electronics Letters, Vol. 16, No. 12, June 1980, pp.459-460.

White, P.M., private communication, 1980.

Williams, R.E. and Shaw, D.W. Graded channel FETs: improved linearity and noise figure. IEEE Trans. Electron Devices, Vol. ED-25, p.600, 1978.

Willing, H.A. and De Santis, P. Modelling of Gunn domain effects in GaAs MESFETs. Electronics Letters, Vol. 13, No. 18, pp.537-539, 1977.

Willing, H.A., Rauscher, C. and De Santis, P. A technique for predicting large signal performance of a GaAs MESFET. 1978 IEEE MTT-S International Microwave Symposium Digest, Ottawa, Canada, pp.132-134.

Wolf, P. Microwave properties of Schottky-barrier field effect transistors. IBM J. Research and Development, Vol. 14, pp.125-141, March 1970.

CHAPTER 4
Material Requirements and Fabrication of GaAs FETs

1. INTRODUCTION

The subject of material growth for GaAs FETs and the fabrication of those devices on the material thus prepared is far too extensive to be covered in detail in this book. Indeed the subject warrants a book in its own right. However, a review is given in this chapter of the most popular methods of material growth. These are namely, vapour phase epitaxy, liquid phase epitaxy, ion implantation and molecular beam epitaxy. Fabrication of MESFETs can be divided broadly into two techniques – firstly using recessed channels together with gate lift-off technology and secondly the so-called self-aligned gate technology. Such fabrication techniques can be adopted for both mesa and planar technologies depending on whether the FETs are isolated from the bulk semi-insulating (SI) GaAs by an etched step or by an isolation implant respectively.

2. MATERIAL REQUIREMENTS

The four material technologies that have been used to produce high quality epitaxial material are: chemical vapour deposition (CVD) based either on chloride chemistry (Knight et al, 1965) or organo-metallic chemistry (Manasevit et al, 1969), molecular beam epitaxy (Cho et al, 1977), liquid phase epitaxy (Rosztoczy et al, 1974) and ion implantation (Stephens et al, 1978). Well over 800 papers have been written on the subject of GaAs and InP growth since 1960, with a sharp increase in the interest in ion implantation since 1975 due to the increased emphasis in the use of this method of producing active regions for IC exploitation.

A. EPITAXIAL LAYERS

Considering the diagram of the GaAs FET in Fig. 4.1, a GaAs buffer layer is first grown onto the GaAs semi-insulating (SI) substrate. This layer has a high resistivity (approximately 10^7 ohm cm) and contains very few mobile electrons. Consequently it acts effectively as an extension of the semi-insulating substrate but protects the subsequently grown active layer from any deleterious effects due to the bulk substrate which would otherwise occur.

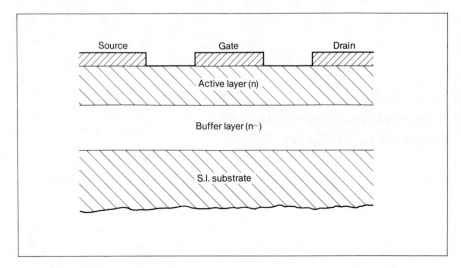

FIG. 4.1. Schematic of Buffer Layer Approach to MESFET Fabrication

The buffer layer and active layer are usually grown in a continuous
epitaxial growth run. The incorporation of the buffer improves the
sharpness of the carrier density profile and helps to maintain a high
electron mobility right up to the layer–substrate interface. Fig.
4.2(a) shows a typical buffered epitaxial n–type layer having a peak
carrier concentration of approximately 6×10^{16} cm^{-3} and shows that
the carrier concentration drops to below 10^{15} cm^{-3} 0.25μm further
into the material than the epitaxial layer 'knee' which occurs at a
depth of approximately 0.55μm. Fig. 4.2b shows the marked difference
that can be obtained between the mobility profile of a non–buffered
and buffered layer. These profiles were obtained by measuring
successively etched away surfaces using a Van der Pauw Hall mobility
technique (Van der Pauw, 1958). Such buffer layers markedly improve
the noise figure, gain and other properties of FETs.

There are two crucial parameters relevant to the active layer – the
mobility of the active layer as a function of depth as well as the
uniformity of the doping and thickness across the wafer. These factors affect the transconductance of the FET device to be fabricated
on the active layer particularly as the drain current is reduced by
applying negative gate voltage. This is because as negative gate
voltage is increased the depletion region starts to extend beyond the
knee of the level carrier concentration. If the 'tail' of the
profile, i.e. the region in which the carrier level is decreasing is
poor, i.e. the rate of change of carrier level with depth is low,
the transconductance of the device will be correspondingly low. Also,
if this 'tail' characteristic is not constant over the wafer then
the g_m will change from device to device. It is also desirable to

maintain the steepness of the tail right down to the SI background level such that the g_m of the device is maintained at a high level under bias currents considerably less than I_{DSS} (typically at low noise bias). The transconductance of the device has a strong influence on both the noise figure and gain of the FET, as we have seen in Chapter 2, when the device is biased for minimum noise figure.

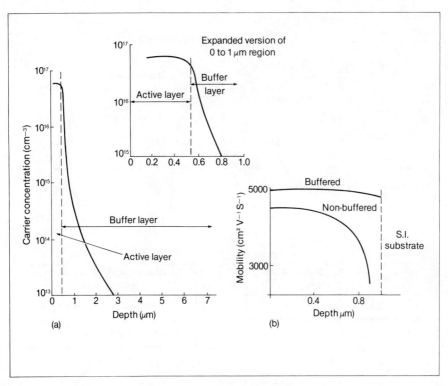

FIG. 4.2.(a). Doping Profile of FET Structure
      (b). Mobility Profiles of GaAs Epitaxial Layers (after Butlin et al)

Fig. 4.3 shows the transfer characteristics of two FET devices (Butlin et al, 1976) where device A has a buffer layer and the transconductance is maintained at a high level to very low current values. Device B has no buffer layer and it can be seen how the transconductance drops as the drain current is decreased. The noise figure of device A was 2.5 dB at 8 GHz whilst the noise figure of device B was 3.3 dB at the same frequency. The uniformity of active layer doping level and the layer thickness will have a marked effect on the yield of the wafer when processed for FETs since these parameters control the spread in drain saturation current I_{DSS}, pinch-off voltage, V_P,

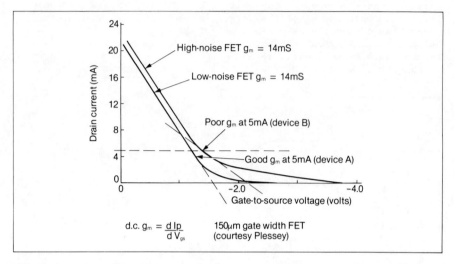

FIG. 4.3. Transfer Characteristics of High and Low Noise Figure GaAs FETs

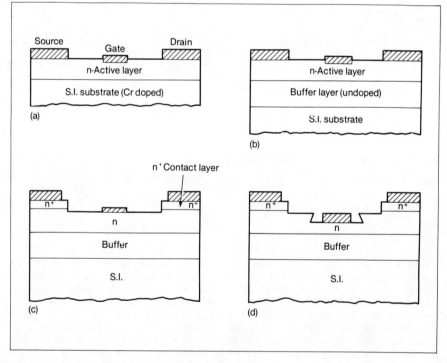

FIG. 4.4. Evolution of GaAs FET Structures

as well as the microwave parameters of the device.

Fig. 4.4. shows, the evolution of material structures used for FETs in various companies worldwide using the vapour phase epitaxy technique. Initial structures were grown on n-type active layers on chromium (Cr) doped substrates (Fig. 4.4a). Noise performance of these devices was poor and soon a buffer layer was introduced (Fig. 4.4b) to lower noise figure and increase gain. Even lower noise figures and higher gains were produced by introducing an n^+ contact layer (Fig. 4.4c) which apart from improving ohmic contact resistance also gives a considerable decrease in source to gate and gate to drain resistance with recessed gate FETs (Butlin et al, 1976) (Fig. 4.4d). It was later discovered that the undoped high-resistivity buffer layers being used were high resistivity due to out-diffusion of unidentified acceptors from the substrate (Cox et al, 1976). Fig. 4.5 shows the evolution of the profiles used to produce low noise FETs since the introduction of the MESFET some 20 years ago (Di Lorenzo, 1977).

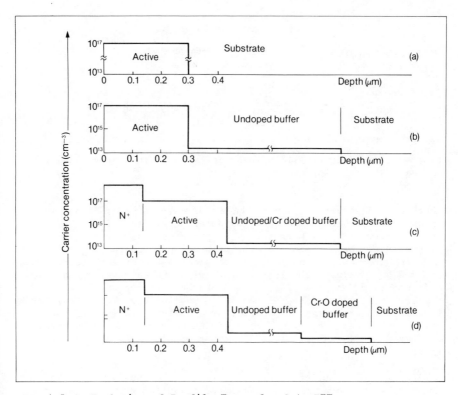

FIG. 4.5.   Evolution of Profile Types for GaAs FETs

Fig. 4.5(a) shows the basic device profile the performance of which
is aided by the introduction of buffer layers (Fig. 4.5(b) and (c)).

Referring again to Fig. 4.2(a) the background level of the layer
drops below 10^{13} cm^{-3} which allows the growth of a high resistivity
buffer as required. If the thickness of the buffer is increased by
extending the buffer layer growth time, the profile of Fig. 4.6
results where the majority of the buffer layer has the correct resistivity but that closest to the active layer has a higher carrier
level of approximately 8 x 10^{14} cm^{-3}.

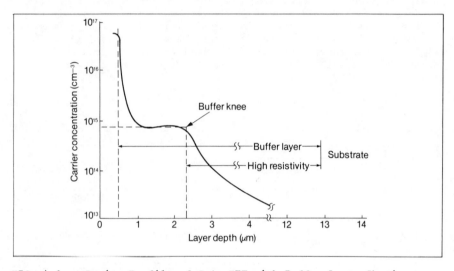

FIG. 4.6. Doping Profile of GaAs FET with Buffer Layer Showing
Knee (After Di Lorenzo)

The high resistivity region is due to out-diffusion of acceptors from
the substrate which compensate the donor population of the buffer in
the vicinity of the substrate. Thus for a good high-resistivity
buffer layer the acceptor level should be approximately equal to the
uncompensated background level at the active-buffer interface. Thus
undoped buffer layer quality can vary with the GaAs ingot used and
even within a GaAs ingot. Thus undoped buffer properties are not
reproducible without careful condition monitoring. Tuck et al (1979)
have reported a series of radiotracer experiments which suggested
that Cr was the acceptor and more recent secondary-ion mass spectrometry results (Evans et al, 1979; Huber et al, 1979) have confirmed
this.

Although doped buffer layers suffer from the disadvantage that the
dopant that is added to the system during buffer growth is an undesirable impurity if it reaches the active layer, the advantage is that

variable substrate effects can be dramatically reduced (Fairman et al 1979). Chromium doping of epitaxial GaAs to form SI layers was first reported by Mizumo, Kikuchi and Seki (1971). Resistivities in excess of 10^8 Ω cm were obtained. Kato et al (1979) reported results with MESFETs fabricated on Cr doped GaAs having similar performances to those made using undoped buffers.

Drukier et al (1975) have also reported improvements in gain and power output of MESFETs when Cr-doped buffers were used. Fairman (1979) has recently produced semi-insulating layers using a halide transport technique with final sheet resistances for the buffers in excess of 10^9 ohms per square exhibiting excellent thermal stability.

The effect of a Cr doped buffer layer of the correct thickness can be compared to undoped buffer layers by reverse biasing the substrate with respect to the buffer layer. This can be achieved by using a mercury probe Schottky contact (Bonnet et al, 1980). The mercury contact (Fig. 4.7) is reverse biased with respect to the mercury ring.

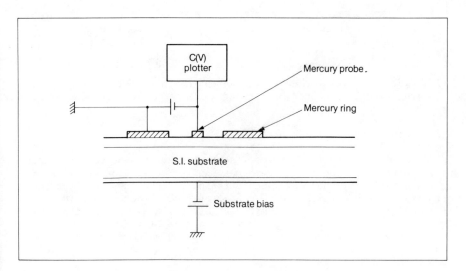

FIG. 4.7. Reverse Biased Substrate C(V) Measurement

The carrier concentration is measured at 0 volts and -25 volts, for example, and for the undoped layer where a space charge develops at the interface between the active layer and the buffer the carrier profile changes (Fig. 4.8(a)). It may be seen from Fig. 4.8(b) that the effect is reduced considerably for a 4μm thick chromium-doped buffer layer. The main advantage of the chromium-doped buffer layer is to suppress substrate effects provided the buffer is greater than 3μm thick. Referring to Fig. 4.5(d) the latest technique reported by

Di Lorenzo et al (1979) is to also incorporate an undoped buffer layer between the Cr doped buffer and the active layer to isolate the active layer from Cr traps.

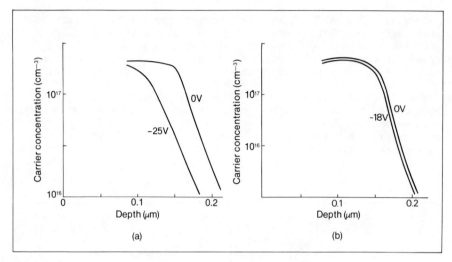

FIG. 4.8.   Carrier Concentration Versus Depth on FET Structures with (a) 0.5μm Undoped Buffer and (b) 4μm Chrome Doped Buffer

(i) Vapour Phase Epitaxy

Epitaxy refers to the growth of a material on a substrate which in this particular case is the same material. Preparation of films of solid material by chemical vapour transport techniques involves the transfer of the material to be deposited from a source zone in the form of gaseous species of limited stability into a deposition zone where these gaseous products are made to react (by controlling the temperature of the zone) to form the compound which is deposited on the desired substrate. Various techniques have been developed which may be divided into:-

1. The use of GaAs as the source with some transport agent, and

2. The use of Ga and/or As with other elements and a carrier gas.

Chloride Transport Process. In the chloride transport technique GaAs is made to react with HCl gas at a temperature of approximately 800°C in a closed quartz tube so that gallium monochloride and elemental arsenic are produced, thus:

$$4 \text{ GaAs} + 4 \text{ HCl} \rightarrow 4 \text{ GaCl} + \text{As}_4 + 2\text{H}_2$$

These products are transported along a temperature gradient in a furnace and GaAs is deposited at the cool end, thus

$$6GaCl + As_4 \ \rightleftarrows \ 4GaAs + 2GaCl_3$$

Moest and Shupp (1962) have deposited epitaxial GaAs using such a process with film thicknesses up to 2μm with thickness variation less than 10%.

The Arsenic Trichloride Process

In the $AsCl_3$-Ga-H_2 open flow system a reaction first takes place between $AsCl_3$ and H_2 being accomplished at temperatures greater than $425^{\circ}C$ so that HCl gas is formed, thus:

$$4AsCl_3 + 6H_2 \rightarrow 12HCl + As_4$$

HCl is made to react with a gallium source at approximately $800^{\circ}C$ resulting in GaCl which is transported to the deposition zone where GaCl and arsenic react to form GaAs, thus

$$2Ga + 2HCl \rightarrow 2GaCl + H_2 \quad \text{(source zone)}$$

$$6GaCl + As_4 \ \rightleftarrows \ 4GaAs + 2GaCl_3 \quad \text{(deposition zone)}$$

Knight et al (1965) used an open tube flow system schematically shown in Fig. 4.9 and this system has become very popular (Eddolls,

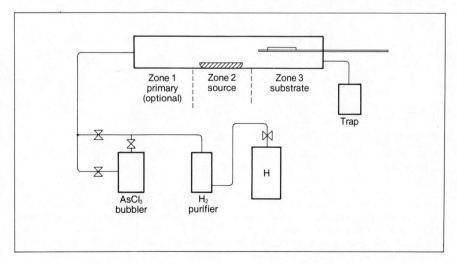

FIG. 4.9. Typical Arrangement for a Single Bubbler $AsCl_3$-Ga-H_2 Open Flow VPE System

1966; Bobb et al, 1966; Wolfe et al, 1970; Cairns et al 1968; Nozaki et al, 1974). The equipment is usually made from high purity quartz (for high temperature zones) and pyrex for the auxiliary systems.

High purity material is readily grown with such a system where the background level of undoped layers grown (e.g. for buffer layers) is found to decrease with increasing $AsCl_3$ molecular fraction (Cairns et al, 1968).

Thermodynamic calculations of this type of transport system indicate that the most probable source of background contamination of undoped layers is Si from the quartz walls and other hardware transported as chlorosilanes.

As may be appreciated from Fig. 4.9 the $AsCl_3$ gas is passed through a single bubbler but considerable versatility can be produced by introducing a second $AsCl_3$ bubbler (Cox et al, 1976; Hewitt et al, 1976). A schematic diagram of this system is shown in Fig. 4.10.

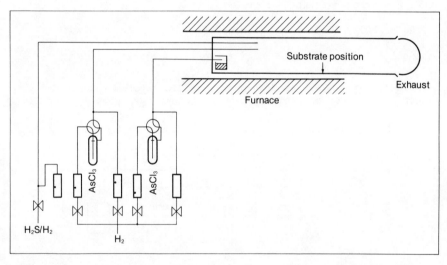

FIG. 4.10. Typical Arrangement for a Two-Bubbler $AsCl_3$-Ga-H_2 VPE System

$AsCl_3/H2$ from the main bubbler passes over the gallium source. The flow from the second bubbler is introduced into the reactor between the source and the substrate at a high flow rate to produce an in-situ etch of the substrate prior to epitaxy thus improving the interface. This flow is also used at a slower rate during the growth of the undoped buffer. H_2S/H_2 mixture is admitted to the reactor for doping of the active and n^+ contact layer if the latter is required. Flow rates in production equipments, for high material uniformity, are controlled by mass flow controllers and the control of gas valves, furnace temperatures etc. is microprocessor controlled. Such actions have led, for example, to large substrates of 40 cm^2 being grown with a standard deviation of 1% or less being obtained for the saturated drain current of processed FETs (Komeno et al, 1979).

The Metal Alkyl-Hydride or Organometallic Technique

Over the past few years a number of workers (Ito et al, 1973; Hallais et al, 1978; Rai-Choudbury, 1969) have developed an alternative vapour system using metal alkyl vapours and the group V hydrides for many III-V compounds. The basic reaction is an irreversible pyrolysis which takes place on a heated substrate, thus:

$$Ga(CH_3)_3 + AsH_3 \rightarrow GaAs + 3CH_4$$

This is the analogue of the silane reaction used to produce epitaxial silicon.

The alkyl system has a number of advantages over the chloride transport system. As all the reactants and dopants can be in the vapour phase and only the substrate is heated the equipment is much simpler and is therefore capable of faster turn around time. Because adding small amounts of gases in the gas stream can be readily achieved the GaAs stoichiometry can be varied. Since the system requires a lower growth temperature than the $AsCl_3$-Ga-H_2 system, dopant diffusion is lower and n to semi-insulating interfaces are sharper. GaAs is doped n-type by adding small amounts of hydrogen sulphide into the gas stream of the basic reactor shown in Fig. 4.11. Fig. 4.12 shows the doping level as a function of sulphur temperature used to generate the H_2S and the H_2S partial pressure for a number of GaAs layers (Bass 1975).

Morkoc (1979) has reported 1.5μm gate length FETs fabricated on organometallic grown layers with noise figures of 3 dB and associated gains of 5.5 dB at 8 GHz.

(ii) Liquid Phase Epitaxy

All the techniques discussed so far involved deposition of the compound from the vapour phase. Liquid phase epitaxy has also received considerable attention (Holger et al, 1966; Morkoc et al, 1979) and both undoped and Cr doped high resistivity layers have been grown using this technique. Low doped layers grown by LPE have high resistivity regions near the substrate, presumably caused by acceptor compensation as in VPE growth. These high resistivity regions have been successfully used for MESFET structures by Nanishi, Takahei and Kuroiwa (1978) and Kim et al (1979) who found that the active layer mobility and device performance improved with the use of buffer layers as with VPE.

110

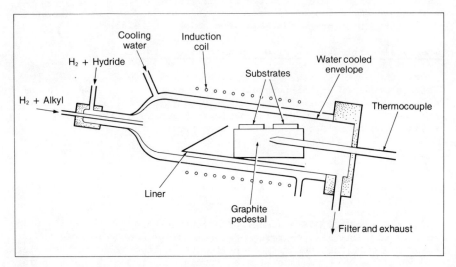

FIG. 4.11.   Reactor for the Epitaxial Growth of GaAs by the
Organometallic Process

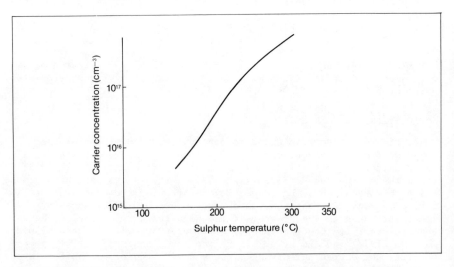

FIG. 4.12.   Dependence of n-type Doping on Sulphur Temperature

In LPE a saturated solution of GaAs in gallium is brought in contact with the substrate and allowed to cool slowly whence the GaAs is deposited on the substrate. A simple equipment is shown in Fig. 4.13.

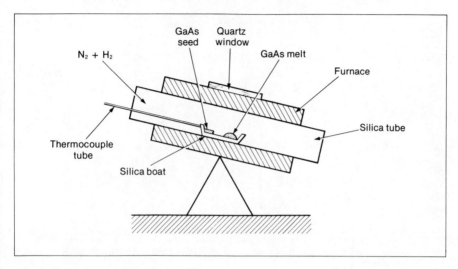

FIG. 4.13.   Typical Equipment for Liquid Phase Epitaxy of GaAs
   (After Bolger)

It consists of a silica furnace containing a silica boat which has a GaAs "seed" at one end and the gallium-GaAs mixture at the other. The Ga-GaAs mixture is brought to a temperature of 850°-900°C upon which GaAs dissolves in gallium and a saturated solution is formed. A N_2 + H_2 atmosphere inside the furnace is maintained to avoid oxidation of the Ga. The molten gallium-GaAs solution is made to run onto the GaAs substrate by tilting the furnace and the furnace temperature lowered to allow crystallization of the GaAs on the substrate. Although LPE is simple in principle it requires very critical control of the conditions to produce reproducible layers and for this reason it has not become popular as a production source of GaAs.

(iii) Molecular Beam Epitaxy

   The rapid development in microwave FETs fabricated on molecular beam epitaxial (MBE) material has come about because this method provides well controlled film thickness, steep doping profiles, good composition and an exceedingly smooth surface (Luscher, 1977).

   The basic MBE process achieves epitaxial growth in an ultra high vacuum (UHV) environment through the reaction of multiple molecular beams of differing flux density and chemistry with a heated single crystal substrate. The process is schematically illustrated in

Fig. 4.14 which shows the essential elements for MBE of GaAs.

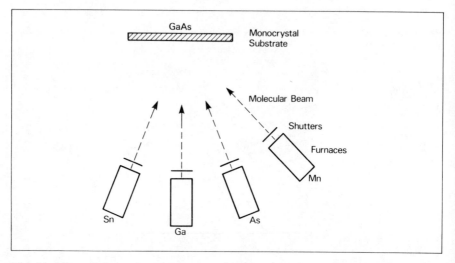

FIG. 4.14.   Schematic Diagram of a MBE System for the Growth of
   Doped GaAs

Each furnace contains a crucible containing one of the constituent
elements. For example, Sn is used as an n-type dopant whilst Mn is
used as a p-type dopant for FET buffer layers. The temperature of
each furnace is such that the vapour pressures of the materials are
high enough for free evaporation of thermal energy molecular beams to
take place. The furnaces are so arranged that the central portions
of the beam flux distributions intersect at the substrate. Choice of
furnace and substrate temperatures as well as control of the beams by
shutters between the furnaces and the substrate allow epitaxial films
of the desired chemical composition to be produced. Growth rate is
determined by the rate of arrival of the group III elements while the
condition of stoichiometry is satisfied simply by growing in an excess
flux of the group V elements.

One of the distinguishing characteristics of MBE is the low growth
rate - approximately 1μm per hour or one monolayer per second. The
molecular beam at the substrate can therefore be readily modulated in
monolayer quantities in times below one second. Because MBE takes
place in UHV 'in situ' analytical instrumentation is possible such as
mass spectrometry, Auger spectroscopy for surface analysis and secondary ion mass spectrometry.

Epitaxial films for both low noise (Bandy et al, 1979; Cho et al,
1976; Cho et al, 1977) and power FETs (Cho et al, 1977; Wataze et al,
1978) have been grown by MBE. Noise figures of 1.5 dB with 15 dB

associated gain at 8 GHz have been reported by Bandy et al where the
material was grown using Sn as the dopant. Wataze et al have reported power FETs having power outputs of over 4W with linear gains of
5.4 dB at 8 GHz using Si doped MBE epitaxial layers. Hysteresis is
often present in the d.c. characteristics of FETs grown by other techniques without a high resistivity buffer layer between the Cr doped
substrate and the channel. This hysteresis, attributed to the alter-
nate filling and emptying of interface states, is not observed in FETs
fabricated on MBE material indicative of a higher quality interface.
MBE has some important attractions for future microwave transistor
devices such as the GaAlAs FET and other structures requiring highly
doped n and p-type layers. For these reasons MBE is likely to play a
more important role in the future in ICs rather than in discrete FET
production and certainly there are no FET production facilities which
use MBE to grow layers, at the present time.

B. LIQUID ENCAPSULATED CZOCHRALSKI GROWTH

In 1962 a simplification in the growth of GaAs was made by the use
of liquid encapsulation techniques (Metz et al, 1962). In this method
the melt from which the crystal is pulled is immersed under an inert
encapsulant which confines the volatile constituents and simplifies
the equipment. GaAs is now grown routinely using the LEC method
(Caruso et al, 1972; Au Cain et al, 1979; Ware, 1979). The LEC
process essentially requires a Czochralski growth furnace which can
be pressurised to approximately 1 atmosphere with nitrogen used as
the ambient gas.

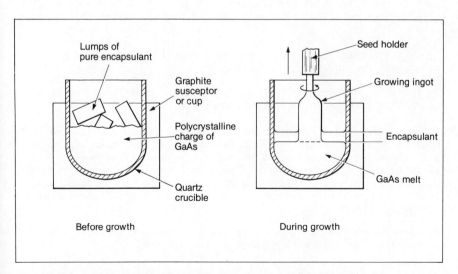

FIG. 4.15.   The Liquid Encapsulated Czochralski Process

The LEC process is represented in Fig. 4.15 where the crucible and susceptor assembly are charged with polycrystalline GaAs and encapsulant lumps prior to heating. Also shown is an illustration of the assembly during growth.

The fabrication of the crucibles and susceptors is important to the purity of the ingot. The encapsulant for the growing ingot must be inert, transparent, have a low melting point, be of high purity and be less dense than the melt. For these reasons boric oxide (B_2O_3) has become popular as an encapsulant.

After the charge is melted at $1240^{\circ}C$ a cooled 'seed' is lowered into the melt through the encapsulant and it is then slowly withdrawn at a carefully selected temperature at which nucleation and single crystal growth occur on the seed. Since the withdrawn seed and ingot are coated in B_2O_3 the evaporation of arsenic from the ingot is minimized.

C. ION IMPLANTATION

Because of the increasing use of GaAs FETs in microwave systems and the introduction of GaAs integrated circuits requiring large areas of uniform material, high yield and reproducible material preparation methods have become a primary concern. A major breakthrough has been accomplished in this respect by developing ion implantation into bulk-grown substrate material as a viable approach for preparing the very thin active layers.

Ion implantation is a process whereby controlled amounts of chosen foreign species can be introduced into the near surface regions of the GaAs in the form of an accelerated beam of ions to form active layers with defined profiles. The principal advantages of ion implantation are:

(i) The total amount and purity of material implanted can be accurately controlled and monitored.

(ii) The concentration of impurities as a function of depth can be controlled by means of the ions' energy.

(iii) Implanted ions enter the surface usually as a well directed beam so that a very high lateral definition of the doped region can be achieved using conventional lithographic masking techniques. This leads to the lack of any need to produce a mesa structure for FETs since either the SI substrate produces isolation or a proton, say, implant can be used for device to device isolation in GaAs integrated circuits.

(iv) The process is very versatile and a single implantation facility can be used with a range of ion species.

(v) The process can be controlled automatically with a large throughput rate giving uniformities determined by the host material.

The large energy transfer of the ions as they give up their energy
to the atoms of the target material occur as elastic recoils which
disrupt the crystalline structure where the distribution of damage
along the ion track varies considerably with the mass of the ion. A
region of disorder occurs but this can fortunately be restored by
thermal annealing.

In order to employ ion implantation into semi-insulating GaAs substrates for device fabrication it is important that the doping produced by implantation not be affected by changes in the substrate. This
leads to a need to qualify the substrate material. Some SI substrates
convert to either p-type or n-type following high temperature annealing even in the absence of implanted ions. One qualification test
consists of implanting the qualification sample with argon or krypton,
then annealing at, typically 800 to 900°C for 15 to 30 minutes. A
qualified SI substrate does not convert after this test.

One of the major problems of ion implantation into GaAs is that the
material begins to dissociate at the commonly used anneal temperatures
of 800°C to 1000°C. To prevent problems caused by dissociation, it is
common practice to use an encapsulant such as SiO_2 (Foyt et al, 1969),
Si_3N_4 (Harris et al, 1972), Al_2O_3 (Chu et al, 1973), AlN (Pashley et
al, 1975) or Al (Sealy et al, 1974). Capless annealing has also
become a popular method where an arsenic overpressure is maintained
to prevent decomposition of the GaAs.

Development of horizontal Bridgman and gradient freeze technology
for LED applications has not been found particularly advantageous for
the growth of SI GaAs for FETs. The resulting electrical yields for
ion implantation into Bridgman grown material is low. GaAs ingots
grown in the <100> direction by liquid encapsulated Czochralski
methods have shown considerable promise. Workers as early as 1965
(Mullins et al, 1965) have shown the virtues of the B_2O_3 encapsulation
technique. Weiner, Lassota and Schwartz (1971) were the first to show
the relationship between silicon contamination and the thermal stability of undoped semi-insulating GaAs. Recently Swiggard and Henry
(1977) and Rumsby (1979) have demonstrated the merits of the LEC
method. The dissociation of the volatile As from the GaAs melt, which
is contained in a crucible, is avoided by encapsulating the melt in an
inert molten layer of boric oxide (B_2O_3) and pressurising the chamber
with a non-reactive gas, such as nitrogen or argon, to counterbalance
the As dissociation pressure.

With the recent introduction of the Melbourn high pressure LEC
puller (manufactured by Metals Research, England) 'in'situ' compound
synthesis can be carried out from the elemental Ga and As components.
The attainment of low background doping levels has led to the growth
of highly uniform 'undoped' crystals of up to 75 mm diameter.

(i) Carrier Concentration and Mobility

Successful ion implantation has resulted from the use of ^{28}Si, ^{32}S,
Se and ^{29}Si for n-type layers. The choice of ion species depends to a

certain extent on the implant energy available and the fluence chosen.

Figs. 4.16 and 4.17 show, for example, the carrier concentration and mobility profiles of two implanted samples, one using 200 keV ^{32}S and the other using 70 keV ^{28}Si at a much higher fluence (Lui 1980).

Post implant annealing was done at 825°C for 20 minutes in both cases using capless annealing. The profile of Fig. 4.16 shows a maximum carrier concentration of 2.4×10^{17} cm^{-3} with a mobility which varies from 3000 cm^2 V^{-1} s^{-1} at the surface to over 4000 cm^2 V^{-1} s^{-1} toward the SI substrate. For the higher dose example, the average mobility and peak carrier concentration of the wafer are 1800 cm^2 V^{-1} s^{-1} and 1.8×10^{18} cm^{-3}.

The Hall mobility of the implanted layer is given by

$$\mu_H = \frac{R_S}{\rho_S} \qquad \qquad 4.1$$

where R_S is the Hall coefficient and ρ_S is the sheet resistance. The Hall coefficient is given by

$$R_S = 10^8 \frac{\Delta V_{24}}{BI_{13}} \qquad \qquad 4.2$$

where I_{13} is the current, B is the magnetic flux density applied perpendicularly to the surface of the sample and ΔV_{24} is the voltage change with and without the magnetic field. The subscripts correspond to the four ohmic contacts (Van der Pauw, 1958). The sheet resistance, ρ_S is given by

$$\rho_S = \frac{\pi}{2\ln 2} \left(\frac{V_{34}}{I_{12}} + \frac{V_{23}}{I_{41}}\right) F \qquad \qquad 4.3$$

where F is a geometrical correction factor. Using Equation 4.1 the sheet carrier concentration N_S is given by

$$N_S = \frac{1}{qR_S} \qquad \qquad 4.4$$

where q is the electronic charge (1.6×10^{-19} coulomb). The activation efficiency for an implanted and annealed sample is given by

$$\eta = \frac{N_S}{N_S'}, \qquad \qquad 4.5$$

where N_S' is the fluence used.

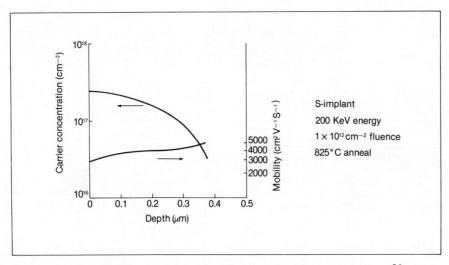

FIG. 4.16. Carrier Concentration and Mobility Profile of a ^{32}S
Implanted Sample (after Liu et al)

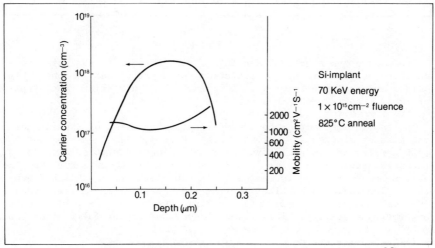

FIG. 4.17. Carrier Concentration and Mobility Profile of a ^{28}Si
Implanted Sample (after Liu et al)

Carrier concentration can be approximately determined using

$$N_m = \frac{N_S}{\sqrt{2\pi}\ \Delta R_p} \qquad\qquad 4.6$$

where ΔR_p is the standard deviation of the projected range. Table 4.1 shows for example the range and standard deviation on that range for ^{32}S and ^{28}Si ions implanted into GaAs at particular energies.

TABLE 4.1. Projected Range and Straggle Statistics for ^{32}S and ^{28}Si Ions Implanted into GaAs

^{32}S into GaAs			^{28}Si into GaAs		
Energy (KeV)	Projected Range, Rp (μm)	Standard Deviation ΔR_p (μm)	Energy (KeV)	Projected Range, Rp (μm)	Standard Deviation, ΔR_p (μm)
10	0.0102	0.0078	10	0.0111	0.0105
50	0.0395	0.0194	50	0.0441	0.0221
100	0.0758	0.0337	100	0.0861	0.0383
200	0.1515	0.0578	200	0.1732	0.0646
300	0.2279	0.0778	300	0.2582	0.0855
400	0.3029	0.0942	400	0.3421	0.1024
800	0.5919	0.1423	800	0.6549	0.1488
1000	0.7219	0.1580	1000	0.7941	0.1638

From the above equations and data it is possible to evaluate the mobility versus carrier concentration for a particular implant. Fig. 4.18 shows the result for a ^{28}Si implant where a carrier concentration of 10^{17}/cc corresponds to a mobility of 4500 cm^2 V^{-1} s^{-1}.

The equivalent carrier concentration as a function of implant dose can also be plotted at a given energy (Fig. 4.19). It may be seen that carrier concentration varies almost linearly with fluence for fluences between 2.5×10^{12} and 2×10^{13} for Si implants and that the carrier concentration increases at a much slower rate at high fluences. Also there is no electrical activation at fluences lower than 10^{12} cm^{-2} for 200 keV implants the limit depending to a certain extent on substrate quality. This cut-off behaviour is not fully understood.

(ii) Cr-Redistribution

Conventional (i.e. Si$_3$N$_4$ or SiO$_2$) as well as capless annealing techniques can induce Cr depletion in the region near the GaAs surface.

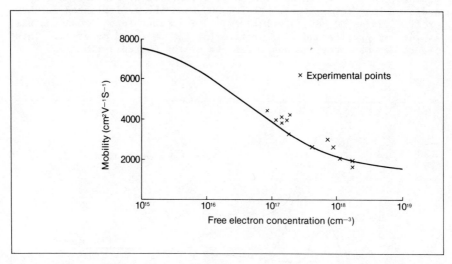

FIG. 4.18. Mobility Versus Carrier Concentration for Si Implanted
into GaAs

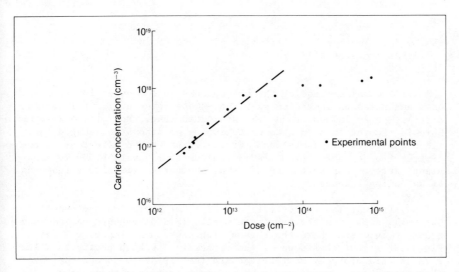

FIG. 4.19. Carrier Concentration as a Function of Implantation Dose
for Si Implantation into GaAs

In this depleted region the Cr concentration may drop below the residual donor level allowing them to contribute to the electrical profile of the implant. Secondary ion mass spectrometry (SIMS) has been used to determine the atomic redistribution of sulphur and chromium, for example, in ^{32}S implanted SI GaAs as a function of fluence, energy and annealing temperature. Fig. 4.20 shows the as-implanted sulphur profile for a 300 keV, 4 x 10^{15} cm^{-2} ^{32}S implant into SI GaAs as well as the ^{32}S profile and ^{52}Cr profile for the annealed material. This figure demonstrates the basic results of the Cr redistribution.

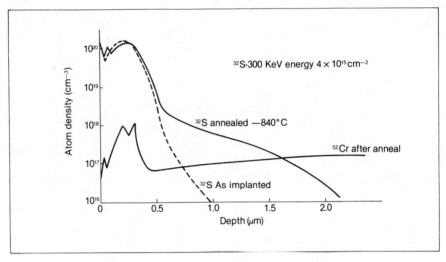

FIG. 4.20. Thermal Redistribution of S and Cr in SI GaAs
(after Evans et al)

After thermal processing the Cr depletes in the outer 2μm of the GaAs and piles up in the encapsulant - GaAs interface. Cr shows two distinct peaks in and around the projected range, R_p, of the S implant. This is thought to be due to Cr taking up positions in damaged lattices due to the implantation. Fig. 4.21 shows the effect of Cr redistribution as a function of ^{32}S fluence at a constant 300 keV implant energy. Concentration of Cr at the surface is also a function of annealing temperature.

Thus, the carrier concentration profile of implanted and annealed GaAs will depend on the substrate used. This dependence can be very dramatic in cases where there is a high background impurity level. Fig. 4.22 illustrates some of the effects. The high purity bulk substrates were grown by Hewlett Packard (For, private communication). This material had very low impurity concentration with no Cr added to produce the semi-insulating properties. Profiles obtained using high

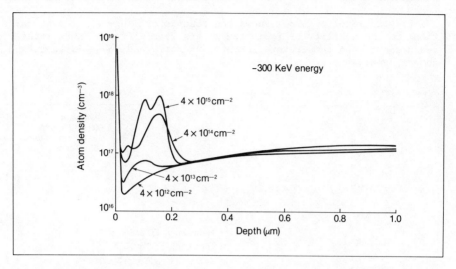

FIG. 4.21.   Thermal Redistribution of Cr as a Function of ^{32}S
Implantation Fluence

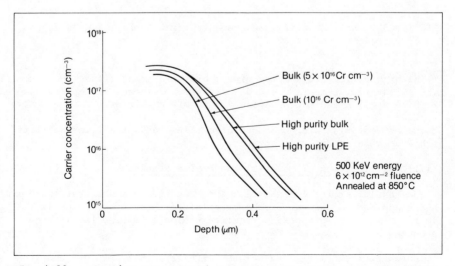

FIG. 4.22.   Carrier Concentration Profiles for Various Se Implanted
GaAs Samples

purity LPE substrates are seen to be very similar. However, ingots containing the Cr content shown exhibit a marked difference the effect being greater than that predicted by the level of the Cr concentration due to the Cr pile-up in the implant damage region as explained above.

Fig. 4.23 (Stolte, 1980) shows the results of long-term investigations on the mobility of implanted wafers from different substrates measured at room temperature and at 77°K for standard Se implants and anneal conditions.

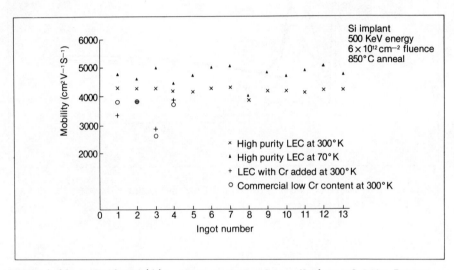

FIG. 4.23.   Hall Mobility Measurements for a Number of GaAs Ingots of Different Types

The high purity LPE samples give mobilities greater than 4000 cm^2 V^{-1} s^{-1} at room temperature and a significant increase when the sample is held at 77°K indicating a low amount of impurities.

The LEC material with Cr added also illustrates the lower mobility achieved with Cr doped substrates.

(iii) Qualification of Material and Device Results

The results of ion implantation into undoped GaAs substrates are now being applied to the use of the material for discrete FET and IC fabrication. The properties required of ion implanted layers are produced by:

1. High purity starting substrates with less than 5×10^{15} cm^{-3} donor and acceptor impurities.

2. Freedom from harmful crystalline defects such as inclusions,

precipitates and segregates.

3. Resistivity greater than 10^7 ohm cm or sheet resistance greater than 10^7 ohms per square.

4. Thermal stability during an 850°C annealing process usually with a Si_3N_4 cap or using an arsenic overpressure capless anneal.

5. The achievement of high activation efficiency of the implanted species.

6. The electrical compensation by deep donors, acceptors or both.

Semi-insulating GaAs is usually qualified by:-

1. Thermal annealing where a sheet resistance greater than 10^7 ohm per square is required following an anneal at 850°C.

2. Ensuring active ion implantation at a 300 keV, 3×10^{12} cm^{-2} Se dose, for example where

   a). the implanted layer shows an activation greater than 80% of the dose.

   b). correlation with the range statistics used (e.g. Table 4.4)

and c). uniformity in the above properties over the length of the ingot (usually by taking slices from the two ends of the ingot).

As was stated earlier the prime reasons for exploiting ion implantation into GaAs is to produce high uniformity, large area wafers leading to high yield processing of discrete FETs and perhaps more importantly the newer GaAs ICs. In this respect the use of 'undoped' substrates produces by far the best results which are illustrated by the pinch-off voltage variations achieved, for example, by Rockwell (Zucca et al, 1980) shown in Fig. 4.24. These particular results, showing a standard deviation of 85 mV in a mean of 1.18V, indicate the potential of ion implantation since such a standard deviation indicates a variation of only 50Å in the depth at a particular electron concentration over the entire wafer.

Table 4.2 shows some of the results achieved at Plessey (Sanders et al, 1980) using selenium implants into chromium doped substrates of various manufacturers. The implants used produced single Gaussian profiles. Table 4.3 shows the same electrical measurements using silicon as the implant species where it may be noticed that the resultant sheet resistance is much more reproducible, Hall mobilities are some 10% higher than the Se implants and the activation is also consistently higher. Table 4.4 summarises some of the measurements of silicon implants into undoped substrates showing that the mobilities are consistently 10% higher than the corresponding silicon implants into chromium doped material. The reproducibility of activity is also good.

TABLE 4.2. Electrical Measurements of Selenium Implants into Various Cr Doped Ingots

Ingot Manufacturer	Se Implant		Peak Concentration (cm^{-3})	Hall Measurements		
	Dose (cm^{-2})	Energy (KeV)		Sheet Resistance $(\Omega/\square)$	Mobility $cm^2\ V^{-1}\ s^{-1}$	% Activity
Metals Res.	4×10^{12}	380	1.7×10^{17}	1298	3311	36
Metals Res.	4×10^{12}	380	1.4×10^{17}	764	3571	57
Sumitomo	5×10^{12}	380	2×10^{17}	337	4118	90
Metals Res.	6×10^{12}	380	2.5×10^{17}	498	3217	65
Metals Res.	8×10^{12}	380	2.4×10^{17}	295	3403	78
Metals Res.	8×10^{12}	380	4×10^{17}	426	3178	58

TABLE 4.3. Electrical Measurements of Silicon Implants into Various Cr Doped Ingots

Ingot Manufacturer	Si Implant		Peak Concentration (cm^{-3})	Hall Measurements		
	Dose (cm^{-2})	Energy (KeV)		Sheet Resistance $(\Omega/\square)$	Mobility $(cm^2\ V^{-1}\ s^{-1})$	% Activity
Monsanto	6×10^{12}	240	2×10^{17}	285	4055	90
Sumitomo	8×10^{12}	240	3×10^{17}	262	3319	90
Metals Res.	7×10^{12}	240	2.5×10^{17}	318	4042	69
Metals Res.	7×10^{12}	240	2.6×10^{17}	296	3936	77
Sumitomo	6×10^{12}	240	2.7×10^{17}	311	4066	82

TABLE 4.4. Electrical Measurements of Si Implants into Various Undoped Ingots

Ingot Manufacturer	Implant		Peak Concentration (cm^{-3})	Hall Measurements		
	Dose (cm^{-2})	Energy (KeV)		Sheet Resistivity $(\Omega/\square)$	Mobility $(cm^2\ V^{-1}\ s^{-1})$	% Activity
Metals Res.	6×10^{12}	240	2.3×10^{17}	336	4144	75
"	"	"	2.1×10^{17}	376	4310	64
"	"	"	2.1×10^{17}	447	4037	58
"	"	"	1.8×10^{17}	372	4323	65
"	"	"	1.6×10^{17}	358	4174	70

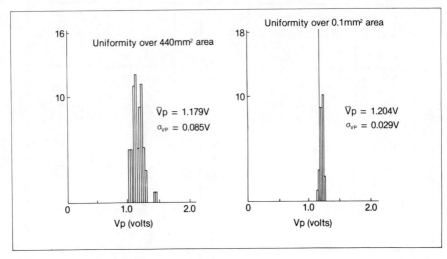

FIG. 4.24. Histograms of Pinch-off Voltages of GaAs FETs on Ion-Implanted Wafers

Many FET devices have been made using undoped semi-insulating substrates implanted with a single room temperature silicon implant of 6×10^{12} ions cm^{-2} at a 240 keV energy. Resulting peak carrier concentrations are in the 2×10^{17} cm^{-3} region with the profile peak occurring at approximately $0.2\mu m$ depth and sheet Hall mobility values in excess of 4000 cm^2 V^{-1} sec^{-1}.

Table 4.5 shows a summary of some of the saturated source-drain current measurements (I_{sat}) after source-drain delineation and before gate etching from six implantations taken from three different ingots. The means and standard deviation for I_{sat} from about 200 devices on each wafer over areas greater than 2 cm^2 are shown. The mean I_{sat} values represent a run-to-run variation of $\pm$ 10%. Standard deviations are between $\pm$ 5% and $\pm$ 15%. However over small areas of the order of 10 mm^2 standard deviations are as low as 2.5%.

Device parameter spread can also be assessed by the sample of Fig. 4.25 which shows histogram plots of I_{DSS} from $1\mu m$ gate length devices on ion implanted undoped and Cr doped substrates and from a VPE layer (the sample size of each histogram is about 100 devices).

The d.c. uniformity of FET devices is improving steadily as material characteristics are improved and substrate impurity effects are more fully understood.

Table 4.6 shows typical r.f. performance for devices fabricated on Si implanted substrates showing that they compare very favourably if not better than corresponding epitaxial FETs with the same geometry.

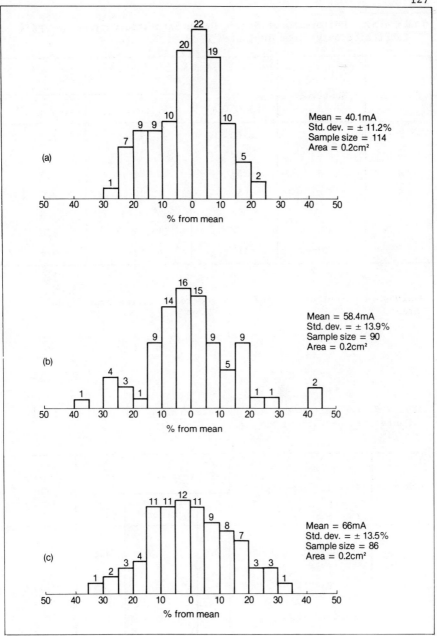

FIG. 4.25.   Histogram of I_{DSS} Measurements from Central Areas of (a) Si Implanted undoped, (b) Si Implanted Cr Doped and (c) VPE Wafers

TABLE 4.5.   Uniformity of Source Drain Saturated Current for FETs
Fabricated on Various Implanted Undoped Substrates

Run No.	Substrate	I_{SAT}		
		Mean (mA)	STD. Dev. (%)*	Area (cm^2)
1	A	60	15	3.7
2	A	58	10	2.5
3	A	69	8	2.7
4	A	65	10	2.0
5	B	62	6.5 (2.4)	3.5
6	C	82	11 (6)	3.4
7	Cr doped	112	8 (5)	0.6

*values in parenthesis refer to standard deviation over 10 mm^2
areas.

TABLE 4.6.   RF Performance of 1μm Gate Length, 300μm Gate Width FETs
Fabricated on Various Materials

Run No.	Material Description	RF Parameters (dB)			Measurement Frequency (GHz)
		MAG	Min. Noise Figure	Assoc. Gain	
1	Si implant, undoped substrate	7.4	2.9	4.6	14
2	Si implant, Cr doped substrate	9.3	2.6	5.4	12.75
3	Standard VPE	8.0	3.0	6.5	10

3. FET FABRICATION TECHNIQUES

The two techniques used for GaAs MESFET fabrication which have gained the most popularity are the self-aligned and the etched-channel technologies both of which can exploit either optical or electron beam lithography to define the gate stripe depending on the transistor's gate length. Electron beam lithography is used generally for gate lengths of less than 0.5μm.

A. SELF-ALIGNED GATE TECHNOLOGY

Fig. 4.26 shows the basic processing steps of the self-aligned gate technology. The first step is that the isolation mesa is defined by etching away the active n-layer until the semi-insulating substrate is reached. The gate metal, say aluminium is then evaporated over the active area as shown in Fig. 4.26(a).

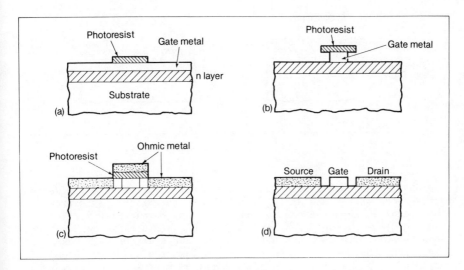

FIG. 4.26. Processing Steps of Self-Aligned GAte Technology

Source and drain areas are defined in photoresist and the exposed gate metal is removed by etching. Over etching is used to undercut the resist as shown in Fig. 4.26(b) to allow the necessary space between gate and drain and gate and source. Gold ohmic contact metallisation, usually In-Ge-Au or Au-Ge-Ni is then evaporated (Fig. 4.26(c)). The resist which is protecting the gate stripe is now covered with this ohmic metallisation but this is conveniently removed by 'floating off' the gold by dissolving away the resist. Thus the remaining thin gate is left situated between source and drain contacts as shown in Fig. 4.26(d).

B. RECESSED CHANNEL TECHNOLOGY

Rather than define the active channel thickness by the thickness of
the n-type epitaxial layer, a thicker layer is grown and the channel
region under the gate is defined by etching. This removes the high
tolerance in thickness required for the epitaxial layer when the
channel region is not etched. Most companies use a preferential etch
that gives a flat bottomed recess. Source and drain contacts are
deposited first as in Fig. 4.27(a) and the gate is defined in photoresist.

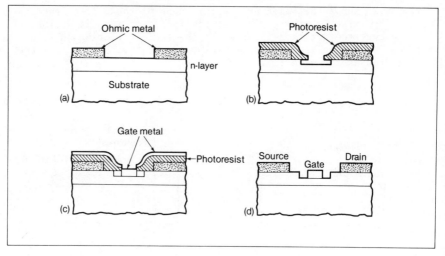

FIG. 4.27. Processing Steps of Etched-Channel Technology

A channel is etched in the GaAs until a specific current is measured
between the source and drain contacts (the so-called I_{SAT} current).
Gate metal is then evaporated and the excess metal (Fig. 4.27(c))
removed by using the 'float-off technique'. This basic method works
equally well with both photolithographic and electron beam resist
exposure techniques and gate lengths as low as 0.25µm have been
produced. Fig. 4.28 is a scanning electron micrograph of a 0.3µm
electron beam exposed FET fabricated using the recessed channel technique by Butlin et al (1978). The SEM clearly shows the channel where
it joins the mesa edge and the gate lying between the source and drain
contacts.

Adaptations of this basic technique are numerous (Vokes et al, 1977;
Ohata et al, 1980; Murai et al, 1977). For example, Murai et al have
reported a technique of intentional side etching of the gate metal to
produce a cross-sectional shape much as shown in Fig. 4.29 where the
effective gate length is 0.5µm but because of the mushroom shape the
gate metal resistance (which, as has been seen in Chapter 2, contributes significantly to the overall noise figure) is reduced.

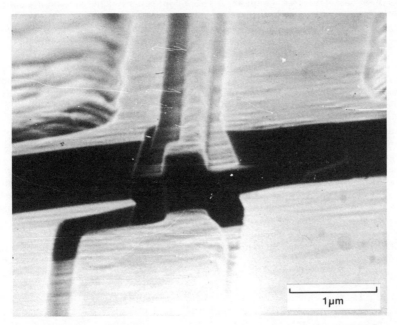

FIG. 4.28.   Photograph of 0.3µm GaAs FET Showing the Mesa Edge Area
(courtesy Plessey Co. Ltd.)

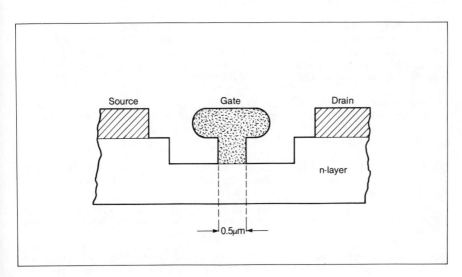

FIG. 4.29.   GaAs FET with Intentionally Side Etched Gate

As has been seen in Chapter 2 the adoption of the recessed channel technology can significantly improve the performance of the small signal GaAs FET since such a recess can decrease the source resistance of the device and thereby improve the noise figure and associated gain. Together with the incorporation of an n^+ contact layer for the source and drain electrodes it is possible to reduce the source resistance by a factor of four over a self aligned FET with a planar geometry. Fig. 4.30. shows the cross sections of several types of low-noise MESFET fabricated using the recessed-gate technique.

In power FETs the breakdown voltage between gate and drain is of importance in determining the r.f. power handling capability. The recessed channel structure improves this breakdown voltage. Fig.4.31 shows a power FET device structure which improves the breakdown voltage. Three regions are identified where high electric fields can occur. The field in region 1 is reduced by smoothing the channel near the drain thus reducing the possibility of breakdown due to avalanching and electron-hole pair generation (Tiwari et al, 1979). The field in region 2 is reduced by recessing the gate. This recess depth has to be correct since if it is too deep local field enhancement will take place lowering the breakdown voltage. The electric fields in region 3 occur because of rising fringing fields towards the active-buffer depletion region through which substantial space charge limited current may be flowing. This leads to current crowding of carriers as they approach the drain region of the channel. By recessing the active layer 0.5 to 1μm away from the drain contact edge and making the thickness under the drain contact edge equal to the breakdown depletion width these effects are avoided.

C. ION IMPLANTED FET PROCESSING

Both mesa isolation and planar isolation techniques can be used with ion-implanted GaAs layers, the latter increasing yield and device reliability since the gate metal does not have to go over a mesa step. With reference to Fig. 4.32 typical processing steps for an ion-implanted planar FET are as follows:

1. The wafer is coated with a layer of silicon nitride which remains on the substrate during subsequent processing. The nitride is removed from the channel region using a plasma etching technique (Rode et al, 1979).

2. A Si^+ ion implantation creates the active channel. The Si_3N_4 acts as an implantation mask, thus ensuring that only the channel is implanted.

3. Activation of the ions is accomplished by annealing using a Si_3N_4 cap.

4. The cap is removed only in the source and drain regions which are defined using photoresist. This same photoresist is used to lift off the ohmic metal (in this case, Ge-Au-Ni-Au). The source and drain ohmic contacts are alloyed to produce low specific contact

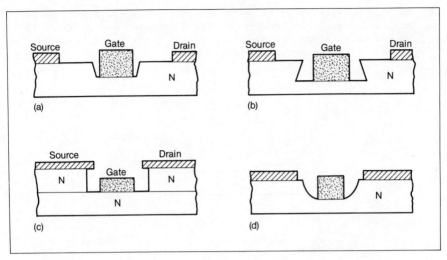

FIG. 4.30.   Cross Sections of Various FET Recessed Channel
Structures

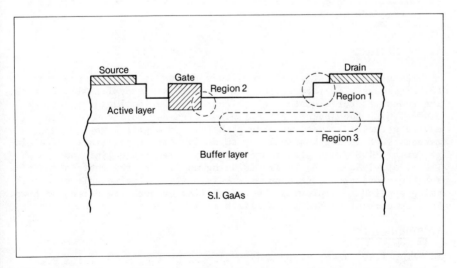

FIG. 4.31.   Power FET Geometry Showing Recessed Gate and Recessed
Channels

resistance.

5. The gate metallisation (in this case for a dual-gate FET) is put down by using a photolithographic technique to open up areas in the remaining Si_3N_4 followed by a resist stage and a float-off process to define the gate stripes resulting in the structure shown in Fig. 4.32(c).

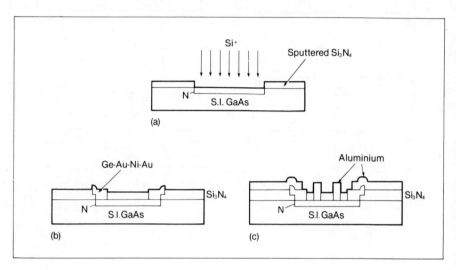

FIG. 4.32. Example of GaAs FET Processing using Ion-Implanted GaAs (after Rode et al)

4. CONCLUSIONS

This chapter has attempted to give a review of material preparation techniques and FET fabrication processes related specifically to GaAs. The reader is referred to the bibliography that follows for further study. With the advent of ion implantation and more complicated device structures 'wet' etching techniques are giving way to 'dry' methods such as ion beam milling to define gate metallisation (Bollinger et al, 1980) and plasma etching for dielectric definition (Tolliver, 1980). The latter is being applied to the fabrication of scratch protection layers on GaAs FETs. Also, considerable effort is being applied to improving the reliability of both low-noise and power FETs by careful choice of metallizations and device design.

5. BIBLIOGRAPHY

Au Coin, T.R., Ross, R.L., Wade, M.J. and Savage, R.O., Solid State Technology, Vol. 22, No. 1, p.59, 1979.

Bandy, S.G., Collins, D.M. and Nishimoto, C.K. Low-noise microwave FETs fabricated by molecular beam epitaxy. Electronics Letters, 12 April 1979, Vol. 15, No. 8, pp.218-219.

Bass, S.J. Device quality epitaxial gallium arsenide grown by the metal alkyl-hydride technique. Journal of Crystal Growth Vol. 31, pp.172-178, 1975.

Bobb, L.C., Holloway, H., Maxwell, K.H. and Zimmerman, E. Journal of Phys. Chem. Solids, Vol. 27, p.1679, 1966.

Bolger, D.E., Franks, J., Gordon, J. and Whitaker, J. Gallium Arsenide and Related Compounds, Institute of Physics Conference, 1966, p.16.

Bollinger, D. and Fink, R. A new production technique: Ion milling. Solid State Technology, November 1980, pp.79-84.

Bonnet, M., Duchemin, J.P., Huber, A.M. and Morillot, G. Low pressure organometallic growth of chromium-doped GaAs buffer layers Semi-insulating III-V Materials, Nottingham 1980, pp.68-75, Shiva Publishing Ltd.

Butlin, R.S., Parier, D., Crossley, I. and Turner, J. Correlation between device characteristics and material quality in low-noise GaAs FETs. Gallium Arsenide and Related Compounds, 1976, Institute of Physics, Conference Series No. 33a, pp.237-245.

Butlin, R.S., Hughes, A.J., Bennett, R.H., Parker, D. and Turner, J.A. J band performance of 300 nm gate length GaAs FETs. Int. Electron Devices Meeting, Washington DC, Dec. 1978, pp.136-139.

Cairns, B. and Fairman, R.D., Journal of the Electrochemical Society, Vol. 115, p.327C, 1968.

Cairns, B. and Fairman, R., Journal of the Electrochemical Society, Vol. 117, p.197C, 1968.

Caruso, R., Di Domenico, M., Verleur, H.W. and Von Neida, A.R. Journal Phys. Chem. Solids, Vol. 33, p.689, 1972.

Cho, A.Y., Di Lorenzo, J.V., Hewitt, B.S. et al, Journal of Applied Physics, Vol. 48, p.346, 1977.

Cho, A.Y., and Ch'en, D.R. GaAs MESFET prepared by molecular beam epitaxy. Applied Phys. Lett, Vol. 28, pp.30-31, 1976.

Cho, A.Y., Di Lorenzo, J.V., Hewitt, B.S., Nichaus, W.C., Schlosser, W.O. and Radice, C. Low noise and high power GaAs microwave field effect transistors prepared by molecular beam epitaxy. J. Applied Physics, Vol. 48, pp.346-349, 1977.

Chu, W.K. et al, Proc. 3rd Int. Conf. on Ion Implantation, Plenum Press, New York, 1973.

Cox, H.M. and Di Lorenzo, J.V. Gallium Arsenide and Related Compounds (St. Louis) 1976, Institute of Physics Conference Series No. 33b, pp.11-12.

Cox, H.M., Di Lorenzo, J.V. and D'Asaro, L.A. 1st GaAs Integrated Circuit Symposium Abstracts, Lake Tahoe, USA 1979.

Cox, H.M. and Di Lorenzo, J.V. Characteristics of an $AsCl_3/Ga/H_2$ two bubbler GaAs CVD system for MESFET applications. Institute of Physics Conference Series No. 33B, pp.11-22, 1976.

Di Lorenzo, J.V. Progress in the Development of low noise and high power GaAs FETs. Proceedings of the 6th Biennial Cornell Electrical Engineering Conference 1977, Cornell University, Ithaca, New York, pp.1-28.

Drukier, I., Camisa, R.L., Jolly, S.T., Huang, H.C. and Narayan, S.Y. Electronics Letters, Vol. 2, p.104, 1975.

Eddolls, D.V., Phys. Status Solidi, Vol. 17, p.67, 1966.

Evans, C.A., Deline, V.R., Sigmon, T.W. and Lidow, A., Applied Physics Conference Series No. 45, p.114, 1979.

Fairman, R.D., Morin, F.J. and Oliver, J.R. Institute of Physics Conference Series No. 45m p.134, 1979.

Fairmam, R.D., Morin, F.J., Oliver, J.R. and Dreon, J.K. High resistivity chromium doped GaAs buffer layers grown by halide transport methods. Proceedings of the Seventh Biennial Cornell Electrical Engineering Conference, Ithaca, New York, 1979, pp.177-187.

For, W.M. - private communication.

Foyt, A.G., Donnelly, J.P. and Lindley, W.T. Efficient doping of GaAs by Se^+ ion implantation. Appl. Phys. Lett. Vol. 14, p.372, 1969.

Hallais, J., Andrew, J.P., Baudet, P. and Beccon-Gibod, D. New MESFET devices based on GaAs-(Ga,Al)As heterostructures grown by organometallic VPE. 7th Int. Symp. on GaAs and related compounds, Institute of Physics Conference Series 1978.

Harris, J.S., Eisen, F.H., Welch, B.K., Haskell, J.D. et al. Influence of implantation temperature and surface protection on tellurium implantation in GaAs. Appl. Physics Lett., Vol. 21, p.601, 1972.

Henry, R.L. and Swiggard, E.M., Institute of Physics Conference Series No. 33b, p.28, 1977.

Hewitt, B.S., Cox, H.M., Fukui, H., Di Lorenzo, J.V., Schlosser, W.O. and Ingelsias, D.E. Low noise GaAs MESFETs: fabrication and performance. Institute of Physics Conference Series No. 33a, pp.246-254, 1976.

Huber, A.M., Morillot, G., Merenda, P. and Linh, N.Y. 2nd International Conference on Secondary Ion Mass Spectroscopy, Palo Alto, USA, 1979.

Ito, S., Shinohara, T. and Seki, Y. Journal of the Electrochemical Society, Vol. 120, p.1419, 1973.

Kato, Y., Mori, Y. and Morizane, K. Journal of Crystal Growth, Vol. 47, p.12, 1979.

Kim, C.K., Malbon, R.M., Omoni, M. and Park, Y.S. Institute of Physics Conference Series No. 45, p.305, 1979.

Knight, J.R., Effer, D. and Evans, P.R. The preparation of high purity gallium arsenide by vapour phase epitaxial growth. Solid State Electronics, Vol. 8, pp.178-180, 1965.

Komeno, J., Nogami, M., Shibatomi, A. and Ohkawa, S. Ultra high uniform GaAs layers by vapour phase epitaxy.

Lui, S.G., Douglas, E.C., Wu, C.P. et al. Ion Implantation of sulphur and silicon in GaAs. RCA Review, Vol. 41, June 1980, pp.227-263.

Luscher, P.E. Crystal growth by molecular beam epitaxy. Solid State Technology, Vol. 20, No. 12, pp.43-52, Dec. 1977.

Manasevit, H.M. and Simpson, W.I. Journal of Electrochemical Society, Vol. 116, p.1725, 1969.

Metz, E.P.A., Miller, R.C. and Mazelsky, R. Journal of Applied Physics, Vol. 33, p.2016, 1962.

Mizuno, O., Kikuchi, S. and Seki, Y. Japanese Journal of Applied Physics, Vol. 10, p.208, 1971.

Moest, R.R. and Shupp, B.R. Journal of the Electrochemical Society, Vol. 109, p.1061, 1962.

Morkoc, H., Andrews, J. and Aeki, V. GaAs MESFET prepared by organometallic chemical vapour deposition. Eletronics Letters, 15 Feb. 1979, Vol. 15, No. 4, pp.105-106.

Morkoc, H. and Eastman, L.F. Journal of Crystal Growth, Vol. 47, 1.12, 1979.

Mullins, J.B., Heritage, R.J., Holliday, C.H. and Straughan, B.W. Journal Phys. Chem. Solids, Vol. 26, p.782, 1965.

138

Murai, F., Kurono, H. and Kodera, H. Intentional side etching to achieve low-noise GaAs FET. Electronics Letters, 26 May 1977, Vol. 13, No. 11, pp.316–318.

Nanishi, Y., Takahei, K. and Kuroiwa, K. Journal of Crystal Growth, Vol. 45, p.272, 1978.

Nozaki, T., Ogawa, M., Terao, H. and Watanabe, H. GaAs and Related Compounds Institute of Physics Conf. Series, No. 24, 1974, p.46–54.

Ohata, K., Itoh, H., Hasegawa, F. and Fujiki, Y. Super low noise GaAs MESFETs with a deep-recess structure. IEEE Trans. on Electron Devices, Vol. ED-27, No. 6, pp.1029–1034, June 1980.

Pashley, R.D. and Welch, B.M. Tellurium-implanted n^+ layers in GaAs. Solid State Electronics, Vol. 18, p.977, 1975.

Rai-Choudbury, P. Journal of Electrochemistry Soc., Vol. 116, p.1745, 1969.

Rode, A.G. and Verma, K.B. An ion implanted planar dual-gate GaAs MESFET. Proceedings of the Seventh Biennial Cornell Electrical Engineering Conference, Ithaca, New York, 1979, pp.249–256.

Rocztoczy, F.E., Goldwasser, R.E. and Kinoshita, J. Gallium arsenide and Related Compounds, 1974, Institute of Physics Conference Series No. 24, pp.37–45.

Rumsby, D. IEEE Workshop on Compoud Semiconductors for Microwave Materials and Devices, Atlanta, USA, 1979.

Sanders, I.R., Peake, A.H. and Surridge, R.K. Selenium ion implanted GaAs MESFETs. Semi-insulating III-V Materials, Nottingham 1980, pp.349–352, Shiva Publishing Ltd.

Sealy, B.J. and Surridge, R.K. A new thin film encapsulant for ion-implanted GaAs. Thin Solid Films, Vol. 26, p.119, April 1974.

Stephens, K.G. and Sealy, B.J. Microelectronics Journal, Vol. 9, No. 2, p.13, 1978.

Stolte, C.A. The influence of substrate properties on the electrical characteristics of ion implanted GaAs. Semi-insulating III-V materials, Nottingham 1980, pp.93–99, Shiva Publishing Ltd.

Tiwari, S., Woodard, D.W. and Eastman, L.F. Domain formation in MESFETS - effect of device structure and materials parameters. Proceedings of Seventh Biennial Cornell Electrical Engineering Conference, Ithaca, New York, 1979, pp.237–248.

Tolliver, D.L. Plasma processing in microelectronics - past, present and future. Solid State Technology, November 1980, pp.99–105.

Tuck, B., Adegboyega, G.A., Jay, P.R. and Cardwell, M.J.   Institute
of Physics Conference Series No. 45, p.114, 1979.

Van der Pauw, L.J.   A method of measuring specific resistivity and
Hall effect of discs of arbitrary shape.   Phillips Research Reports
Vol. 13, p.1, 1958.

Vokes, J.C., Barr, W.P., Dawsey, J.R., Hughes, B.T. and Schrubb,
S.J.W.   A low noise FET amplifier in coplanar waveguide.   Proc.
1977 IEEE International Microwave Symp. pp.185-186.

Ware, R.M.   Int. Conf. on Crystal Growth, ICCG-5, Boston, 1979.

Wataze, M., Mitsui, Y. et al.   High power GaAs FET prepared by molecular beam epitaxy.   Electronics Letters, 23 Nov. 1978, Vol. 14,
No. 24, pp.759-760.

Weiner, M.E. Lassota, D.T. and Schwartz, B.   Journal Electrochem.
Soc. Vol. 118, p.301, 1971.

Wolfe, C.M. and Stillman, G.E.   GaAs and Related Compounds, Proc.
3rd Intern. Symposium (Institute of Physics Conf. Series No. 9,
London 1970) p.3.

Zucca, R., Welch, B.M., Asbeck, P.M., Eden, R.C. and Long, S.I.
Semi-insulating GaAs - a user's view.   Semi-insulating III-V
Materials, Nottingham 1980, pp.335-345, Shiva Publishing Ltd.

CHAPTER 5
The Design of
Transistor Amplifiers

1. INTRODUCTION

As has been seen in the previous chapters modern GaAs FET devices are
capable of producing noise figures of less than 3 dB at 18 GHz with
almost 10 dB associated gain whilst at the lower frequencies noise
figures of less than 0.5 dB are possible. Power FETs are now capable
of producing 15 watts up to X-band in Class A operation with promise
of considerably more power in the pulsed mode of operation.

The GaAs FET offers in its simplest form a device having somewhat
more convenient impedances for matching than the bipolar device and
it is also capable of exhibiting much superior performance above
approximately 4 GHz than the bipolar transistor. This chapter deals
with the theory and design of amplifiers using GaAs FETs.

2. LOW NOISE/SMALL SIGNAL AMPLIFIERS

A. S-PARAMETERS

The measurements which are usually made on a three terminal device
are those where one of the terminals is effectively earthed. In the
case of a microwave GaAs FET it is usual to measure the device in
common source configuration, i.e. where the source electrodes are
bonded to r.f. ground. This arrangement usually provides the design
engineer with the most stable arrangement and low feedback capacitance between drain and gate. Alternative configurations such as
common gate and common drain are dealt with later.

The gain parameters of a microwave transistor can be completely
specified by a set of 2 port parameters, the so-called 'scattering or
S-parameters'. Other descriptions of the transistor such as H or Y
parameters require open- and short-circuited measurements that are
difficult to establish accurately at microwave frequencies.

Because S-parameters are measured using travelling waves it is
essential to specify the plane of reference of the measurements with
respect to the device. Usually FET data sheets specify the reference

TABLE 5.1. Typical S-Parameters for GAT6 Chip at I_{DSS}, V_{DS} = 5 volts

Frequency GHz	S_{11}		S_{21}		S_{12}		S_{22}	
	Mag	Phase	Mag	Phase	Mag	Phase	Mag	Phase
2	.95	−38	2.95	150	.02	75	.725	−15
3	.92	−48	2.77	140	.02	74	.725	−17
4	.89	−60	2.6	130	.02	78	.725	−21
5	.85	−72	2.41	120	.02	83	.725	−26
6	.825	−84	2.25	110	.025	90	.725	−31
7	.8	−96	2.09	100	.03	97	.73	−36
8	.775	−107	1.96	90	.035	104	.735	−42
9	.75	−118	1.84	81	.04	110	.74	−47
10	.73	−128	1.73	73	.045	114	.75	−52
11	.71	−136	1.64	64	.053	115	.755	−58
12	.695	−145	1.55	56	.06	115	.765	−62
13	.68	−153	1.47	47	.065	113	.77	−68
14	.665	−161	1.4	40	.073	110	.78	−72
15	.655	−168	1.33	33	.08	106	.79	−77
16	.64	−175	1.26	27	.09	100	.8	−81
17	.63	−180	1.2	22	.1	93	.805	−86
18	.62	174	1.14	16	.11	86	.815	−90

planes at the end of wire bond lengths furthest from the device in the case of bare chip transistors and at the package lead to substrate interface in the case of packaged devices.

The S-parameters of a GaAs FET whether chip or package have to be measured accurately over a large range of frequencies.

Since errors can occur in measurement due to equipment and also the test jigs used to hold the devices, all measurements made on FETs should be obtained using an automatic network analyser with post measurement processing to extract errors due to the equipment and others due to causes such as microstrip transmission line losses, coaxial to microstrip transition errors and dispersion with frequency.

Commercial FET data sheets are the result of a relatively large number of S-parameter measurements on a particular FET type allowing 'typical' S-parameters to be generated. Since the maximum available gain of the device and the minimum noise figure are produced at different d.c. bias conditions at least two sets of S-parameters are needed. Also the S-parameter data for the device biased for maximum output power is often required. (This occurs at approximately 50% of the saturated drain current for a GaAs FET).

Table 5.1 shows typical S-parameters for a FET chip device (Plessey GAT6) over 2 to 18 GHz whilst Table 5.2 also shows the same data over 4 to 16 GHz for a packaged GAT6. It may be seen that the package produces a considerable rotation in the phase of the S-parameters.

Bare chip devices should be used where the ultimate in noise figure or gain-bandwidth product is used. Packaged devices have the advantage of being pretested by manufacturers, being usually supplied in hermetically sealed enclosures and being more easily handled. Offset against these advantages are the disadvantages of narrower band and slightly inferior r.f. performance.

Chips or packaged devices should always be used with the grounding and mounting methods used by the manufacturer when measuring the S-parameters.

B. STABILITY OF A 2 PORT

One important parameter in the design of microwave amplifiers is that of stability. An amplifier can be either unconditionally or conditionally stable. A circuit is unconditionally stable if its input and output resistances are positive for passive terminations. If the amplifier is conditionally stable then either the input or output resistances are negative.

The measured S-parameters of a transistor enable the maximum available gain (MAG) of the device to be determined.

The transducer power gain, G_T, of a transistor can be calculated as:

TABLE 5.2. Typical S-Parameters for a P103 Packaged GAT6 at I_{DSS}, V_{DS} = 5 volts

Frequency GHz	S_{11}		S_{21}		S_{12}		S_{22}	
	Mag	Phase	Mag	Phase	Mag	Phase	Mag	Phase
4	0.865	−78	2.37	112	.04	43	0.74	−35
5	0.84	−94	2.24	99	.04	37	0.72	−44
6	0.81	−112	2.14	85	.05	37	0.71	−53
7	0.78	−113	2.04	70	.05	34	0.692	−62
8	0.76	−150	1.94	55	.05	37	0.68	−70
9	0.74	−164	1.85	44	.06	47	0.666	−76
10	0.724	−178	1.77	30	.06	46	0.654	−84
11	0.71	166	1.68	20	.07	49	0.644	−95
12	0.695	150	1.6	6	.07	44	0.634	−100
13	0.685	134	1.53	−4	.08	43	0.626	−114
14	0.672	121	1.47	−15	.09	37	0.618	−125
15	0.664	110	1.4	−25	.11	38	0.615	−136
16	0.66	100	1.34	−34	.125	34	0.61	−149

$$G_T = \frac{|S_{21}|^2 (1 - |\Gamma_s|^2)(1 - |\Gamma_L|^2)}{|(1 - S_{11}\Gamma_s)(1 - S_{22}\Gamma_L) - S_{12}S_{21}\Gamma_s\Gamma_L|^2}$$

When the generator and load, Γ_S and Γ_L are conjugately matched to the two FET ports the gain can be maximised. Assuming $S_{12} = 0$ then we have:

$$G_{MAX} = \frac{|S_{21}|^2}{(1 - |S_{11}|^2)(1 - |S_{22}|^2)}$$

where $\Gamma_S = S_{11}*$, $\Gamma_L = S_{22}*$, $S_{11}*$ and $S_{22}*$ being the complex conjugate reflection coefficients, and the device is assumed unilateral. The S-parameters also determine Rollett's stability factor, K (Rollett, 1962).

$$K = \frac{1 + |S_{11}S_{22} - S_{12}S_{21}|^2 - |S_{11}|^2 - |S_{22}|^2}{2|S_{12}S_{21}|}$$

If K is larger than unity, i.e. when the FET is unconditionally stable, an optimum combination of Γ_S and Γ_L can simultaneously match the 2 FET ports to maximise the gain (Bodway, 1967). If K is smaller than unity, the FET is only conditionally stable and Γ_S and Γ_L must be carefully chosen to operate the device in a stable region (Bodway, 1967; Gledhill et al, 1974). Since the FET's stability factor, K, will change with frequency it is possible to plot the regions of stability onto a Smith Chart at each frequency.

The input and output reflection coefficients with arbitrary source and load terminations are given by:-

$$\Gamma_{IN} = S_{11} + \frac{S_{21}S_{12}\Gamma_L}{1 - S_{22}\Gamma_L}$$

and

$$\Gamma_{OUT} = S_{22} + \frac{S_{21}S_{12}\Gamma_S}{1 - S_{11}\Gamma_S}$$

If we set $|\Gamma_{IN}|$ equal to unity, a boundary is established beyond which the device is unstable and then

$$\left| S_{11} + \left| \frac{S_{21}S_{21}\Gamma_L}{1 - S_{22}\Gamma_L} \right| \right| = 1$$

This equation can be changed in form to give the solution as a circle whose radius is given by

$$\gamma_L = \frac{S_{12} S_{21}}{|S_{22}|^2 - |\Delta|^2}$$

and whose centre C_L is given by

$$C_L = \frac{(S_{22} - \Delta S_{11}{}^*)^*}{|S_{22}|^2 - |\Delta|^2}$$

where $\Delta = S_{11} S_{22} - S_{12} S_{21}$

and the origin of the Smith Chart is at $_L = 0$.

  The area either inside or outside the circle represents a stable
operating condition. If $|S_{11}|<1$ and $|\Gamma_{IN}|<1$ then the shaded region
in Figs. 5.1(a), 5.1(b) will enable stable gain to be realised.
In Fig. 5.1(b) the stability circle encloses the origin of the Smith
Chart and hence its interior represents the stable region. In Fig.
5.1(a) the origin is outside the circle and the exterior of the
circle represents the stable region.

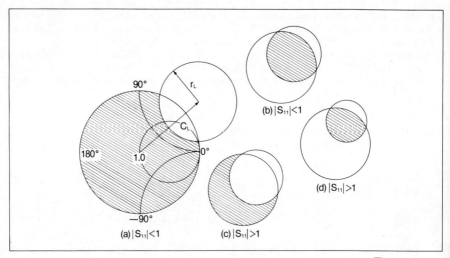

FIG. 5.1.   The Four Different Regions of Stability in the Γ Plane

  If $|S_{11}|>1$, the origin (i.e. at $\Gamma_L = 0$) represents an unstable
point. Fig. 5.1(c) is the case where the stability circle encircles
the origin and its exterior is the stable region. For the case where
the circle does not encircle the origin, its interior represents the
stable region as in Fig. 5.1(d).

  For unconditional stability we must ensure that the magnitude of
the vector C_L (the distance from the centre of the Smith Chart to the

centre of the stability circle (Fig. 5.1(a)) minus the radius of the stability circle r_L is greater than one. Table 5.3 lists the stability results for a packaged Plessey GAT6 from 4 to 16 GHz. The device is conditionally stable at all these frequencies where the stability circles are given in terms of their location in magnitude and angle on the Smith Chart and the radius of the circle with an indication of the area of stability.

For the case where K<1 the maximum stable gain or MSG of the device can be calculated as

$$\text{MSG} = \left| \frac{S_{21}}{S_{12}} \right|$$

In general,

$$\text{MAG} = \left| \frac{S_{21}}{S_{12}} \right| (K - \sqrt{K^2 - 1})$$

Thus MAG equals MSG when K is 1 and the MSG is the gain that can be obtained from the GaAs FET when the input and output reflection coefficients fall on the boundaries of the instability regions.

C. TRANSDUCER POWER GAIN

In the design of small signal amplifiers, a transducer power gain as defined, for example, in the Hewlett Packard Application Note 154 (1972), can be calculated for the device terminated in arbitrary load and generator impedances, Z_L and Z_S.

The transistor power gain is given by $\quad G_T = |S_{21}|^2$

We can write

$$\Gamma_s = \frac{Z_S - Z_O}{Z_S + Z_O}$$

$$\Gamma_L = \frac{Z_L - Z_O}{Z_L + Z_O}$$

When S_{12} is sufficiently small to be neglected (see Table 5.1 at the lower frequencies), the device is defined to have a unilateral transducer power gain

$$G_U = \frac{|S_{21}|^2 (1 - |\Gamma_s|^2)(1 - |\Gamma_L|^2)}{|1 - S_{11} \Gamma_s|^2 |1 - S_{22} \Gamma_L|^2}$$

$$G_U = G_S \cdot G_O \cdot G_L$$

TABLE 5.3. Stability Circles for a Packaged GAT6 having the S-parameters of Table 5.2

Frequency GHz	INPUT PLANE				OUTPUT PLANE				K	MSG dB
	Location MAG.	ANG.	Radius	Stable Region	Location MAG.	ANG.	Radius	Stable Region		
4	1.21	88	0.29	Outside	1.61	59	0.76	Outside	.67	17.7
5	1.22	103	0.26	Outside	1.51	63	0.59	Outside	.79	17.5
6	1.23	122	0.31	Outside	1.42	70	0.54	Outside	.68	16.3
7	1.31	122	0.31	Outside	1.51	77	0.52	Outside	.97	16.1
8	1.24	156	0.26	Outside	1.35	79	0.38	Outside	.87	15.9
9	1.23	168	0.29	Outside	1.32	82	0.39	Outside	.76	14.9
10	1.26	-179	0.28	Outside	1.34	88	0.37	Outside	.9	14.7
11	1.26	-166	0.30	Outside	1.35	95	0.39	Outside	.84	13.8
12	1.30	-150	0.30	Outside	1.38	100	0.38	Outside	1	13.6
13	1.32	-137	0.33	Outside	1.41	110	0.42	Outside	.97	12.8
14	1.36	-125	0.37	Outside	1.45	120	0.46	Outside	.97	12.1
15	1.41	-117	0.44	Outside	1.50	127	0.54	Outside	.89	11.05
16	1.46	-109	0.50	Outside	1.57	138	0.62	Outside	.88	10.3

The first term is related to the transistor used and remains invariant throughout the amplifier design. The other two terms, however, are not only related to the S-parameters S_{11} and S_{22} but also to the load and source reflection coefficients. These latter two quantities are used to control the design of the amplifier in terms of gain slope compensation etc.

The unilateral transducer gain can be divided into three independent gain blocks.

where

$$G_S = \frac{1 - |\Gamma_s|^2}{|(1 - S_{11}\,\Gamma_s)|^2}$$

$$G_O = |S_{21}|^2$$

and

$$G_L = \frac{1 - |\Gamma_L|^2}{|(1 - S_{22}\,\Gamma_L)|^2}$$

D. CIRCLES OF CONSTANT UNILATERAL GAIN

Considering equation 5.12a we have that for $\Gamma_S = S_{11}{}^*$, $G_S = G_{S\,max}$

and for $\Gamma_S = 1$, $G_S = 0$

Thus for any arbitrary value of G_S there is a value of Γ_S which lies on a circle. The output gain term in Eq. 5.12c is of a similar form and thus a similar set of circles can be generated. Such circles are called CONSTANT GAIN CIRCLES and can be generated from the FET S-parameters since

$$d_i = \frac{g_i\,|S_{ii}|}{1 - |S_{ii}|^2\,(1 - g_i)}$$

$$R_i = \frac{\sqrt{1 - g_i}\,(1 - |S_{ii}|^2)}{1 - |S_{ii}|^2\,(1 - g_i)}$$

where

$$g_i = G_i(1 - |S_{ii}|^2) = \frac{G_i}{G_{imax}}$$

where G_i is the constant gain represented by the circle, d_i is the distance from the centre of the Smith Chart to the centre of the constant gain circle along the vector $S_{11}{}^*$ and R_i is the radius of a particular circle (Hewlett Packard Application Note 154 (1972)).

150

Thus gain circles enable the design engineer to choose a matching topology which, for example, will allow matching over a range of frequencies to produce constant gain or indeed gain of a specified slope with frequency. Fig. 5.2 is an example of the constant gain circles for a Plessey P103 packaged GAT5 at 10 GHz.

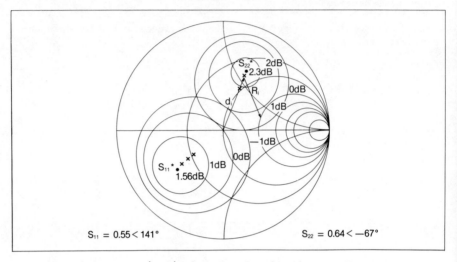

FIG. 5.2. Constant Gain Circles for a Packaged GaAs FET at 10 GHz

E. UNILATERAL FIGURE OF MERIT

Although the assumption that $|S_{12}| = 0$ is often made in initial amplifier design it is desirable to quantify the error involved in doing this.

The transducer power gain G_T, from Eqs. 5.1 and 5.12 can be written as

$$G_T = Gu \cdot \frac{1}{|1 - X|^2}$$

where

$$X = \frac{\Gamma_s \Gamma_L S_{12} S_{21}}{(1 - \Gamma_s S_{11})(1 - \Gamma_L S_{22})}$$

The ratio of the true gain to the unilateral gain is bounded by

$$\frac{1}{|1 + 1 \ X \ 1|^2} < \frac{G_T}{Gu} < \frac{1}{|1 - 1 \ X \ 1^2|}$$

For complex conjugate matching $|S_{11}| < 1$ and $|S_{22}| < 1$

$$\frac{1}{(1 + X')^2} < \frac{G}{Gu} < \frac{1}{(1 - X')^2}$$

where

$$X' = \frac{|S_{11}| |S_{22}| \ |S_{12}| \ |S_{21}|}{(1 - |S_{11}|^2)(1 - |S_{22}|^2)}$$

A typical value for X' at 10 GHz for the GAT5 in Fig. 5.1 is 0.1 giving

$$\frac{1}{(1 + X')^2} \quad \text{equivalent to } -0.8 \text{ dB;} \qquad \frac{1}{(1 - X')^2} \quad \text{equivalent to } +0.9 \text{ dB}$$

Thus the error in gain by assuming that the device is unilateral in any design will be approximately $\pm$ 0.8 dB.

F. VARIATION OF GAIN WITH DRAIN CURRENT AND TEMPERATURE

The variation of maximum available gain of a GaAs FET with drain-source current I_{DS} can be plotted at different frequencies. This enables the design engineer to calculate the gain that can be expected under certain bias conditions. Fig. 5.3 shows the gain v drain current ratio I_{DS}/I_{DSS} for a typical high frequency FET at 8 GHz. It may be seen that below a certain I_{DS}/I_{DSS} ratio there is a sharp decrease in gain due to a rapid decrease in transconductance g_m and consequently of f_t as might also be seen by inspecting the drain current versus drain to source voltage characteristic of the device for gate voltages close to pinch off.

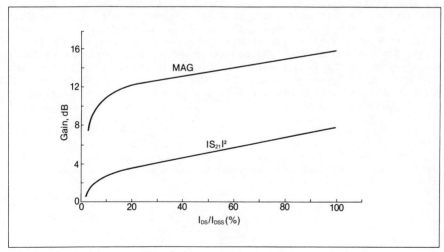

FIG. 5.3.   Gain v I_{DS}/I_{DSS} FOR A GaAs FET Chip at 8 GHz

Fig. 5.4 shows the associated gain change with temperature for a packaged FET device over the temperature range -40 to +80°C where the transistor is biased and matched for minimum noise figure as explained

152

in Chapter 2.

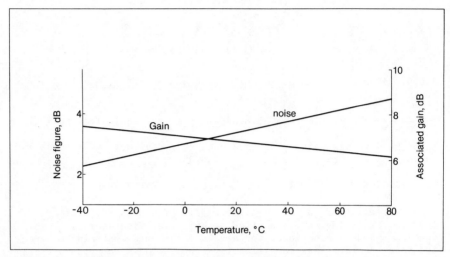

FIG. 5.4.   Typical Associated Gain and Noise Figure Versus
   Temperature for a GaAs FET

G. OPTIMUM LOAD CONDITIONS FOR OUTPUT POWER

In certain cases the FET is designed to operate in a linear (Class A) mode where the objectives are either lowest intermodulation distortion, largest added r.f. power or largest linear output power. In the last case, for example, the d.c. bias is graphically determined from the static drain-current versus drain-voltage characteristic (Fig. 5.5). The bias and load conductance line are chosen to maximise the product of the linear voltage and current swing. The limitations are determined by the maximum d.c. power dissipation, the drain-to-gate breakdown voltage and the positive gate bias ($V_{GS} \simeq 0.5V$) above which appreciable gate current flows. The optimum load conductance is typically much larger than the GaAs FETs output conductance. Consequently the GaAs FET is not matched to the load and does not provide maximum gain. The output load conductance and susceptance determine the optimum reflection coefficient of the load, $_{LP}$. The generator reflection coefficient that provides a complex conjugate match to the input of the device is then

$$\Gamma_{SP} = \left[S_{11} + \frac{S_{12} S_{21} \Gamma_{LP}}{1 - S_{22} \Gamma_{LP}} \right]^*$$

where the associated gain is then determined using Eq. 5.1.

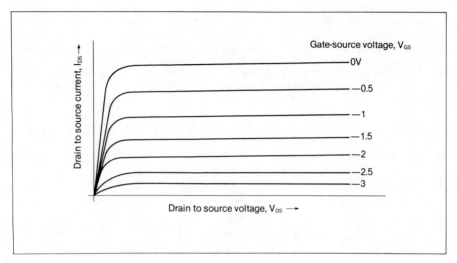

FIG. 5.5.   D.C. Characteristics of a Grounded Source GaAs FET

H. EQUIVALENT CIRCUIT OF THE GaAs FET

It is convenient to use equivalent circuits of the GaAs FET for forming a clear view of microwave characteristics and noise characteristics of the device. The equivalent circuit can be used as the first step in amplifier design. Fig. 5.6a shows an equivalent circuit of the GaAs FET.

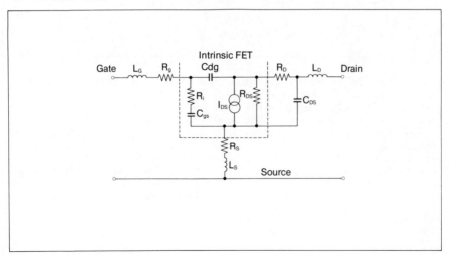

FIG. 5.6a.   Equivalent Circuit of GaAs FET Chip

This equivalent circuit is shown for the device operated in the saturated current region in common-source configuration. In the intrinsic FET model, the elements $(C_{DG} + C_{GS})$ represent the total gate-to-channel capacitance, R_i and R_{DS} show the effects of the channel resistance and I_{DS} defines the voltage controlled current source. The transconductance g_m relates I_{DS} to the voltage across C_{GS}. The extrinsic (parasitic) elements are: R_S, the source resistance, R_D the drain resistance, R_g the gate metal resistance, and C_{DS} the substrate capacitance. Typical element values for a 0.5μm gate length and 300μm gate width FET are shown in Table 5.4.

TABLE 5.4.   Typical Equivalent Circuit Element Values for a 0.5μm
   Gate Length GaAs FET

Intrinsic FET		
R_i	=	6.95Ω
R_{DS}	=	400Ω
C_{gs}	=	0.43 pF
C_{dg}	=	0.01 pF
g_m	=	30 mS
Parasitic Components		
R_g	=	0.8Ω
R_D	=	3Ω
R_S	=	3Ω
C_{DS}	=	0.16 pF
L_G	=	0.2 nH
L_S	=	0.05 nH

Very often the input circuit of an FET can be modelled as a series RC and the output circuit as a parallel RC for the case of a bare chip device. The equivalent circuit of a packaged transistor can also be found where the addition of package wire bond inductances, L_G, L_D and L_S and shunt parasitic capacitances C_{P1}, C_{P2}, C_{P3} and C_{P4} on input and output together with increased feedback capacitances, C_{FB1} and C_{FB2} are usually sufficient to model accurately the device reflection coefficients as shown in Fig. 5.6b. This is covered further in Chapter 8. By adding the current source as an element giving gain to the transistor we have built up a simple equivalent circuit for the FET.

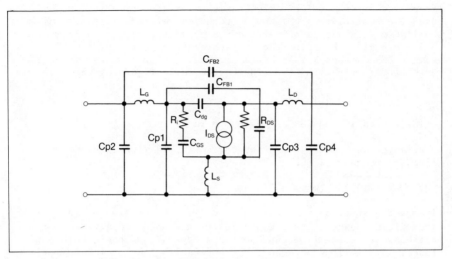

FIG. 5.6b.   Equivalent Circuit of Packaged GaAs FET

3. EXAMPLE OF NARROW BAND AMPLIFIER DESIGN

This example deals with the design of a simple narrow band amplifier to operate between a source and load of characteristic impedance
of 50 ohms, which provides a gain of at least 18 dB with input and
output reflection coefficients of less than 0.1 at 2 GHz.   The
scattering parameters of the transistor used are given in Table 5.1.
The equivalent circuit of the transistor (Liechti et al, 1974)
derived from the S_{11} and S_{22} parameters is given in Table 5.4.   The
transistor is assumed unilateral (i.e. $S_{12} = 0$) for the initial
design.

In order to resonate the reactance of the input circuit and transform the impedance a series inductor is required to resonate with the
0.43 pF series capacitor and a transformer to transform 10.75Ω (R_i +
R_g + R_S) to 50Ω.   The latter can be achieved using a simple single
section quarter wavelength transformer if the amplifier is to be
realised using transmission lines or a combination of series inductance and shunt capacitance if lumped element matching is to be used.

The scattering parameters of the transistor are

S_{11} = $0.95\angle-38°$

S_{21} = $2.95\angle150°$

S_{12} = $.02\angle75°$

S_{22} = $.725\angle-15°$

The stability factor K is calculated using equation 3.1 to be 0.33 i.e. the transistor is only conditionally stable. The device can be stabilised by the use of a 100 ohm shunt resistor at its output. The use of a resistive stabilization technique will decrease the maximum available gain and also increase the noise figure, the degradation being minimized by using the resistor on the output of the FET. Common lead feedback is a better method to effect an increase in the stability factor without degrading the noise figure of the FET.

A. INPUT MATCHING CIRCUIT

The source impedance for maximum gain is 10.75 + j 145.

First transform the 50Ω generator to 10.75Ω with a quarter wave line of impedance Z, where

$$Z = \sqrt{50 \times 10.75} \quad \text{ohms} = 23 \text{ ohms}.$$

Then add a shunt open circuit quarter wavelength line of low characteristic impedance. This essentially adds shunt capacitance to the matching structure lowering the imaginary part of the matching admittance. A high impedance series transmission line or inductance completes the input matching network. Fig. 5.7a shows the steps in the input matching sequence for the elements shown.

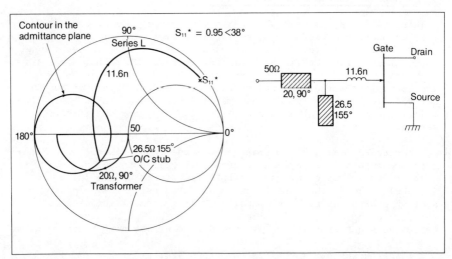

FIG. 5.7a. Input Matching Network for Narrow Band Small Signal Amplifier

Starting at the centre of the Smith Chart, i.e. at 50Ω. The low impedance, λ/4 transformer is used to transfer to a lower real impedance. From this point an open circuit stub adds capacitive reactance enabling the series inductance to transform the impedance to S_{11}*.

The resulting component values are optimised using CAD to give the optimum input VSWR and the values shown in Fig. 5.7a.

B. OUTPUT MATCHING NETWORK

The output matching is considerably simplified because of the use of the stabilising resistor which lowers the magnitude of the output reflection coefficient from 0.725 to 0.28. A 0.5 nH inductance (wire bond) is used between the FET and the resistor to give enough space for the resistor when the circuit is made. The output matching circuit consists simply of a series wire bond (which can be treated as a lumped 1.1 nH inductance) followed by a single section transformer nominally 60 ohms impedance, $\lambda/4$ long.

Referring to Fig. 5.7b, the output source impedance needed is $0.725 \angle 15^{\circ}$. Initially 50 ohms is transformed with a quarter wavelength line of impedance 60 ohms to a higher real impedance point on the Smith Chart. A series inductor of value 1.1 nH is then used followed by the shunt stabilising resistor. Fig. 5.7b shows the sequence of matching of the output of the device.

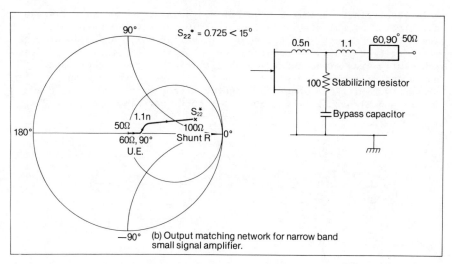

(b) Output matching network for narrow band small signal amplifier.

FIG. 5.7b. Output Matching Network for Narrow Band Small Signal Amplifier

By using Equations (5.4), (5.5) and (5.11) the resulting circuit design gives an input reflection coefficient of 0.01 and an output reflection coefficient of 0.1 at 2 GHz with a gain of 18.9 dB and a stability factor of 1.2. The latter indicates that whatever passive loading is put on the input or output of the amplifier it is not possible to make the circuit oscillate.

4. EXAMPLE OF A BROADBAND AMPLIFIER DESIGN

This example deals with the design of a broadband amplifier to operate over the frequency band 8 to 12 GHz. The amplifier design objective is to give maximum gain consistent with a bandpass characteristic having a gain ripple of less than 0.5 dB over the band of interest. The scattering parameters of the transistor are again given in Table 5.1. These parameters can be interpolated from the table in 500 MHz intervals.

The input and output matching networks are designed to give the best input and output VSWRs over the band. However, because of its bandwidth the VSWRs will only be optimum at certain frequencies and the main design objective is to obtain flat gain. To achieve this the device can be mismatched at the low frequency end of the band to reduce the contribution of G_S and G_L of Eq. 5.12 to the overall amplifier gain. The gain can also be reduced by using absorptive resistive loading preferably on the drain side of the FET.

A. LUMPED ELEMENT DESIGNS - DESIGN WHERE STABILITY IS NOT CONSIDERED

The equivalent circuit of the FET has been given in Table 5.4 and calculation of the stability factor at 10 GHz gives a value of 0.8. Again let us assume that the device is unilateral. A lumped element matching technique will be discussed. Lumped element matching gives inherently broader bandwidth capability since the designer is not dependent on components whose electrical length changes with frequency. Fig. 5.8 shows the input constant gain circles at 8 and 12 GHz and also shows the sequence of input matching used at 8 GHz.

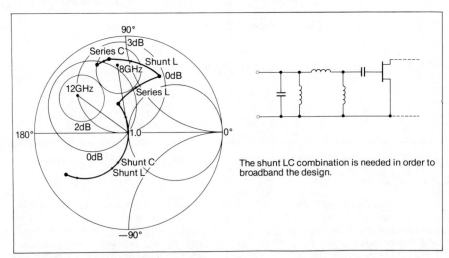

The shunt LC combination is needed in order to broadband the design.

FIG. 5.8.   Input Constant Gain Circles at 8 and 12 GHz and the Matching Sequence of the Network shown at 8 GHz.

The input impedance of the FET will in practice be determined by the output matching conditions when the device is conditionally stable as given by Eq. 5.4. A schematic of the lumped circuit is shown in Fig. 5.9(a) together with its performance shown in Fig. 5.9(b).

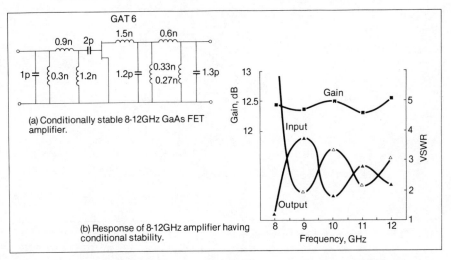

(a) Conditionally stable 8-12GHz GaAs FET amplifier.

(b) Response of 8-12GHz amplifier having conditional stability.

FIG. 5.9. (a) Conditionally Stable 8-12 GHz GaAs FET Amplifier
(b) Response of 8-12 GHz Amplifier having Conditional Stability

As may be appreciated from Fig. 5.8 it is essential in the design of broadband matching networks to ensure that the network is capable of synthesising the necessary changes in impedance such that a wanted gain slope compensation and gain ripple can be met. It is often necessary to optimise the initial design values on a computer to minimise gain ripples etc. Such an optimisation will also account for the non-unilateral behaviour of the transistor.

B. DESIGN WHERE STABILITY IS CONSIDERED

A shunt resistor at the output of the FET can be used to increase the stability factor. A 500 ohm resistor can be used in this particular case to produce a stable gain condition with a K = 1.7 at 10 GHz. This lowers the amplifier gain by approximately 3 dB over the maximum available gain with no resistor. Fig. 5.10 shows the output matching sequence again using lumped elements where in this example the gain slope compensation needed (approximately 3.8 dB per octave) is produced by reactively mismatching the FET at low frequencies but matching at the top end of the band. Figs. 5.11a,b show the complete matching networks together with the amplifier response. It is noticed, because the input network has been chosen to give 0 dB gain slope, that the input VSWR is much more constant than the output VSWR which is low at the high end of the band becoming progressively larger

160

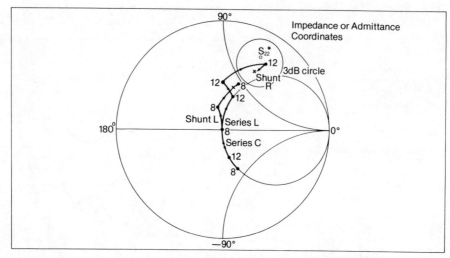

FIG. 5.10.   Output Matching Sequence for Stable Amplifier

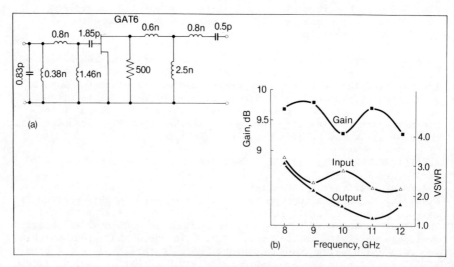

FIG. 5.11.(a)   Lumped Element 8-12 GHz GaAs FET Amplifier
     (b)   Response of Lumped Element 8-12 GHz GaAs FET Amplifier

towards the low end, thus mismatching the output of the FET and lowering its gain.

Stabilisation can also be produced by using common lead inductive feedback, which is a popular method when noise figure is of importance. It is also possible to stabilise the previous example by using a balanced amplifier approach or using input and output isolators where the reflected energy, due to $|S_{11}|$ and $|S_{22}|$ not being zero, is dumped into a load other than the source load.

C. DISTRIBUTED DESIGNS

Since many amplifier designs are based on synthesis techniques requiring lumped elements it is often desirable to convert lumped element designs to a close distributed element equivalent by using the following equations which are based on the use of π equivalent circuits for transmission lines

$$\omega L = Z_o \sin \theta$$

and

$$\omega C = \frac{1}{Z_o} \tan \frac{\theta}{2} \qquad \text{for } \theta < 90°$$

where L is the inductance, C is the capacitance, Z_o is the characteristic impedance of the distributed elements and θ their phase length. It is also possible to design matching networks directly from transmission line theory as was done in the narrow-band amplifier design previously.

Considering the conversion of the circuit of Fig. 5.11 it is worth noting that a good approximation is that inductors can be replaced by 120Ω characteristic impedance short circuit stubs or unit elements and capacitors can be considered as 25Ω characteristic impedance open circuit stubs. This approximation is particularly useful when implementing the resultant network in microstrip on alumina of 0.635 mm thickness since the two impedance levels correspond to the generally accepted extremes of line widths which are conveniently reproduced. Fig. 5.12 shows the result of converting inductors and capacitors to transmission lines. Note in Fig. 5.12 that a small amount of inductance has been left at the FET ports allowing wire-bonding to take place in the realisation. Fig. 5.12 also shows the resultant response of the circuit.

5. DESIGNING AN AMPLIFIER FOR OPTIMUM NOISE FIGURE

A. INTRODUCTION

In many applications such as for receiver preamplifiers the amplifier design objective is minimum noise figure rather than maximum gain. The main task is to synthesise the optimum source admittance for minimum transistor noise figure at the FET input plane.

162

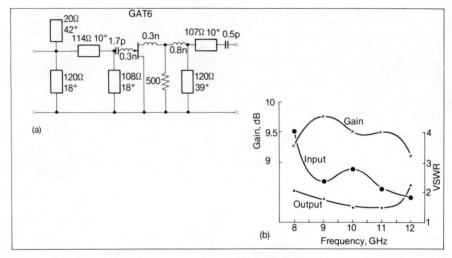

FIG. 5.12(a). Distributed Element 8-12 GHz GaAs FET Amplifier
        (b). Response of Distributed 8-12 GHz GaAs FET Amplifier

The noise performance of GaAs FETs is superior to that of bipolar
transistors above approximately 4 GHz. For example the Plessey GAT6
device exhibits chip gains of 10 dB with a noise figure of 2 dB at
12 GHz. A theoretical treatment of GaAs FET noise performance has
been given by Hewitt et al (1976) and the bias dependence of the FET
device noise figure is predicted by the noise theory of Pucel et al.
This theory predicts that the minimum noise figure is obtained at low
drain currents of approximately 15% of the saturated drain current.
This is more fully covered in Chapter 2.

B. CONSTANT NOISE FIGURE CIRCLES

In many amplifier designs the overall noise figure of the amplifier
is important since this defines the signal to noise ratio at the
output of the amplifier.

If we consider an infinite cascade of transistors all operated at
their minimum noise figure d.c. bias conditions, then the lowest
cascaded noise figure (often referred to as the noise measure, M) is
given by

$$M = \frac{(F - 1)}{(1 - 1/G)}$$

where each stage has a noise figure F and associated gain, G.

In general, for an arbitrary source reflection coefficient Γ_S, the noise figure is determined by the equation

$$F = F_{min} + 4R_N \frac{|\Gamma_s - \Gamma_o|^2}{(1 - |\Gamma_s|^2)|1 + \Gamma_o|^2}$$

where Γ_o is the optimum reflection coefficient of the source which produces the minimum noise figure F_{min} and R_N is the equivalent noise resistance of the device. Eq. 5.20 has the form of a circle. To determine a family of noise figure circles the noise figure parameter N_i is defined where

$$N_i = \frac{F_i - F_{min}}{4R_N} \cdot |1 + \Gamma_o|^2$$

where F_i is the value of the desired noise figure circle and Γ_o, F_{min} and R_N are as previously defined. The radius of the circle is given by

$$R_{Fi} = \frac{1}{1 + N_i} \sqrt{N_i^2 + N_i(1-|\Gamma_o|^2)^2}$$

and the centre by

$$C_{Fi} = \frac{\Gamma_o}{1 + N_i}$$

From these equations we see that $N_i = 0$ when $F_i = F_{min}$ and the centre of the F_{min} circle with zero radius is located at Γ_o on the Smith Chart. The equivalent impedance to Γ_o is often referred to as Z_{OPT}, the optimum source impedance for minimum noise figure.

The value of R_N can be found by measuring the noise figure when the source impedance is 50 ohms so that $\Gamma_s = 0$ and

$$R_N = (F_{\Gamma s = 0} - F_{min}) \frac{|1 + \Gamma_o|^2}{4|\Gamma_o|^2}$$

Fig. 5.13 shows the noise circles at 8, 10 and 12 GHz for a GAT6 FET chip device. All these plots show Z_{OPT} at a point where the gate wire bond is attached to the 50 ohm transmission line. Fig. 5.14 shows a plot of R_N against frequency for the same device.

C. NOISE MODELLING

It is useful to have the concept of a noise model when designing amplifiers which need to have optimum noise tracking over their frequency band. A device model can be derived that has an input impedance equivalent to the complex conjugate (Z_{OPT}^*) of the optimum noise impedance, Z_{OPT}.

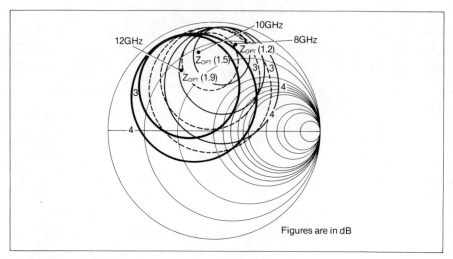

FIG. 5.13.   Constant Noise Circles

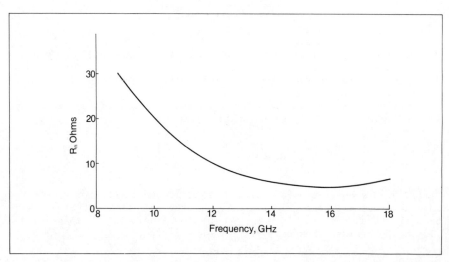

FIG. 5.14.   Frequency Sensitivity of R_n for a Plessey GAT6 Chip FET

For an FET connected to a network at its input that presents Z_{OPT} to the FET, the input impedance at the driving source terminal (50Ω) is not power matched, i.e. not 50Ω. However, if the device is re-placed by an equivalent noise model which represents $Z_{OPT}*$ over a frequency range, the input impedance is now 50Ω. This technique enables networks to be derived that provide optimum noise tracking. From the plots of Z_{OPT} that have been measured or extracted from a data sheet, $Z_{OPT}*$ can be found and an equivalent RLC model derived that fits this $Z_{OPT}*$ data. This RLC model can be computer optimised. This RLC model is used to provide the impedance necessary to derive a minimum noise matching network which can be combined with the S_{21}, S_{12} and S_{22} of the FET.

6. EXAMPLE OF BROADBAND AMPLIFIER DESIGNED FOR MINIMUM NOISE FIGURE

It is desired to design a low noise broadband 8-12 GHz amplifier for optimum noise performance over the band together with a flat gain response using the device whose S-parameters are given in Table 5.5 and whose noise data is given in Fig. 5.13.

TABLE 5.5. Typical S-Parameters for a GAT6 Chip at I_{DS} = 10 mA, V_{DS} = 5 volts (Minimum Noise Figure Conditions)

Frequency GHz	S_{11} Mag	S_{11} Ang	S_{21} Mag	S_{21} Ang	S_{12} Mag	S_{12} Ang	S_{22} Mag	S_{22} Ang
4	0.93	-43	1.66	134	.039	64	0.79	-24
5	0.9	-54	1.59	125	.042	62	0.79	-29
6	0.88	-64	1.52	115	.045	61	0.78	-34
7	0.85	-75	1.46	106	.048	60	0.78	-39
8	0.83	-85	1.4	97	.05	60	0.77	-44
9	0.81	-93	1.34	90	.052	61	0.77	-49
10	0.79	-101	1.29	80	.055	63	0.77	-54
11	0.77	-110	1.24	72	.057	65	0.77	-60
12	0.75	-116	1.19	64	.059	67	0.77	-65
13	0.74	-123	1.14	55	.061	69	0.77	-70
14	0.72	-130	1.1	47	.063	71	0.77	-75
15	0.7	-137	1.06	38	.066	72	0.77	-81
16	0.68	-144	1.03	31	.068	71	0.77	-85

A. INPUT MATCHING

The equivalent noise input impedance can be approximated with a capacitance C_{im} in series with a resistance R_{im} in series with an inductor L_{im} with good accuracy over the desired frequency range. The values C_{im} = 0.23 pF, R_{im} = 12Ω and L_{im} = 0.41 nH are found by fitting the measured input impedances to those predicted. It is desired to produce a Tchebyshev impedance matching network to match the generator impedance Z_0 to the noise input impedance Z_{OPT}. Tchebyshev prototypes (Matthaei et al, 1964) can be selected to maximise the bandwidth and provide a good match at the band edges.

In order to establish the number of resonators required to synthesise Z_{OPT}^* with a specified accuracy between the lower and upper band edges a maximum noise figure increase from the optimum values has to be specified. Let us assume that 0.6 dB excess noise figure can be tolerated. For example, at 10 GHz by inspecting the constant noise figure circles it can be seen that a worst case VSWR of approximately 2:1 can be tolerated with respect to Z_{OPT}^*. In synthesising Z_{OPT}^* with a Tchebyshev impedance matching network the bandwidth can be computed using the VSWR versus load decrement curves plotted in (Matthaei et al, 1964, Fig. 4.09-3).

The network shown in Fig. 5.15 will provide wide-band input impedance matching. Details on the use of inverters in network synthesis, shown in Fig. 5.15, can be found in Matthaei et al.

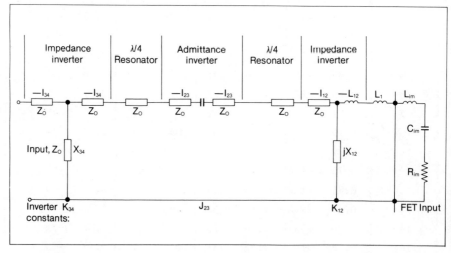

FIG. 5.15. Wideband Impedance Matching Network for Minimum Noise Figure

The FET's equivalent noise input impedance is resonated with the series inductance L_1 and coupled to a network consisting of 2 quarter wavelength resonators. In this technique the resistance $R_{im} = 12\Omega$ can be considered the load resistance and $(L_1 + L_{im}) - C_{im}$ the last resonator of a 3 stage bandpass filter coupling R_{im} to the source resistance Z_0.

By choosing the resonator line impedances to be equal to Z_0 the inverter constants of Fig. 5.15 are given by (Matthaei et al, 1964)

$$K_{12} = Z_0 \left(\frac{\pi \omega\, R_{im}}{4\, g_1\, g_2\, Z_0\, \delta} \right)^{\frac{1}{2}}$$

$$J_{23} = \frac{\pi \omega}{4\, Z_0} \left(\frac{1}{g_1\, g_3} \right)^{\frac{1}{2}}$$

and
$$K_{34} = Z_0 \left(\frac{\pi \omega}{4\, g_3\, g_4} \right)^{\frac{1}{2}}$$

where ω is the relative bandwidth:
$$\omega = \frac{(f_U - f_L)}{f_0}$$

where f_0 is the centre frequency $= \dfrac{f_U + f_L}{2}$ and where f_u and f_L are the upper and lower amplifier band limits.

δ is the decrement of the load impedance at f_0 given by
$$\delta = \frac{(2\pi\, f_0\, R_{im}\, C_{im})}{\omega}$$

g_1, g_2, g_3 and g_4 are the normalized element values of the Tchebyschev impedance matching prototype for the decrement δ (Matthaei et al, 1964).

1. Series Inductance

$$L_1 = \frac{1}{(2\pi\, f_0)^2\, C_{im}} - L_{im}$$

2. Impedance Inverters

$$X_{iK} = \frac{K_{iK}}{1 - \left(\dfrac{K_{iK}}{Z_0} \right)^2}$$

$$\ell_{iK} = \frac{\lambda}{4\pi} \tan^{-1} \left(\frac{2\, X_{iK}}{Z_0} \right)$$

and
$$L_{12} \approx \frac{K_{12}}{2\pi f_0}$$

3. <u>Admittance Inverter</u>

$$B_{23} = \frac{J_{23}}{1 - (J_{23} Z_0)^2}$$

$$\ell_{23} = \frac{\lambda}{4\pi} \tan^{-1}(2B_{23} Z_0)$$

where λ is the wavelength of the transmission line at f_0.

The synthesis for the X-band amplifier with

f   = 8 to 12 GHz

R_{im} = 12Ω

C_{im} = 0.23 pF

L_{im} = 0.41 nH (including 0.12 nH of wire bond)

is as follows

$$\omega = \frac{(12-8)}{10} = 0.4$$

$$\delta = \frac{(2 \times 10 \times 10^9 \times 12 \times 0.23 \times 10^{-12})}{0.4}$$

$$\delta = \underline{0.433}$$

For a decrement δ, Fig. 4.09-3 of (Matthaei et al, 1964) indicates that an insertion loss maximum of 0.85 dB will be achieved with n = 2 with a corresponding Tchebyschev ripple of 0.4 dB. However, this will not fall within the criterion of 0.6 dB increase in noise figure. Thus with n = 3 we have an insertion loss maximum of 0.6 dB and a corresponding ripple of 0.2 dB.

For the decrement δ = 0.43 and n = 3 the normalized element values of the Tchebyschev impedance-matching prototype can be found from Fig. 4.09-7 of (Matthaei et al, 1964).

g_1 = 2.3, g_2 = 0.68, g_3 = 1.5 and g_4 = 0.53.

1. <u>Series Inductance</u>

$$L_1 = \frac{10^9}{(2\pi \times 10^{10})^2 \times 0.23 \times 10^{-12}} - (0.41-0.12) \text{ nH}$$

$$= \underline{0.81} \text{ nH}$$

2. Impedance Inverters

$$X_{12} = \frac{K_{12}}{1 - \left(\dfrac{K_{12}}{Z_0}\right)^2}$$

where

$$K_{12} = Z_0\left(\frac{\pi \times 0.4 \times 12}{4 \times 2.3 \times 0.68 Z_0 \times 0.43}\right)^{\frac{1}{2}} = 16.74$$

$$\therefore \frac{18.85}{2\pi \times 10^{10}} \text{ nH, i.e. } \underline{0.3 \text{ nH}}$$

the equivalent shunt inductor being

$$\therefore \quad \ell_{12} = \left(\frac{\lambda}{4\pi}\right) \tan^{-1}\left(\frac{2X_{12}}{Z_0}\right) = 0.05\lambda \ (18.5°)$$

and

$$L_{12} \approx \frac{K_{12}}{2\pi f_0} = \underline{0.27 \text{ nH}}$$

Thus the first inductor is $(0.8-0.27)$ nH = $\underline{0.53 \text{ nH}}$

$$X_{34} = \frac{K_{34}}{1 - \left(\dfrac{K_{34}}{Z_0}\right)^2}$$

where

$$K_{34} = Z_0 \sqrt{\frac{(0.4\pi)}{4 \times 1.5 \times 0.53}}$$

$$\therefore \quad X_{34} = 52,$$

the equivalent shunt inductor being $\dfrac{52}{2\pi \times 10^{10}}$ = $\underline{0.83 \text{ nH}}$

and

$$\ell_{34} = \left(\frac{\lambda}{4\pi}\right) \tan^{-1}\left(\frac{2X_{34}}{Z_0}\right) = 0.09\lambda$$

which is equivalent to a phase length of $26.3°$.

3. Admittance Inverters

$$B_{23} = \frac{J_{23}}{(1 - J_{23} Z_0)^2}$$

where

$$J_{23} = \frac{0.4\pi}{4 Z_0}\left(\frac{1}{0.68 \times 1.5}\right)^{\frac{1}{4}} = 0.0062$$

therefore $B_{23} = 13.1 \times 10^{-3}$ the equivalent series capacitor being

$$\frac{13.1 \times 10^{-3}}{2\pi \times 10^{10}} = \underline{0.21 \text{ pF}}$$

$$\ell_{23} = \frac{\lambda}{4\pi} \tan^{-1}(2 \times B_{23} \times Z_0) = 0.073\lambda$$

and $l_{12} + l_{23} = (0.05 + 0.073)\lambda = 0.123\lambda$

Thus, the resonator length is

$(0.25 - 0.123)\lambda = \underline{0.127\lambda}$

and $l_{23} + l_{34} = (0.073 + 0.09)\lambda = 0.163\lambda$, with the corresponding resonator length of:

$(0.25 - 0.163)\lambda = \underline{0.087\lambda}$

B. OUTPUT MATCHING NETWORK

In this example the transistor has been matched for minimum noise figure over the band so that the device will exhibit a gain slope of approximately -3 dB/octave, a figure found by measurement for the particular FET. An ideal output network then has to couple all available power to the load at the upper band edge, f_u and provide a frequency dependent attenuation that compensates the FET's gain slope for $f < f_u$.

A network suitable for this is shown in Fig. 5.16.

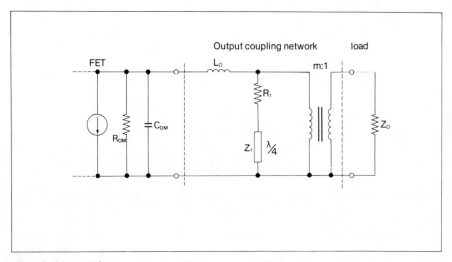

FIG. 5.16.   Wideband GaAs FET Output Matching Network

The output impedance of the transistor R_{om} (in this particular case equal to 250 ohms), in parallel with C_{om} (in this case 0.17 pF) is resonated with the series inductance L_o at the upper frequency, f_u. The transformer matches the resulting resistance to the load Z_o.

At f_u, no power is dissipated in the resistor R_1 since it is connected in series with a short-circuited stub being $\lambda/4$ at f_u. Below f_u the power transferred to the load decreases since the impedance of the resonated FET output changes rapidly and the shunt branch with the resistor becomes increasingly lossy. The resistance R_M and the transmission line impedance Z_M are chosen to give the best approximations to a 3 dB/octave slope.

The complete circuit is shown in Fig. 5.17.

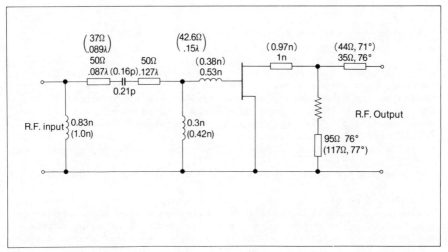

FIG. 5.17. Single Stage Low Noise X Band GaAs FET Amplifier With Absorptive Output Matching

The value of the resonating inductor L_o is found at the upper band edge frequency of 12 GHz to be:

$$L_o = \frac{1}{(2\pi \times 12 \times 10^9)^2 \times 0.17 \times 10^{-12}}$$

$$= 1 \text{ nH}$$

The Q of the FET output is given by

$$Q = 2\pi f_u R_{om} C_{om}$$

$$= 2\pi \times 12 \times 10^9 \times 250 \times 0.17 \times 10^{-12}$$

$$= 3.2$$

Leichti et al (1974) have shown that for such a Q value

$$\frac{R_1}{R_{om}} = 0.21, \text{ i.e. } R_1 = 52.5\Omega$$

$$\frac{Z_1}{R_{om}} = 0.38, \text{ i.e. } Z_1 = 95\Omega$$

and the transformer ratio $m^2 = (\frac{R_{om}}{Z_o}) \times 0.1$

$$m = 0.707$$

i.e. a quarter wavelength resonator of impedance 35Ω of length 0.21λ at the centre frequency of 10 GHz.

The accuracy of the design using the unilateral model has been checked using computer analysis and optimization with the FET represented by its S-parameters and optimum noise reflection coefficients etc., using the COMPACT program. The resulting element values are shown in Fig. 5.17 in brackets. It may be appreciated that the initial element values are close to those finally adopted after optimization and this circuit gives a theoretical noise figure which is 0.6 dB higher than the optimum noise figure at 8 GHz. Worst case input and output VSWR's were 3.5:1 and the gain was 8.8 dB $\pm$ 0.25 dB.

It should be stressed that the above example is only one method that can be used to synthesise networks suitable for low noise amplifier realisation. The choice of a particular means of matching into a device is often dictated by the available technology and the device selected. For packaged devices, for example, it is desirable to add a short length of line to the packaged FET S-parameters to act as a soldering area. Such action moves the matching load impedance of the device and this should be assumed at the beginning of the design.

7. COMPUTER-AIDED DESIGN PRACTICE

As we have seen in the last example the use of a computer optimization routine is of great advantage in producing a final circuit design suitable for realisation. The computer should not be used as a replacement for good microwave engineering but an addition to the microwave engineer's range of aids to design.

A. GENERAL FORMAT OF MICROWAVE CAD PROGRAMS

There are a considerable number of microwave circuit analysis and optimization routines in use today ranging from specially written 'in-house' packages to wide-ranging 'bureaux' type programs (e.g. COMPACT version 5.1).

All programs of the type mentioned have several items in common.

Circuit Elements. Circuit elements are input into the program as separate building blocks and are cascaded within the program using transmission matrices. Elements may be lumped or distributed and may be varied by the program in its optimisation mode. Feedback elements, series or parallel, nodal analysis etc. may also be features of the program. S-parameters and optimum noise data are usually input to the program as separate blocks of data or are stored in a separate data file and 'called-up' by the main working program.

The circuit is described using codes or abbreviations.

Frequency Input. There are usually two ways of entering frequency. Firstly as a series of separate frequencies or as a band with constant intervals.

Noise Figure Data. This is usually specified in terms of the optimum noise figure, the optimum reflection coefficient and either the noise figure at $\Gamma_S = 0$ (i.e. 50Ω) or the noise resistance R_N. There is also a weighting factor associated with noise figure as explained later.

One and Three Port S-parameters. It is often found possible to input single port data such as that used with a diode and also three port data such as that associated with a dual-gate FET. In the latter case the three-port S-parameters of which there are nine elements are converted to 2 port parameters with a specified load on the third port.

Negative Components. Many programs allow the user to subtract element values, thus allowing the designer to 'strip-off' certain parts of a circuit to reveal the S-parameters of an 'imbedded' device. This is useful in obtaining the chip parameters of FETs, for example.

Optimization. In order to affect optimization the program needs to have an error function input to which it attempts to converge with the circuit description available. Usually the total error function, which is a sum of the sum of the squares of the individual error functions is weighted.

The total error function may be written as

$$E = \frac{1}{N} \sum_{F1}^{F2} (W_1 \cdot |S_{11}|^2 + W_2 \cdot |S_{22}|^2 + W_3 \cdot |(S_{21} - G)|^2$$
$$+ W_4 \cdot |F_{OPT} - F_{ACT}|^2)$$

where S_{11}, S_{22} and S_{21} are the scattering parameters of the resultant overall circuit; F_1 is the first and F_2 the last frequency of interest; N is the number of frequencies, W_1 to W_4 are the weighting factors; G is the desired amplifier gain and F_{OPT} and F_{ACT} are the optimum and actual noise figures.

Sensitivity Analysis and Mapping. Certain programs allow the user to analyse his circuit design in terms of its sensitivity to changes in component values. This can be done using a Monte Carlo analysis (e.g. COMPACT Version 5.1) where specified components are altered in some random fashion within specified limits thus simulating, for example, production tolerances on transistor noise figures and associated gains. It is also possible to use some programs in a 'search' mode which will indicate the elements in a particular circuit which are the most sensitive or indicate whether the circuit is capable of performing the task asked of it.

A few rules worth bearing in mind when using computer aided analysis and optimization of microwave circuits are

1. Use conventional matching theory with a Smith Chart or synthesis techniques to get a starting circuit for optimization.

2. Avoid large numbers of frequencies and variables. This will result in excessive computer time and cost. It is often possible to select the most sensitive elements by a few trial analyses.

3. A stability analysis should always be achieved before an optimization run is performed. The addition of resistive loading or feedback will often produce a stable condition after inspection of the stability circles such that optimization will be successful.

4. Do not try and optimize a complicated circuit such as a multistage amplifier before analysing the individual circuits.

5. The selection of weighting factors is important and attention should be paid to the gain and noise figure parameters of the transistors used.

8. NETWORK SYNTHESIS

The use of the Smith Chart and resulting calculations is somewhat time consuming. Passive network synthesis, has been used in the design of filters and can be applied to the design of interstage input and output matching networks for microwave amplifiers where the networks operate between unequal impedance levels and include parasitic elements associated with the active devices.

Synthesis routines (Skwirzynski, 1971; Saal et al, 1958) have been in use for many years and more recently have been extended to include the synthesis of networks of prescribed gain, bandwidth, ripple, slope and impedance transformation. Mellor et al (1975) has produced a synthesis routine on which is based a commercial synthesis computer program, AMPSYN.

An example of the technique is given in the following:

Example: An interstage network for a 2 stage FET amplifier is to be designed to operate between a 135 ohm source and a 10.75 ohm load with a 6 dB/octave gain slope and a 0.4 dB ripple over the 6 to 12 GHz frequency range. A 300Ω shunt resistor is used at the output of the first FET to stabilize the device. The specification for good input and output match implies a frequency response that is flat over the passband of the amplifier.

The interstage matching networks must provide a positive-sloped gain with frequency (Fig. 5.18e) to compensate the transistor's roll-off (Fig. 5.18c) and give an overall flat gain.

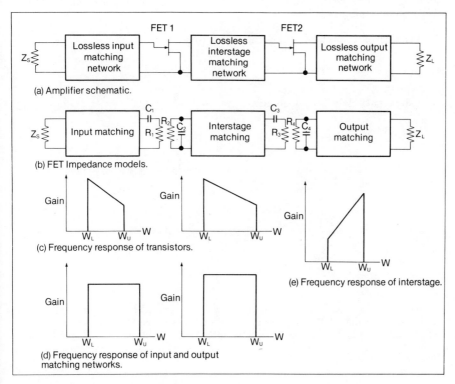

FIG. 5.18.   Typical Amplifier Design Problem for Network Synthesis

The procedure to be described in this example is based on a method of synthesis of matching networks of arbitrarily specified passband slope with provision for inclusion of parasitic elements. The parasitic elements in this particular case are the output capacitance of a GAT6 namely 0.17 pF and the input capacitance namely 0.31 pF.

1) In general two resistors coupled by a lossless ladder network having all transmission zeros at zero and infinite frequency are called trapless filters. The insertion loss for such filters can be

expressed as

$$IL = \frac{\text{power available from source resistance}}{\text{power delivered to load resistance}}$$

$$= \frac{a_o + a_2\omega^2 + a_4\omega^4 + \cdots\cdots\cdots + a_{2N}\omega^{2N}}{\omega^{2J}} \qquad 5.38$$

where N is the order of the network, i.e. the numbers of inductors and capacitors before impedance transformation and J is the number of high pass elements.

2) From this insertion loss function form

$$|\rho_1(\omega)|^2 = 1 - 1/IL \qquad 5.39$$

3) Substitute $s = j\omega$ to obtain

$$|\rho_1(s)|^2 = \rho_1(s)\rho_1(-s) \qquad 5.40$$

4) Obtain the poles and zeros of the resulting expression in the usual manner (Mellor et al, 1975).

5) Form

$$\rho_1(s) = \pm 1 \frac{\displaystyle\prod_{i=1}^{N}(s-Z_i)}{\displaystyle\prod_{i=1}^{N}(s-\rho_i)} \qquad 5.41$$

where the poles of $\rho_1(s)$ are the poles of $|\rho_1(s)|^2$ that are in the left hand plane (LHP) and the zeros of $\rho_1(s)$ are chosen from the zeros of $|\rho_1(s)|^2$ that are in the LHP or the RHP (right hand plane).

6) Form

$$Z_1(s) = R_s \frac{1 + \rho_1(s)}{1 - \rho_1(s)} \qquad 5.42$$

where R_s is the source resistance.

7) Determine the network element values by forming the appropriate expansion of $Z_1(s)$ about zero and infinity.

In order to absorb the parallel capacitance at the FET drain and also the series capacitance at the second stage FET gate we require a network of the form shown in Fig. 5.19a.

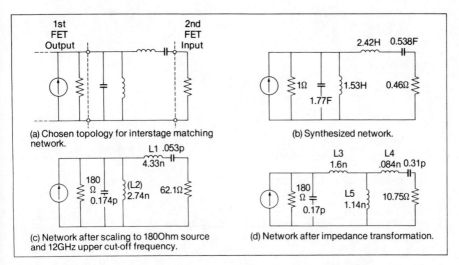

(a) Chosen topology for interstage matching network.

(b) Synthesized network.

(c) Network after scaling to 180Ohm source and 12GHz upper cut-off frequency.

(d) Network after impedance transformation.

FIG. 5.19.   Synthesis Routine for FET Amplifier Interstage Matching

This is a network of N = 4; J = 1 type. Insertion loss functions approximating 6 dB/octave gain slope are easily obtained from flat insertion-loss functions. The flat insertion-loss function is first normalised to an upper cut off frequency of 1 rad/sec and then divided by ω^2. This operation results in a sloped insertion loss of exactly the same ripple and passband as the flat insertion loss case. The order of the IL functions remains unchanged (i.e. at 4) while the number of transmission zeros at d.c. (J) is increased by 1.

1) The insertion loss function is found using the methods of Mellor et al (1975) and is given by:-

$$IL = \frac{0.8 - 5.8\omega^2 + 19.9\omega^4 - 23.9\omega^6 + 10.1\omega^8}{4}$$

2) $|\rho_1(\omega)|^2 = \dfrac{0.8 - 5.8\omega^2 + 18.9\omega^4 - 23.9\omega^6 + 10.1\omega^8}{0.8 - 5.8\omega^2 + 19.9\omega^4 - 23.9\omega^6 + 10.1\omega^8}$

Substituting $\omega = s/j$ gives

3) $|\rho_1(s)|^2 = \dfrac{0.8 + 5.8s^2 + 18.9s^4 + 23.9s^6 + 10.1s^8}{0.8 + 5.8s^2 + 19.9s^4 + 23.9s^6 + 10.1s^8}$

4) The poles of $\rho_1(s)^2$ are

$$S = \pm 0.19 \pm j0.48 \;)$$
$$) \quad \text{from the numerator}$$
$$S = \pm 0.19 \pm j1.01 \;)$$

The zeros of $|\rho_1(s)|^2$ are given by

$$S = \pm 0.18 \pm j0.51 \;)$$
$$) \quad \text{from the denominator}$$
$$S = \pm 0 \pm j0.98 \quad)$$

5) Thus

$$\rho_1 = \frac{(-1)(s+j0.98)(s-j0.98)(s-0.18+j0.51)(s-0.18-j0.51)}{(s+0.19+j0.48)(s+0.19-j0.48)(s+0.19+j1.01)(s+j0.19-j1.01)}$$

where the zeros of ρ_1 are chosen to be in the right hand plane. Multiplying out we obtain

$$\rho_1 = \frac{-s^4 + 0.36s^3 - 1.2s^2 + 0.36s - 0.28}{s^4 + 0.74s^3 + 1.4s^2 + 0.5s + 0.28}$$

$$Y_1 = \frac{1}{Z_1} = \frac{1-\rho_1(s)}{(1+\rho_1(s))} = \frac{2s^4 + 0.38s^3 + 2.7s^2 + 0.14s + 0.56}{1.1s^3 + 0.21s^2 + 0.86s}$$

$$= 1.77s + \frac{1}{1.53s} + \cfrac{1}{\left(2.42s + \cfrac{1}{\frac{1}{0.538s}} + 0.46\right)}$$

The synthesised network is then as given in Fig. 5.19b. The network is scaled to 135 ohm source and 12 GHz upper cut-off frequency to give Fig. 5.19c.

The shunt-series connection of inductors is used to transform the found impedance level of 62 ohms to the wanted level of 10.75 ohms. The resulting TEE network is given in Fig. 5.19d where the new component values are given by

$$L_3 = (1-n)L_2$$

$$L_4 = n^2(L_1+L_2) - nL_2$$

$$L_5 = nL_2 \quad \text{where } n^2 = 10.75/62$$

Thus, $n = 0.416$

$$L_3 = 1.6 \text{ nH}$$

$$L_4 = 0.084 \text{ nH}$$

$$L_5 = 1.14 \text{ nH}$$

and the synthesis is completed by scaling the series capacitor from 0.0528 pF to 0.3 pF.

Now the output capacitance of the 1st FET can be absorbed into the
0.174 pF shunt capacitance whilst the 0.31 pF input capacitance of
the 2nd FET is absorbed exactly by the series capacitance thus yield-
ing the circuit of Fig. 5.19e.

This method is obviously tedious especially where the IL function
is to be calculated and the zeros and poles of the network identified.
However, such a synthesis technique is readily amenable to CAD
programming.

This example is therefore completed by designing a complete 2 stage
amplifier to operate over the 8 to 12 GHz band using the input network for the low-noise 1st stage as previously defined followed by
the presently designed interstage network and completed with the
design of an output stage network using the synthesis procedure
described.

The GAT6 transistor chip on which the above example has been based
has an approximate 4 dB/octave gain slope under maximum gain conditions. Thus a two stage network will have an 8 dB/octave gain slope.
However, the input matching network has been designed for minimum
noise figure and it may be shown by measurement that the available
gain of the FET when matched for minimum noise figure exhibits a
slope of approximately 3.2 dB/octave. Thus the output matching
network has only to provide $-(3.2 + 4-6)$ dB/octave gain slope, i.e.
approximately 1.2 dB/octave slope where the 2nd stage is biased for
maximum gain.

Fig. 5.20a shows the final complete circuit of the two stage amplifier together with its frequency response (Fig. 5.20b).

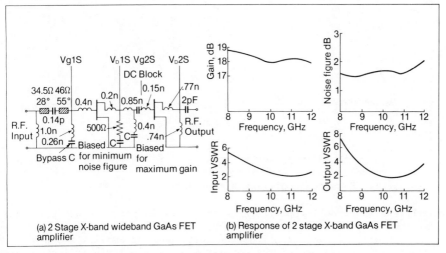

(a) 2 Stage X-band wideband GaAs FET amplifier

(b) Response of 2 stage X-band GaAs FET amplifier

FIG. 5.20. 2 Stage X-Band Wideband GaAs FET Amplifier

9. THE USE OF SINGLE ENDED AND BALANCED AMPLIFIERS

The examples of amplifier design which have been given indicate
that it is possible to design more than one stage of amplification to
achieve gain flatness over a specified bandwidth. However, for
multistage amplifiers where gains of 30 dB or greater are required
the use of such "single-ended" realisations can lead to difficulty
in producing the required gain flatness etc. This is because of the
mismatches occurring between gain stages causing gain and phase
variations. In the single-ended case it is possible to significantly
reduce such gain ripples by using interstage as well as input and
output isolators. These devices can produce significant loss which
will increase the overall noise figure of the unit. For integrated
amplifiers such isolators can also be comparatively large unless they
are integrated into the amplifier medium, e.g. the use of planar
isolators with a microstrip amplifier.

An alternative approach which is often used in commercial amplifiers is the adoption of balanced amplifier techniques. Fig. 5.21
illustrates the concept. Two amplifier units of the same performance
are arranged between the output and input ports of quadrature 3 dB
couplers. The theory of balanced amplifiers has been given by
Kurokawa (1965) and it may be shown providing the two amplifiers are
matched in their characteristics that the input VSWR and output VSWR
are predominantly dependent on that of the couplers.

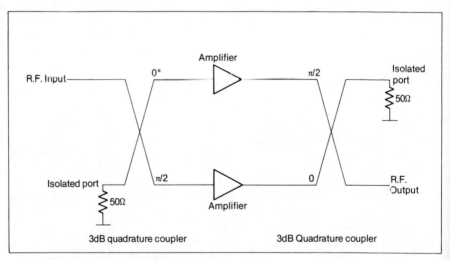

FIG. 5.21. Balanced Amplifier Arrangement

If the individual amplifiers in the balanced pair are not perfectly
matched at certain frequencies then a signal in the 0° arm of the
coupler will be reflected from the corresponding amplifier and a
signal in the $\pi/2$ arm of the coupler will be similarly reflected from

its amplifier. On travelling through the coupler after reflection the signals will again be phased 0 or $\pi/2$ and the reflected power is 'dumped' into the isolated port (Fig. 5.21).

In a microstrip realisation for octave band-amplifiers one of the most popular 3 dB couplers is the so-called Lange coupler which achieves the coupling factor required by using interdigitated $\lambda/4$ fingers (Lange, 1969). Another coupler which is used for more narrow band applications is the 'branch-line' coupler (Reed et al, 1956).

The overall noise figure of the amplifier is the same as the equivalent single-ended amplifier with the addition of the loss of the input coupler. However, the amplifier is able to handle an increase of 3 dB in input r.f. power for the same compression characteristic and the third order intermodulation performance is increased by 9 dB assuming perfect phase matching between the 2 parallel and matched amplifier chains.

Fig. 5.22 shows the microstrip realisation of a balanced amplifier which is fabricated with chip FETs on an alumina substrate medium.

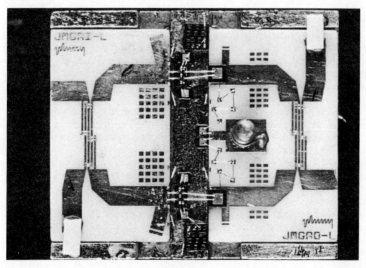

FIG. 5.22. Example of Wideband Balanced Amplifier (5 to 10 GHz)

The balanced amplifier has the further advantage that if for some reason one of the FETs fails in a unit then the amplifier will still operate albeit with a reduced performance.

Balanced amplifiers enable the design engineer to produce a matched 'gain block' which is able to be cascaded with other matched 'blocks' such as other amplifiers, attenuators, detectors etc.

A major advantage of the balanced configuration is that of stability. As has already been demonstrated stable gain can be assured in a single-ended amplifier by using a stabilising resistor (or reactance in the source to ground connection). However, such action reduces the available gain of the transistor and does not allow the minimum noise measure to be achieved in a multistage amplifier. The use of balanced units ensures that the input and output of each FET stage 'sees' an impedance close to 50 ohms. Indeed the stability factor of such a balanced stage can be an order of magnitude higher than its single-ended equivalent, depending on the VSWRs and isolation of the quadrature couplers used (Kurokawa, 1965).

The disadvantage of using the balanced approach is that the unit uses double the number of active devices in a single-ended unit, is larger and consumes more d.c. power. However the advantages to be gained from the balanced approach will usually outweigh the disadvantages.

10. VARIATIONS IN AMPLIFIER PERFORMANCE

A. TRANSISTOR VARIATIONS AND CIRCUIT SENSITIVITY

Transistor parameters which are given in manufacturers' data sheets are only typical sets of data collected from a large number of measurements. In order to allow the use of any transistor of a particular type and also to allow for circuit realisation tolerances it is often necessary to include some means of adjustment in the designed matching circuits.

For amplifiers realised in a planar transmission medium such as microstrip or triplate the use of adjustable open and short circuited series transmission line elements can often compensate for variations in the FET gate-source and drain-source capacitance due to process variations in channel doping density and variations in the input capacitance and gate resistance due to gate length variations.

Fig. 5.23a and b show a typical spread of S-parameters and Z_{OPT} data for a transistor together with a corresponding circuit which was designed to compensate for such variations without affecting the performance of the amplifier (Estabrook et al, 1978).

B. VARIATIONS IN AMPLIFIER PERFORMANCE WITH TEMPERATURE

Fig. 5.4 has shown the variation in the associated gain of a single gallium arsenide FET over the temperature range -40 to $+80^{\circ}C$. Fig. 5.4 also shows the effect of temperature on the noise figure of a single transistor. For a multistage amplifier, therefore, substantial gain changes can occur with temperature as well as variations in the overall amplifier noise figure. Because of small variations in the S-parameters with temperature the gain ripple over the frequency band of interest may also alter its character. Thus for certain applications, particularly military, an amplifier must be constructed

which has integrated compensation to maintain gain to within specified
limits over a certain range of temperatures.

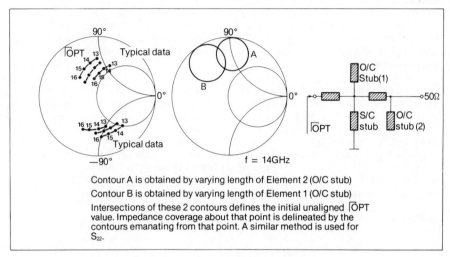

FIG. 5.23. Typical Method for Adjustment of Amplifier Performance
to Account for Transistor Variations

This can be achieved using three different techniques.

1. By varying the bias on single-gate FETs.

2. By using the AGC facility of a dual-gate FET by varying the bias on
the second-gate.

3. By using a PIN diode attenuator.

With all three methods a temperature sensing element is needed such as
a thermistor or diode, a circuit enabling the gain change 'law' of the
amplifier to be realised and a drive circuit.

11. DESIGNING AN AMPLIFIER FOR A SPECIFIED LINEAR OUTPUT POWER

So far in this chapter we have described the design of maximum gain
and low noise figure amplifiers. There are many cases where it is
also desirable to be able to specify the r.f. power handling capabilities of the amplifier. This is usually achieved using the definition of 1 dB gain compression point where large signal gain is 1 dB
lower than the small signal linear gain of the device. For a GaAs
FET the 1 dB gain compression point for a small signal device is that
measured at a specified drain to source voltage (e.g. V_{DS} = 5V) and a
gate to source voltage which optimises the output power (usually such
that I_{DS} = 50% I_{DSS}).

184

Small signal FETs usually have 1 dB gain compression points of +10 to +13 dBm but certain small signal FETs will give much higher output powers, e.g. a Plessey GAT4/021 will give r.f. output powers of +20 dBm at 8 GHz.

If two signals at frequencies f_1 and f_2 are input to a FET device the third order intermodulation products occur at frequencies $2f_1-f_2$ and $2f_2-f_1$ where the products $2f_1+f_2$ and $2f_2+f_1$ are assumed to be out-of-band.

By observing the 3rd order intermodulation products (IP) on a spectrum analyser, for example, it is possible to plot the graph of 3rd order product power output versus input power. If this plot is compared with the fundamental output power plot a point of intersection occurs where the 3rd order IP amplitude equals the fundamental signal amplitude. This point is referred to as the 3rd order intercept point (Fig. 5.24). The 3rd order intercept point allows the designer to predict the 3rd order intermodulation product level at any r.f. input power.

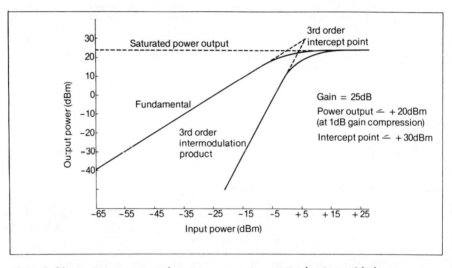

FIG. 5.24.   Intercept Point Response for a Typical Amplifier

An amplifier is designed for optimum output r.f. power handling using the same matching techniques already discussed. However, the S-parameters that are used in the design should be those measured with the transistor operating at $\frac{1}{2}I_{DSS}$. For amplifiers having output powers up to +13 dBm this technique of design is considered valid but for higher power amplifiers the small signal S-parameters become invalid since the transistor impedance levels change when subjected to larger r.f. powers. This is more fully covered later in this chapter.

12. THE USE OF FEEDBACK, COMMON-GATE AND SOURCE-FOLLOWER GaAs FET CONFIGURATIONS IN AMPLIFIER DESIGN

A. FEEDBACK AMPLIFIERS

The use of negative feedback in bipolar transistor microwave amplifier stages is now a universally adopted design approach. The technique allows the designer to control the amplifier's performance and
to design a configuration which allows the use of transistors with
variable parameters.

Negative feedback can control the input and output impedances of a
single-ended stage without resource to balanced techniques. Negative
feedback produces an amplifier capable of multi-octave bandwidths and
gain variations held to a few tenths of a dB. A negative feedback
amplifier circuit is shown in Fig. 5.25a.

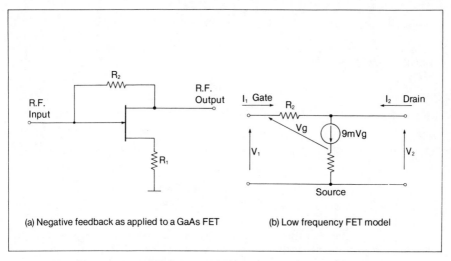

(a) Negative feedback as applied to a GaAs FET   (b) Low frequency FET model

FIG. 5.25.   Negative Feedback FET Amplifier

For simplicity the d.c. bias components and coupling capacitors are
excluded. Representing the FET by its low frequency model (Fig.5.25b)
the results of an analysis in terms of S-parameters is as follows
(Cooke, 1978).

$$S_{11} = S_{22} \cfrac{1 - \cfrac{g_m Z_0^2}{R_2(1 + g_m R_1)}}{1 + \cfrac{2Z_0}{R_2} + \cfrac{g_m Z_0^2}{R_2(1 + g_m R_1)}}$$

$$S_{21} = \cfrac{1 - \cfrac{2Z_0}{R_1} + \cfrac{2Z_0}{R_2}}{1 + \cfrac{2Z_0}{R_2} + \cfrac{g_m Z_0^2}{R_2(1 + g_m R_1)}}$$

and

$$S_{12} = \cfrac{2Z_0}{R_2\left(1 + 2\cfrac{Z_0}{R_2} + \cfrac{g_m Z_0^2}{R_2(1 + g_m R_1)}\right)}$$

where g_m is the transconductance of the FET and Z_0 is the characteristic impedance of the source and load. For the condition that $S_{11} = S_{22} = 0$, it may be shown from Eq. 5.43 that

$$1 + g_m R_1 = \frac{g_m Z_0^2}{R_2}$$

By substituting Eq. 5.46 into Eq. 5.44 and 5.45 expressions for S_{21} and S_{12} are found to be

$$S_{21} = \frac{(Z_0 - R_2)}{Z_0}$$

and

$$S_{12} = \frac{Z_0}{R_2 + Z_0}$$

and

$$R_1 = \frac{Z_0^2}{R_2} - \frac{1}{g_m} \text{ ohms}$$

Thus, unless the transconductance of the GaAs FET is large enough S_{11} and S_{22} cannot equal zero with a positive value of R_1. The minimum required transconductance required occurs when $R_1 = 0$, and

$$S_{21} = |1 - g_m Z_0|$$

Thus, for example if a 10 dB amplifier gain is required with a characteristic impedance of 50 ohms a transistor with a g_m of 83 mS is required with a feedback resistor of 207Ω.

Single-stage amplifiers using the described method of feedback have been fabricated and show extremely wide bandwidths from virtually d.c. to over 8 GHz. Input and output VSWR's can be maintained to less than 2:1 up to approximately 5 GHz, with a stability factor always greater than 1 (Ulrich, 1978).

More recently feedback amplifiers which also feature a degree of

conventional matching have been designed and fabricated. Both small
signal and power FET amplifiers have been produced using such a technique which was described by Niclas et al (1980).

The parasitic elements of a GaAs FET restrict the amplifier bandwidth capability. Minimization of these parasitics together with the
use of controlled negative feedback design has enabled such amplifiers to be produced covering less than 500 MHz to 14 GHz. The
frequency dependence of the feedback is controlled by two inductors,
one in series with the feedback resistor and the other in series with
the output of the GaAs MESFET. In addition to the feedback circuitry
simple matching networks have been employed to improve the input and
output reflection coefficients. The GaAs FET used was designed with
major emphasis on reducing its parasitics. Figure 5.26a shows the
circuit of a 0.5 to 6 GHz amplifier having the response of Fig.5.26b.

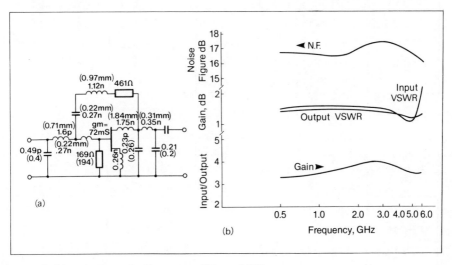

FIG. 5.26.   0.5 to 6 GHz GaAs FET Amplifier

B. COMMON-GATE AND SOURCE-FOLLOWER CONFIGURATIONS

A GaAs FET is most commonly used in common-source operation since
the device is designed to give its maximum gain and minimum noise
figure under this circuit condition.

However considerable advantage can be attained in terms of input
and output reflection coefficients by using common-gate and common-
drain (source-follower) configurations respectively. Table 5.6 shows
the S-parameters for the common gate and source-follower connected
GAT6 and may be compared directly with Table 5.1. It may be seen
that up to 6 GHz the input and output reflection coefficients of the
respective configurations are low indicating that these configurations
would be suitable as the input and output stages of an amplifier.

TABLE 5.6.   Common Gate and Common Drain Connected GAT6 S-Parameters

Frequency GHz	S_{11}		S_{21}		S_{12}		S_{22}	
	MAG	ANG	MAG	ANG	MAG	ANG	MAG	ANG
2	.27	−175	1.26	−9.9	.116	23.6	.91	−7
3	.26	−176	1.25	−13.9	.124	27.8	.92	−8
4	.26	−175	1.25	−18.2	.139	32.1	.94	−10
5	.27	−173	1.25	−22.8	.159	35.6	.96	−13
6	.28	−173	1.25	−27.9	.178	36.3	.98	−16
7	.29	−171	1.27	−33.8	.196	36.5	1.02	−20
8	.3	−172	1.29	−40.2	.218	36.8	1.07	−24
9	.33	−173	1.32	−47.1	.232	36.4	1.13	−28
10	.36	−175	1.35	−54.4	.244	36.1	1.2	−33
11	.36	−178	1.41	−63.3	.263	36.3	1.29	−39
12	.38	178	1.49	−73.8	.251	37.5	1.38	−42

Frequency GHz	S_{11}		S_{21}		S_{12}		S_{22}	
	MAG	ANG	MAG	ANG	MAG	ANG	MAG	ANG
2	.98	−15	1.22	−7.8	.202	73.0	.32	167
3	.98	−20	1.2	−10.9	.267	69.6	.3	158
4	.97	−25	1.18	−13.9	.343	64.2	.28	152
5	.95	−31	1.17	−17.1	.419	58.0	.26	145
6	.94	−37	1.16	−20.4	.492	52.4	.25	138
7	.93	−43	1.14	−23.9	.567	46.5	.23	129
8	.92	−49	1.14	−28.0	.635	40.8	.21	120
9	.91	−56	1.13	−31.8	.7	35.3	.2	110
10	.9	−62	1.13	−35.4	.757	29.7	.18	100
11	.91	−68	1.12	−40.5	.811	25.0	.17	83
12	.88	−75	1.09	−44.4	.882	20.8	.19	63

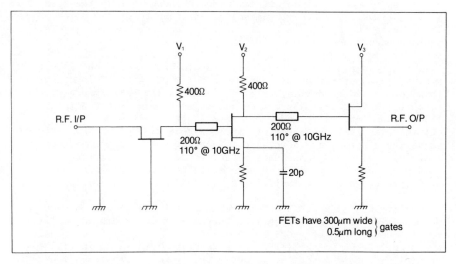

FIG. 5.27(a).   Common Gate/Common Source/Source-Follower Amplifier

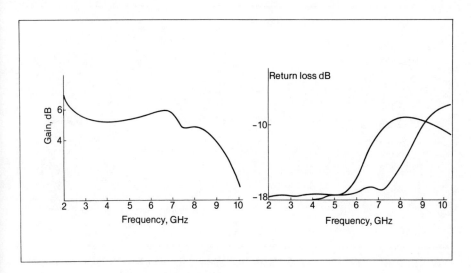

FIG. 5.27(b).   Performance of Circuit in (a)

It should also be noted that $|S_{11}|$ for the source follower configuration is very close to 1, whilst $|S_{12}|$ is large compared with a common source connection. It may also be shown that in the common gate configuration the optimum source admittance for low noise figure is close to 50Ω particularly at the low frequency end for a FET with a transconductance of 20 mS.

In order to provide flat gain over a wide bandwidth the circuit of Fig. 5.27a is of a typical type where a common gate FET is connected to a common source device. Fig. 5.27b shows the performance of the circuit of Fig. 5.27a where the devices used were of the GAT6 type (i.e. 300μm gate width and 0.5μm gate length). In order to flatten-out the gain it is necessary to include some simple matching between stages such as series inductance. Such circuits using impedance matching over broad bandwidths by active devices are particularly suitable where very small components are needed such as in the case of integrated circuits (Suffolk et al, 1980).

Considerable advances are being made in the use of common gate and common drain (source follower) connected FETs in broadband amplifiers for frequencies up to 4 GHz or so where monolithic circuit techniques are used. Decker et al (1980) have described promising initial results of a common source, common gate, common drain cascade used as an IF amplifier covering the 500 to 1000 MHz band. The input impedance of the IF amplifier, shown in Fig. 5.28a is high allowing it to be used directly after a FET mixer. The measured gain of the circuit is shown in Fig. 5.28b and is compared with the predicted gain which takes into account the additional parasitics and losses of the resistors and capacitors in the circuit.

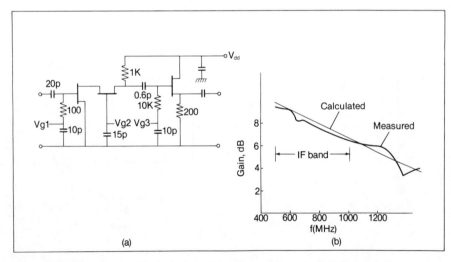

(a)   (b)

FIG. 5.28(a) IF Amplifier Using Cascode Connected FETs and Source-follower Output
(b) Gain of IF Amplifier

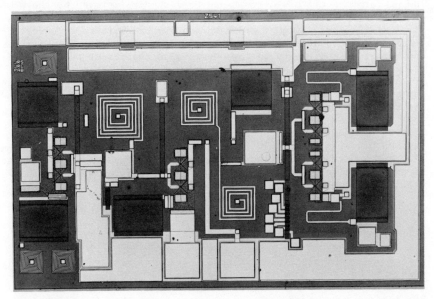

FIG. 5.29(a). Low Noise Common Gate, Common Source, Source Follower
S-Band Active Splitter

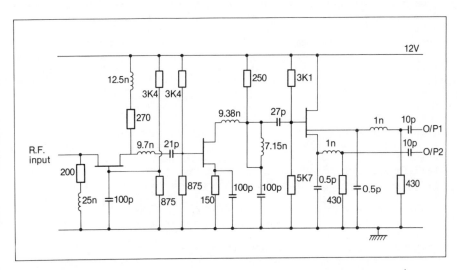

FIG. 5.29(b). Circuit Design for Actively Matched L.N. Preamp/
Splitter

The performance of this circuit can be markedly improved by the use of simple inductive tuning between stages and more recently Suffolk et al (1980) have designed wideband bandpass amplifiers to cover the 2 to 4 GHz frequency range which have 19 dB gain, 3 dB noise figures and very low sensitivity to component value changes of the order of $\pm$ 20% from their mean value. Fig. 5.29a shows the chip realisation of this circuit showing the use of 'spiral' inductors, thin film bias resistors, GaAs FETs, overlay capacitors etc. whilst Fig. 5.29b shows the basic amplifier circuit.

As the parasitic resistances and reactances of GaAs FETs are reduced by using monolithic fabrication techniques the use of common-gate and source follower configurations will become more common-place at higher frequencies. This is also true of very wideband feedback amplifiers covered in another part of this chapter.

13. POWER AMPLIFIERS

A. INTRODUCTION

High power, high efficiency amplification of microwave power using GaAs FETs has been demonstrated over the past few years at frequencies in C, X and Ku band (Macksey et al, 1976; Tserng, 1979). Power FET devices are commercially available which enable linear Class A powers of several watts to be achieved up to X band. A power FET is essentially a multicell structure where the GaAs material and the fabrication techniques used are optimised for higher breakdown voltage than a small signal FET.

GaAs power FETs are very attractive solid state amplifiers. In comparison with silicon bipolar devices, Schottky barrier power GaAs FETs have higher maximum frequency of oscillation, low noise figures and large gains. Above 4 GHz they have higher efficiencies, do not exhibit secondary breakdown, are self-ballasting and have inherently higher input impedance. A typical power GaAs FET is shown in Fig. 5.30a.

B. D.C. CHARACTERISTICS

The device shown in Fig. 5.30a has a total gate width of 2400µm (compared to typically 300µm for a small signal device) comprising 4 unit cells each 600µm wide. Devices with well matched cell characteristics are required since such devices yield good power-added efficiencies (over 30%) and good thermal characteristics.

The maximum output power obtainable from a GaAs FET in class A operation with full gate modulation is determined from the characteristic of Fig. 5.30b. The maximum output voltage and current swings give output power for a sinusoidal input of

$$P_{OUT} = \frac{I_F (V_B - V_S)}{8}$$

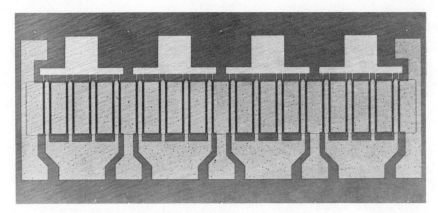

FIG. 5.30(a).   2 Watt Power GaAs FET

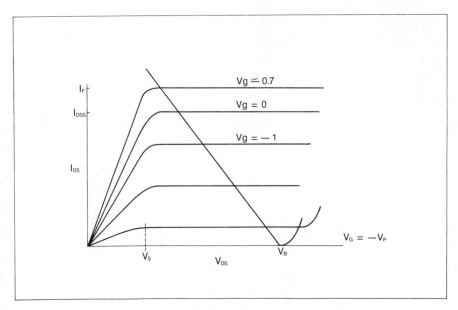

FIG. 5.30(b).   I-V Characteristics of a Power FET Showing the Load
    Line

where I_F is the value of I_{DS} corresponding to a forward gate bias of
0.7V (for an aluminium to GaAs Schottky barrier), V_B is the source-
drain voltage at which breakdown occurs in the pinch-off region and
V_S is the saturation or 'knee' voltage of Fig. 5.30b usually between
1.5 and 2 volts. The optimum load resistance is

$$R_L = \frac{(V_B - V_S)}{I_F}$$

From Equations 5.51 and 5.52 the requirements for high output power
may be seen. Large values of I_F are obtained by placing in parallel
a number of FET 'cells' to give a large total gate width. The number
of gates put in parallel is limited by the minimum load resistance
which can be presented to the device by equation 5.52.

The breakdown voltage V_B is determined principally by the avalanche
breakdown of the gate to drain diode. These are related by the
equation

$$V_B = V_A - V_P$$

where V_p is the pinch-off voltage.

C. R.F. CHARACTERISTICS OF POWER GaAs FETs

(i) S-Parameters

The small signal scattering parameters of power FETs are usually
supplied on manufacturers' data sheets. This enables the design
engineer to obtain a device bandwidth capability and also indicates
the regions of instability of the device. The low frequency S-
parameters of a power FET usually indicate, below 2 GHz, that the
regions of instability are large particularly for the output of such
a device so that it is necessary to use bias networks, for example,
that present an impedance to the device close to 50 ohms. Table 5.7
shows the small signal S-parameters for a 1 watt power FET from 2 to
12 GHz in 1000 MHz steps. The Table shows the maximum available gain
of the device as well as the Rollett stability factor, K. It is
apparent that the gain of a power FET is substantially lower than that
of a small signal device, a result of the device being optimised for
its power handling performance.

(ii) Large Signal Parameters

The small signal S-parameters of a power FET, although useful
design aids, are not so useful for accurate prediction of the large
signal gain and power output of a device either at a spot frequency
or over a bandwidth. The measurement of high power S-parameters is
difficult. The measurement of the source and load impedances for the
device operating at its 1 dB gain compression point are more easily
measured and together with the small signal S-parameters give sufficiently accurate information for design. These impedances are
measured by matching the device using for example, stub tuners at the

r.f. power levels of interest and then separately measuring the tuner impedance using a conventional small signal network analyser. The required impedance is the conjugate of the measured stub impedance.

TABLE 5.7. Typical S-Parameters for a Plessey PGAT1000 Power FET (P105 Package)

Frequency GHz	S_{11}		S_{21}		S_{12}		S_{22}	
	MAG	ANG	MAG	ANG	MAG	ANG	MAG	ANG
2	0.9	−149	3.15	56.8	.035	11.0	0.33	−86.0
3	0.82	−171	2.57	33.5	.035	−4.0	0.38	−104
4	0.86	168	2.11	9.0	.036	−20.7	0.44	−121
5	0.82	145	1.71	−13.1	.038	−30.1	0.50	−142
6	0.79	126	1.38	−39.0	.039	−42.8	0.55	−171
7	0.72	107	1.17	−64.0	.042	−53.0	0.58	166
8	0.70	88	1.07	−90.0	.052	−61.1	0.60	140
9	0.70	68	0.88	−114	.054	−75.0	0.65	125
10	0.71	50	0.784	−138	.060	−87.1	0.70	117
11	0.70	25	0.65	−158	.065	−102	0.72	105
12	0.70	12	0.60	−174	.074	−117	0.75	95

Fig. 5.31a shows the result of doing this for the large signal impedance at the output of a Plessey PGAT1000 FET. The output power of power GaAs FETs largely depends on the output being matched at the large r.f. signal levels. Fig. 5.31b shows the effect of increasing the r.f. power level into a 250 mW power GaAs FET from 10 dBm to 26 dBm. The latter input power is larger than would be used in practice but the trends shown in Fig. 5.31b may be used to modify the remaining small signal S-parameters. For example, $\angle S_{21}$ is virtually unchanged. However both the magnitude and angle of S_{22} change significantly.

For operation over medium bandwidths (i.e. 20% or so) it is desirable to calculate the variation in output power with frequency. This can be achieved by plotting the large signal gain circles or constant output power circles at discrete frequencies over the band of interest and superimposing on these the matching circuit impedance chosen for the amplifier.

The large signal output impedance is usually used in the design of the output matching network together with an $|S_{21}|$ which reflects the decrease in gain under large signal conditions. The small signal S_{11} and S_{12} can be used during design of the matching networks with final circuit adjustment being performed after the amplifier has been con-

196

constructed.

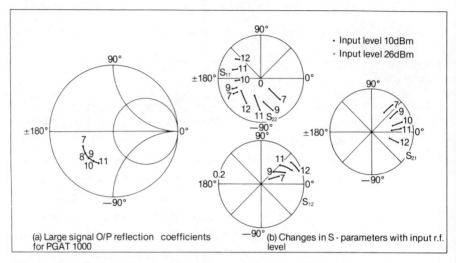

FIG. 5.31. Large Signal S-Parameters of GaAs FET Devices

D. CIRCUIT TOPOLOGIES FOR MATCHING POWER FETs

To determine the suitable circuit topologies for microwave GaAs FET amplifiers in microstrip form, for example, consideration must be given to the bandwidth, output power and frequency of operation.

If high output power is required, a large gate width device will have to be used. This will result in low input and output impedance and the circuit designer will have to choose a circuit topology that can match to these low impedances within the physically realisable limit of the transmission line. One useful approach is the lumped element impedance matching techniques already discussed with reference to small signal amplifiers.

The impedance matching circuits that can be used for power GaAs FET amplifier design include the following:

(i) Quarterwave Impedance Transformers

For moderate bandwidth applications (10 to 20%) a single section of quarterwave impedance transformers may be used. A packaged device is most often used for power amplifiers (since to bond up a power FET is an intricate operation) and if the device impedance appears inductive, either a capacitive shunt stub or a series capacitor may be used for resonance prior to the necessary impedance transformation to the source or load.

(ii) Matching Networks Based on Low Pass Filter Prototypes

For the broadband amplifier design, the lumped element equivalent circuit model of the FETs may be used to compute the input/output Q's of the device. These Q values are then used to determine the complexity of the network (the 'n' value) and the prototype 'g' values (Matthaei et al, 1964).

Using the coupled-resonator matching technique Leichti (1974) has designed a broadband amplifier. The broadband amplifiers with edge-coupled transmission line sections as described by Tserng (1979) can be considered a modified version of this broadbanding technique, since the single-section edge-coupled line can be represented as a series impedance transformer with two open-circuited shunt stubs connected to each end of the transformer.

(iii) Lumped-Element Impedance Matching Networks

Lumped-element impedance matching techniques can also be used for power FET amplifier design. For narrow to moderate bandwidths, a simple design procedure using a Smith Chart will be adequate. GaAs power FET amplifiers using lumped LC elements have been realised (Tserng et al, 1978). For wideband applications the initial element values of the lumped element, low pass Tchebyschev impedance matching network can be obtained from the work of Matthaei (1964). A low-pass to bandpass transformation can be made to obtain the correct element values for the desired frequency of operation.

14. NARROW BAND POWER FET AMPLIFIER DESIGN

This example describes the design of a 9 to 10 GHz power amplifier. The amplifier is to provide 20 dB gain at 32 dBm output power with three cascaded FET amplifier stages.

The optimum large-signal circuit impedances for three FET power devices with various gate widths are shown in Fig. 5.32a. The FETs have gate widths of 300μm, 1200μm and 2400μm to be used in the first, second and third stages respectively. The plot shows the effect of more cells lowering the impedance of the FET. The simple distributed matching networks shown in Fig. 5.32b, are designed to give output powers of 120 mW, 500 mW and 1.5 watts with 9 dB, 7 dB and 5 dB of gain respectively. The 1 dB bandwidth was 1 GHz with power-added efficiencies of 23 to 30%. Based on the circuit topologies shown in Fig. 5.32 an integrated, 3 stage amplifier with 50 ohm interstage transmission lines was designed. The performance of the overall amplifier is 20 dB gain with +12 dBm and an output power of 1.6W is obtained at 9.5 GHz with a power added efficiency of 30%.

15. BROADBAND POWER FET AMPLIFIER DESIGN

The use of distributed matching networks for broadband power amplifiers follows the same synthesis routines as for small signal amplifiers. Fig. 5.33 shows for example the use of a particular circuit

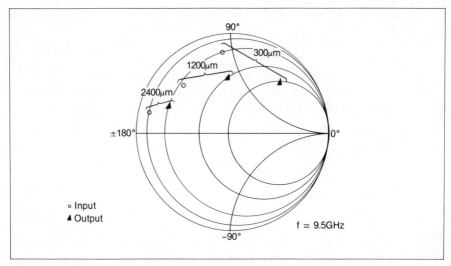

FIG. 5.32(a).　Input and Output Circuit Impedances for Power
GaAs FETs

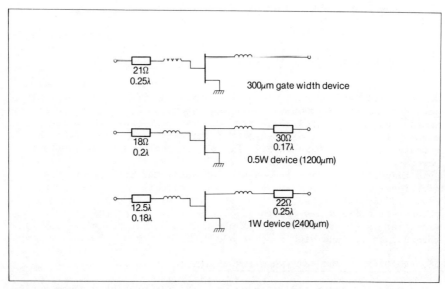

FIG. 5.32(b).　Matching Circuits for 100 mW FET, 0.5W FET and
1W FET

topology for broadband matching of a 600μm gate width device over the frequency band 6-12 GHz giving 100 mW output power. Performance is shown in Fig. 5.33.

One of the most useful methods of broadband matching power FETs is to use 'internal' matching within a microwave package using wire bonds and chip capacitors. Such a circuit is shown in Fig. 5.34 with a set of performance curves in Fig. 5.34. It is seen that output powers of the order of 2 to 3 watts with 5 to 6 dB gain can be achieved with 1 dB bandwidths of approximately 2 GHz over the 7 to 10 GHz frequency range. Power-added efficiencies generally fall into the 20 to 30% range (Honjo et al, 1979; Mitsui et al, 1980).

16. DESIGNING AN AMPLIFIER FOR MAXIMUM SPURIOUS FREE DYNAMIC RANGE

The spurious free dynamic range of an amplifier is defined by the equation

Spurious Free Dynamic Range, $D = {}^2/3 (P_1 - P_0 - 10\log B - F) dB$

where P_1 is the input intercept point (obtained by subtracting the amplifier gain from the output intercept point).

P_0 is the effective input noise power with no signal (-114 dBm/MHz).

B is the system noise bandwidth in MHz (controlled by the amplifier bandwidth).

F is the amplifier noise figure in dB.

Thus for example, if a FET amplifier has a gain of 10 dB with an output intercept point of 25 dBm, a bandwidth of 1000 MHz and a noise figure of 6 dB the spurious free dynamic range is

$D = {}^2/3\{(25-10)-(-114)-10\log 1000-6\}dB$

$= \underline{62 \text{ dB}}$

i.e. the range over which the amplifier will not have any 3rd order intermodulation products greater than the noise floor is 62 dB.

For any specified amplifier bandwidth the parameters available enabling an increase in the dynamic range of the amplifier are the overall noise figure and the output intercept point.

Thus, the use of power FETs with their high 3rd order intercept point and low-noise small signal preamplifier FETs will enable large dynamic ranges to be achieved. In order to accurately predict the distortion products of a power FET amplifier considerable attention has been paid to the use of power FET models for predicting gain-compression, harmonic and intermodulation distortion (Tucker et al, 1977; Willing et al, 1978). Non linear circuit-type device models

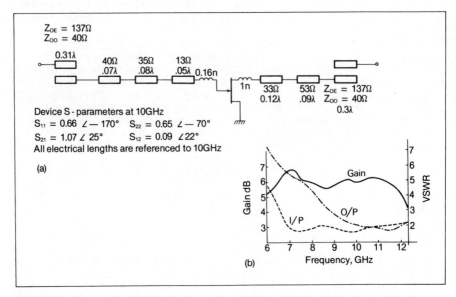

FIG. 5.33(a). Broadband Power GaAs FET Amplifier for 6 to 12 GHz Operation. (b) Gain-Frequency and VSWR Characteristic of 6-12 GHz Power FET Amplifier

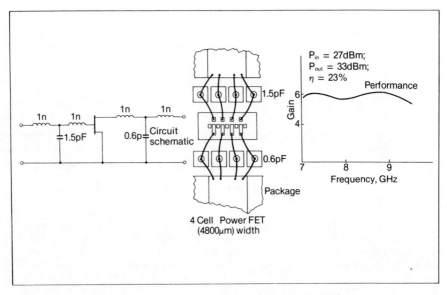

FIG. 5.34. Broadband Power FET with Internal Matching

have been derived which indicate that the third order intermodulation distortion of many GaAs power FETs does not follow a 3:1 slope. Indeed often the intermodulation products do not follow a monotonic relationship with the input power of the test tones (Fig. 5.35). This is covered further in Chapter 3.

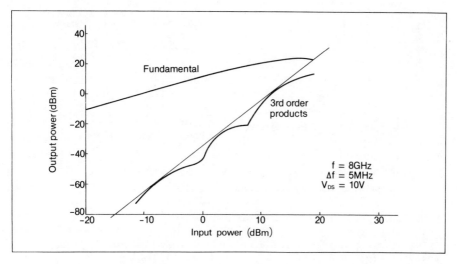

FIG. 5.35.   Measured Values of Output Power Level and IMD Products for a Power GaAs FET

17. POWER COMBINING TECHNIQUES

Often the design of power FET amplifiers is limited to the available r.f. power handling capability of the FETs available. Also, 1 dB gain compression point output powers are at single frequencies whilst the wideband operation power outputs are significantly below these levels. Wide gate devices have low input and output impedances which are difficult to match over wide bands. In addition gain is at a premium at the higher frequencies.

In a search for a solution circuit designers have been employing various types of hybrid power combiners to combine the output of several lower power amplifiers that incorporate more easily matched narrow gate width GaAs FETs. The main disadvantage with the use of hybrid power dividers is that of gain and power losses due to their insertion losses and that effect on overall power added efficiency.

Fig. 5.36 is an example of a 4 way power combiner, where the gain of the driver stages is 6 dB with an efficiency of 20% and the output devices have a gain of 5 dB with the power added efficiencies defined as in Fig. 5.36. As may be seen if circuit losses and combiner losses are both equal to 0.5 dB and the output devices have power added

202

efficiencies of 30% the overall efficiency of the circuit is only 15%.

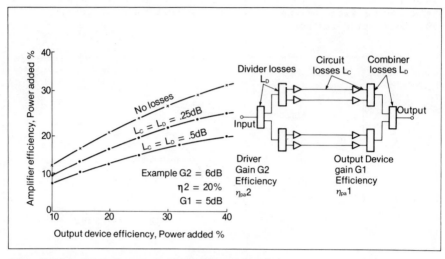

FIG. 5.36. Efficiency of 4 Combined Amplifiers

Two simple planar combining techniques that combine the output powers of four GaAs FETs are shown in Fig. 5.37a,b.

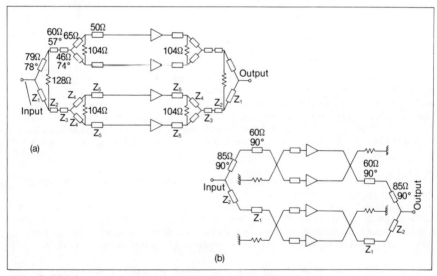

FIG. 5.37(a). Four-way Single-ended Amplifier
(b). Dual Quadrature Hybrid Balanced Amplifier

The first technique makes the use of two-way combiners of the
Wilkinson type (Wilkinson, 1960) while the second approach employs
two hybrids similar to that described by Lange (1969) and a simple
2 way splitter.

Fig. 5.37a shows the use of two-way combiners that have been connected to a branch like network by two cascaded transforming elements.
In order to provide physical access to the single-ended amplifier
submodules, additional line length can be inserted into the combiner
and divider circuits. This is accomplished by means of the elements
Z_2 and Z_3 as well as Z_5. In order to combine the output power of 2
balanced amplifier modules power combiners are used as shown in
Fig. 5.37b. In contrast to the four-way Wilkinson type combiner, the
dual quadrature hybrid combiner has the advantage of the excellent
matching conditions that are typical of balanced amplifiers. As a
result the overall amplifier exhibits very low return loss and a
smooth gain response (Niclas, 1979).

18. THERMAL CONSIDERATIONS IN POWER AMPLIFIER DESIGN

Unlike small signal FET devices attention must be paid to the heat-
sinking of power FET devices to avoid the channel temperatures of the
devices exceeding the manufacturer's specified maximum operating
point. For example, a 3W amplifier having 20 dB of gain using
Plessey PGAT1000 devices over a 1 GHz bandwidth centred at 8 GHz
would have an overall power added efficiency of approximately 15%,
i.e. 6.7W of d.c. power is input to the FETs plus dissipation in
other components such as regulators etc. 10W of heat could be
generated in such an amplifier and this heat must be dissipated as
efficiently as possible. Thus, it is necessary to heat sink the
power FETs effectively. Power FET packages are designed to allow
this to be achieved conveniently, e.g. holes in the package base
allowing the devices to be screwed into a baseplate made of a high
thermal conductivity material such as copper. Power FET data sheets
usually provide a derating curve if the amplifier is to be operated
at temperatures higher than 25°C.

19. PULSED OPERATION OF POWER FETs

With the advances in power FET performance anticipated in the light
of recently reported low noise FETs (Ohata, et al, 1980; Butlin et al,
1978) above 20 GHz, power FETs operating in the 20 to 30 GHz frequency
range can be expected in the near future.

Although the limits of power FET device performance have not been
reached, present performance levels are already sufficiently high to
have created applications in many systems. Schroeder and Gewartowski
(1978) have reported a 2 watt 4 GHz amplifier for radio relay systems
whilst Tserng and Macksey (1980) have developed a 4 watt, 10 GHz
amplifier module suitable for phased array radar applications provided that cost reductions can be achieved. Power combining techniques
are being developed that are resulting in the replacement of TWT's

with their associated bulky power supplies.

Large periphery FET power devices exhibit low input and output impedances which require the special matching techniques discussed in this chapter. The incorporation of matching networks directly on the device carrier has improved power gain performance by minimizing the loss between the device and the circuit (Honjo et al, 1979).

The output power of large periphery devices is limited also by gate-drain breakdown voltage and the channel operating temperature. Continued development in both device and material technology have significantly increased breakdown voltage. Many applications for power FETs, particularly for radars, require pulsed operation of the devices. Significant improvements in output power have been reported by pulsing the gate or drain of the FET whilst the FET is operated at elevated drain voltages.

Pulsing the drain voltage appears to produce the most significant increases in output power. Wade and Drukier (1980) have reported increases in output power from 2.3W to 3.5W with a pulse width of 1.5μsec, a duty factor of 0.5%, for a 6 mm by 1.7μm GaAs FET operating at a V_{DS} of 9V. At a drain voltage of 14V, 5.9W was obtained with 5.2 dB gain. The overall increase in power output from CW to pulsed operation was 4.1 dB. Fig. 5.38a and b show the performance as a function of drain voltage, pulse width and channel temperature. Also a marked improvement in bandwidth is observed probably due to device heating under mismatched conditions during CW operation. Bandwidth improvements of over 3:1 in X band have been reported.

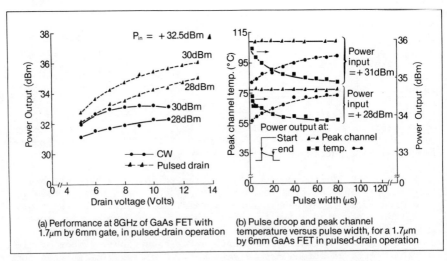

(a) Performance at 8GHz of GaAs FET with 1.7μm by 6mm gate, in pulsed-drain operation

(b) Pulse droop and peak channel temperature versus pulse width, for a 1.7μm by 6mm GaAs FET in pulsed-drain operation

FIG. 5.38.   Pulsed Performance of GaAs FET (Drain Pulsed)

Temple et al (1980) have pulsed the gate of X band FETs to produce
2 dB improvements in power gain when compared to CW operation.
Fig. 5.39 shows the output power improvement versus drain voltage for
a 2µsec pulse width at a duty factor of 20%.

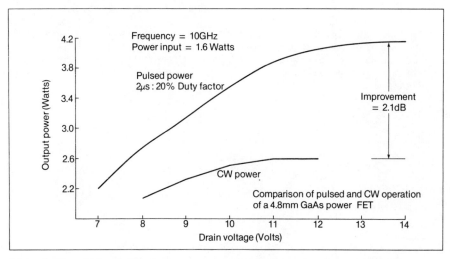

FIG. 5.39. Pulsed Performance of 4.8 mm GaAs FET (Gate Pulsed)

It thus appears that the pulsed mode of power GaAs FET operation
will give significant improvements in performance such that low cost
monolithic power amplifiers, for phased array radar transmitter
applications, will be able to operate at considerably reduced channel
temperatures. This means that not only will less attention have to
be paid to thermal impedance between the chip and its carrier but
that increased reliability will result (Irie et al, 1976).

20. <u>REFLECTION AMPLIFIERS</u>

The GaAs FET is the only three terminal device available in the
frequency range above X band. However the development of low noise
amplifiers above approximately 20 GHz is difficult due to the maximum
available gain per stage. Because of the relatively low associated
gain of the first few stages of a multistage amplifier the overall
noise measure is poor.

Circuit losses, such as those encountered with microstrip at the
higher frequencies, can be substantially reduced by the use of waveguide.

The GaAs FET can be operated as a negative resistance two terminal
device by inducing the correct impedance with feedback. Several
schemes to do this are available including operating the FET in
common drain and terminating the gate in the correct reactive load

206

(Nicotra, 1979). The reflection coefficient measured from the source side of the device Γ_S is then greater than one.

$$\Gamma_s = S_{11} + \frac{(S_{12} \cdot S_{21})}{(1 - S_{22}\Gamma_G)}$$

where Γ_G is the reflection coefficient at the gate of the FET.

As shown in Fig. 5.40 source transforming networks can be designed such that the load matching network, having impedance $R_L + jX_L$, resonates the reactive component X_S' of the FET thus allowing the maximum output power of the FET to be delivered to the load. This is the case when

$$X_L = -X_S'$$

$$R_L < |R_S'|$$

and $G = P_{OUT}/P_{IN} = |(R_S'-R_L)/(R_S'+R_L)|^2$

A circulator is used to provide isolation between the input signal and the output signal.

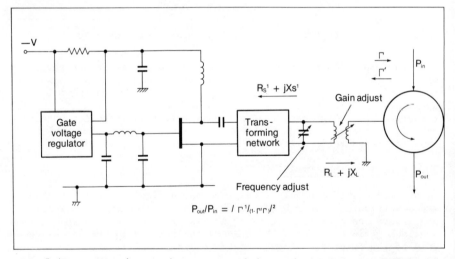

FIG. 5.40. Negative Resistance Amplifier Using Common Drain FET

Tohyama (1979) has recently described the construction of a 23 GHz reflection type amplifier in waveguide which utilises the resonant properties of the sealing ring of a packaged NEC 388 GaAs FET. Fig. 5.41a shows the structural details of the FET package whilst Fig. 5.41b shows the waveguide mount.

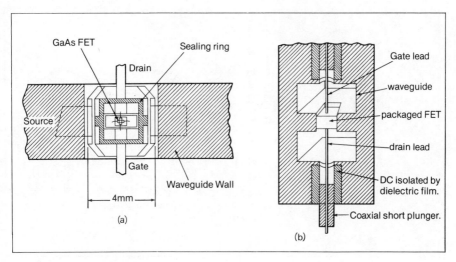

FIG. 5.41(a).   GaAs FET Package Structure
        (b).   Waveguide Mount Structure

In this amplifier the coaxial short circuits on the gate and drain
of the device are adjusted such that the impedance looking into the
drain-source port is negative over a considerable range of frequencies. For example Fig. 5.42a shows the impedance at the drain-source
port when the circuit has been adjusted for operation at 24 GHz.
Fig. 5.42b shows the gain and noise figure of the amplifier over the
23.3 to 24.2 GHz band. A gain of 8 + 1 dB was achieved with a noise
figure between 6 and 7 dB. In the usual transmission type amplifier
circuit the gain of such a packaged 0.5μm gate length FET would be
very low particularly since the package exhibits a resonance around
20 GHz or so.

Broadbanding of reflection amplifiers depends on the ability of the
matching circuit to provide a reactance which is the conjugate of the
reactive part of the FETs input impedance. Normally this is provided
by, for example, resonating the gate-to-source capacitance, C_{GS}, of
the FET with an inductance. In order to broadband such a technique
some of the network synthesis routines described in this chapter can
be adopted.

However, an attractive method of broadbanding is to use the active
reactance compensation technique developed by Aitchison et al (1980).

Fig. 5.43, for example, indicates that by connecting a similar
negative resistance FET device λ/4 away from the FET under consideration the input admittance, Y_L of the FET can be exactly compensated.

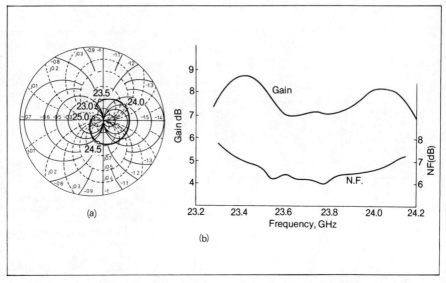

FIG. 5.42(a). Measured Impedance at the Drain—Source Port
    (b). Gain and Noise Figure Characteristics of GaAs MESFET
  Reflection Amplifier in the 23 GHz band

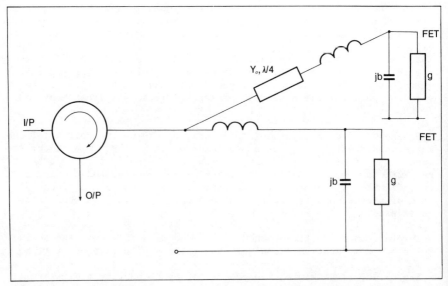

FIG. 5.43. Active Reactance Compensation of GaAs FET Reflection
    Amplifier

If the input admittance,

$$Y_L = g + jb$$

then a perfect quarter-wavelength of characteristic admittance Y_0, terminated in Y_L will have an input admittance

$$Y_{in} = \frac{Y_0^2}{g + jb}$$

$$= \frac{Y_0^2}{g^2 + b^2} (g-jb)$$

If Y_L is resonant at the centre frequency then $b = 0$ and if $g2 \gg b^2$ elsewhere in the band of interest

$$Y_{in} \simeq \frac{Y_0^2}{g^2} (g-jb)$$

Connecting the two loads in parallel gives

$$Y_{TOT} = g + jb + \frac{Y_0^2}{g^2} (g-jb)$$

By choosing $Y_0 = g$

$$Y_{TOT} = 2g$$

21. CONCLUSIONS

This chapter has dealt with the extensive subject of microwave amplifier design with particular reference to the use of gallium arsenide field effect transistors. The descriptions given are by no means complete and the reader is referred to the references as a means to further study.

22. BIBLIOGRAPHY

Aitchison, C.S. and Maclani, K. An investigation into the technique of active reactance compensation to improve the gain-bandwidth product performance of microwave bipolar transistor amplifiers. IEE Proceedings on Microwaves, Optics and Acoustics, June 1980.

Bodway, G. Two port power flow analysis using generalised scattering parameters. Microwave Journal, Vol. 10, pp.61-69, May 1967.

Butlin, R.S., Hughes, A.J., Bennett, R.H., Parker, D.R. and Turner, J.A. J-band performance of 300 nm gate length GaAs FETs. 1978

Int. Electron Devices Meeting, Digest Tech. papers, pp.136–139.

Compact Version 5.1 Compact Engineering Inc., 1088 Valley View Court, Los Altos, Calif. U.S.A.

Cooke, H.F. Microwave FETs – A status report. IEEE International Solid State Circuits Conference Digest of Technical Papers, pp.116–117 (February 1978).

Decker, D.R., Gupta, A.K., Peterson, W. and Ch'en, D.R. A monolithic GaAs IF amplifier for integrated receiver applications. 1980 IEEE MTT-S International Microwave Symposium Digest, May 1980, Washington DC, pp.363–366.

Estabrook, P., Knowne, C.M. and Crescenzi, E.J. A low noise single-ended GaAs Schottky FET amplifier for a 14 GHz satellite communication application. 1978 IEEE MTT-S International Microwave Symposium Digest, June 27–29, Ottawa, Canada, pp.129–131 (78CH1355-7).

Gledhill, C. and Abulela, M. Scattering parameter approach to the design of narrow-band amplifiers employing conditionally stable active elements. IEEE Trans. Microwave Theory and Techniques, Vol. MTT-22, pp.43–48, January 1974.

Hewitt, B. et al. Electronics Letters, Vol. 11, p.309, 1976.

Hewlett-Packard Application Note 154, S-parameter design. pp.25–34, April 1972.

Honjo, K. et al. Broad-band internal matching of microwave power GaAs MESFETs. IEEE Transactions on Microwave Theory and Techniques, Vol. MTT-27, January 1979, p.3.

Irie, T., Nagasako, I., Kohzu, H. and Sekodo, K. Reliability study of GaAs MESFETs, IEEE Trans. Microwave Theory Tech., Vol. MTT-24, pp.321–328, June 1976.

Kurokawa, K., Design theory of balanced transistor amplifiers. Bell System Technical Journal, Vol. 44, No. 10, pp.1675–1798, October 1965.

Lange, J. Integrated stripline quadrature hybrids. IEEE Trans. MTT, Vol. MTT-17, December 1969, pp.1150–1151.

Liechti, C.A. and Tillman, R.L. Design and performance of microwave amplifiers with GaAs Schottky-gate field effect transistors. IEEE Transactions on Microwave Theory and Techniques, Vol. MTT-22, No. 5, pp.510–517, May 1974.

Macksey, H.M., Adams, R.L. and Wisseman, W.R., X-band performance of GaAs power FETs. Electronics Letters, Vol. 12, pp.54–56, Jan. 22, 1976.

Matthaei, G.L., Young, L. and Jones, E.M.T. Microwave filters, impedance-matching networks and coupling structures. New York: McGraw-Hill, 1964.

Matthaei, G.L. Table of Tchebyschev impedance-transforming networks of low-pass filter form. Proc. IEEE, Vol. 52, p.939 (1964).

Mellor, D.J. and Linvill, J.G. Synthesis of interstage networks of prescribed gain versus frequency slope. IEEE Transactions on Microwave Theory and Techniques, Vol. MTT-23, No. 13, December 1975, pp.1013-1020.

Mitsui, Y., Kobiki, M. et al. 10 GHz - 10W internally matched flip-chip GaAs power FET. 1980 IEEE MTT-S International Microwave Symposium Digest, May 1980, Washington DC, pp.6-8.

Niclas, K.B. Planar power combining for medium power GaAs FET amplifiers in X/Ku-bands. Microwave Journal, June 1979, pp.79-84.

Niclas, K.B., Wilser, W.T., Gold, R.B. and Hitchens, W.R. A 350 MHz to 14 GHz GaAs MESFET amplifier using feedback. 1980 IEEE International Solid State Circuits Conference, Digest of Technical Papers CH1490-2/80/0000/0164, pp.164-165.

Nicotra, S. 13 GHz FET negative resistance 0.5W amplifier. Conference Proceedings of the 9th European Microwave Conference, Brighton, England, September 1979, pp.303-307.

Ohata, K., Itoh, H., Hasegawa, F. and Fujiki, Y. Super low noise GaAs MESFETs with a deep recess structure. IEEE Transactions on electron devices, Vol. ED-27, No. 6, June 1980, pp.1029-1034.

Pucel, R., Haus, H. and Statz, H. Signal and noise properties of gallium arsenide microwave field effect transistors. Advances in Electronics and Electron Physics, Vol. 38, pp.195-265.

Reed, J. and Wheeler, G.J. A method of analysis of symmetrical four port networks. IRE Transactions on Microwave Theory and Techniques, October 1956, pp.246-252.

Rollett, J. Stability and power gain invariants of linear two ports. IRE Trans. on Circuit Theory, Vol. CT-9, pp.29-32, March 1962.

Saal, R. and Ulbrick, E. On the design of filters by synthesis. IRE Trans. Circuit Theory, Vol. CT-5, pp.284-327, December 1958.

Schroeder, W.E. and Gewartowski, J.W. A 2W, 4 GHz GaAs FET amplifier for radio relay applications. 1978 IEEE MTT-S International Microwave Symposium Digest, June 1978, Ottawa, pp.279-281.

Skwirzynski, I.F. On synthesis of filters. IEEE Transactions on Circuit theory (Special Issue on Computer-Aided Circuit Design), Vol. CT-18, pp.152-163, Jan. 1971.

Suffolk, J.R., Pengelly, R.S., Cockrill, J.R. and Turner, J.A.
A monolithic S-band image rejection receiver using GaAs FETs.
Presented at the IEEE 1980 GaAs IC Symposium, Las Vegas, U.S.A.

Temple, S.J., Galani, Z., Dormail, J., Healy, R.M. and Hewitt, B.S.
Pulsed power performance of GaAs FETs at X-band. 1980 IEEE MTT-S
International Microwave Symposium Digest, May 1980, Washington DC,
pp.177-179.

Tohyama, H. and Mizuno, H. 23 GHz band GaAs MESFET reflection-type
amplifier. IEEE Trans. Microwave Theory, Tech., Vol. MTT-27, No. 5,
pp.408-411, May 1979.

Tserng, H.W. and Macksey, H.M. Microwave GaAs power FET amplifiers
with lumped element impedance matching networks. 1978, IEEE MTT-S
International Microwave Symposium Digest, June 1978, Ottawa,
pp.282-284.

Tserng, H.Q. Design and performance of microwave power GaAs FET
amplifiers. Microwave Journal June 1979, pp.94-100.

Tserng, H.W. and Macksey, H.M. Ultra wideband medium power GaAs MESFET
amplifiers. 1980 IEEE International Solid State Circuits Conference,
Digest of Technical Papers, pp.166-167.

Tucker, R.S. and Rauscher, C. Modelling the 3rd-order intermodulation
distortion properties of a GaAs FET. Electronic Letters, Vol. 13,
No. 17, pp.508-510 (1977).

Ulrich, E. Use negative feedback to slash wideband VSWR. Microwaves,
October 1978, Vol. 17, No. 10, pp.66-70.

Wade, P.C. and Drukier, I. A 10W X-band pulsed GaAs FET. 1980 IEEE
International Solid State Circuits Conference, Digest of Technical
Papers, pp.158-159.

Wilkinson, E.J. An N-way hybrid power divider. IRE Trans. MTT,
Vol. MTT-8, January 1960, pp.116-118.

Willing, H.A., Rauscher, C. and de Santis, P. A technique for predicting large signal performance of a GaAs MESFET. 1978 IEEE MTT-S
International Microwave Symposium Digest, Ottawa, Canada, pp.132-134
(78CH1355-7).

CHAPTER 6
FET Mixers

1. INTRODUCTION

The GaAs MESFET has gained a well deserved position amongst microwave solid state devices as a low noise amplifying device up to 30 GHz or so. The performance of Schottky barrier FETs as mixers, particularly utilizing low IF frequencies has been somewhat disappointing in respect of noise figure performance although there are undoubted advantages in terms of conversion gain and high compression point over Schottky diode mixers. Extensive studies have been carried out by laboratories worldwide into the applications of both single and dual-gate MESFETs in up and down-conversion circuits. Recently, with the advent of gallium arsenide integrated circuits, new device and circuit techniques have been adopted which are realising low-noise figures with high conversion gains at frequencies up to X-band (Pengelly, 1980). The reason for this latter success has been the realisation of biasing and circuit designs giving optimum performance in respect of signal and local oscillator injection as well as a fuller understanding of FET requirements.

GaAs MESFETs utilize their nonlinearities to produce mixing (Pucel et al, 1975; Sitch et al, 1973) - namely, the $I_{GS} - V_{GS}$ nonlinearity and the $I_{DS} - V_{GS}$ nonlinearity. The first of these effects results from the Schottky barrier between the gate and source and it shows characteristics analogous to Schottky barrier diodes. The second non-linearity is caused by the pinch-off effect described in Chapter 2.

2. THE GaAs FET AS A MIXING ELEMENT

A. LO APPLIED BETWEEN GATE AND SOURCE

Fig. 6.1 shows the simplified equivalent circuit of a chip FET where parasitic components such as wire bond inductances have been neglected. As has been described in Chapter 2, a low noise FET will have been designed to minimize the parasitic resistances R_G, R_S and R_D, where R_G and R_S are the principal contributors of extrinsic noise (Fukui, 1979).

214

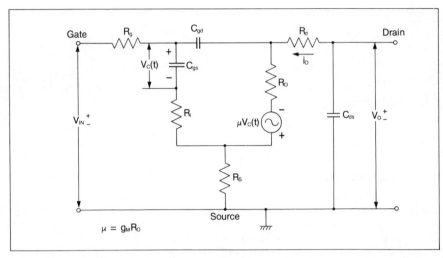

FIG. 6.1.   Small Signal Equivalent Circuit of the GaAs FET used in
    Mixer Analysis

Mixing will occur in a FET if certain of the 'small-signal' elements
can be varied at a periodic rate by a large local oscillator (LO)
signal connected across two of the device terminals.   The connections
which have been investigated by various workers include putting the
LO signal between gate and source and between drain and source.   In a
GaAs MESFET the strongest gate-bias dependent parameter is the transconductance, g_m.   Mixing products due to the parametric 'pumping' of
the source-gate capacitance C_{GS} and the intrinsic resistance R_I
(which can be considered as a charging resistance associated with
C_{GS}) are also present but are neglected in the following analysis
since frequency conversion due to these is inefficient.   Drain resistance is also a strong function of gate-bias and such dependence is
analysed later in this chapter.   For the present analysis this parameter is time-averaged.

A plot of the measured transconductance of a typical 300μm gate
width, 1μm length FET is shown in Fig. 6.2 as a function of applied
voltage between gate and source where the drain-to-source voltage V_{DD}
is above the knee of the $I_{DS}-V_{DS}$ characteristic.

Consider the case where a large LO signal is present across the
gate-to-source of the FET in addition to a fixed bias voltage.   Thus
the transconductance becomes a time varying function $g_m(t)$ with a
frequency equal to that of the LO.   If ω_0 denotes the LO radian
frequency then

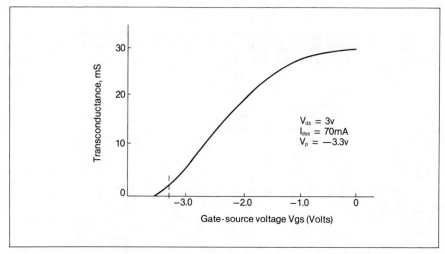

FIG. 6.2. Measured Transconductance as a Function of Gate Bias
for a GaAs MESFET

$$g_m(t) = \sum_{K=-\infty}^{\infty} g_K \, e^{jK\omega_0 t} \qquad\qquad 6.1$$

where

$$g_K = \frac{1}{2\pi} \int_0^{2\pi} g_m(t) \, e^{-jK\omega_0 t} \, d(\omega_0 t) \qquad\qquad 6.2$$

Neglecting the harmonic components of $R_0(t)$ the time-varying voltage
amplification factor $\mu(t)$ can be written as

$$\mu(t) \simeq \bar{R}_0 \, g_m(t) \qquad\qquad 6.3$$

where $\bar{R}_0$ is the time-average drain resistance.

With reference to Fig. 6.1 a signal $v_c(t)$ whose amplitude is much
less than the LO and of radian frequency ω_1 is also present across the
capacitance, C_{GS}. By the nonlinearity of the time-varying transconductance a voltage $\mu(t)v_c(t)$ will appear across the drain-to-source
terminals. This signal will have 'sideband' components at $|n\omega_0 \pm \omega_1|$,
where n is an integer, since

$$\mu(t) = \bar{R}_0 \, (g_0 + g_1 \cos\omega_0 t + g_2\cos2\omega_0 t + \ldots.)$$
$$\times \, (V_0 + V_1 \cos\omega_1 t + V_2 \cos2\omega_1 t + \ldots.)$$

For convenience this analysis will concentrate on the prime use of the MESFET, i.e. as a downconverter, such that the intermediate frequency will be $\omega_2 = |\omega_0 - \omega_1|$ (where $\omega_2 < \omega_1$). The other frequency component of significance is the image frequency ω_3 where $(\omega_3 - \omega_1) = 2\omega_2$.

Fig. 6.3 represents the FET mixer with the LO signal generator impedance represented by Z_0 and the input signal ω_1 provided by a voltage source E_1 with impedance Z_1.

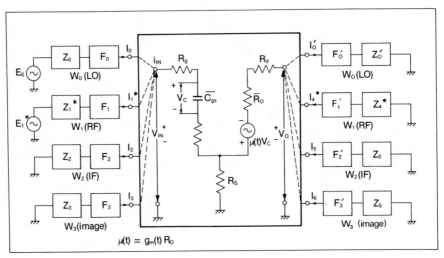

FIG. 6.3.   Diagram of FET Gate Mixer Showing Signal, IF, LO and Image Circuits Used in Analysis

The other sideband components, namely the IF frequency, ω_2, and the image frequency ω_3 are terminated in complex impedances. The rectangles F_K and F_K' are ideal lossless filters having zero impedance at the frequency of interest and infinite impedance at all other frequencies. Of course, the distinct sets of input and output ports are not physically separable in the real FET. Although in this analysis the LO is assumed to be injected between gate and source it can be inserted between source and ground. A port for the LO signal is necessary at the output since a strong LO component is present due to modulation of the drain current at this frequency. The LO frequency is usually terminated in a short circuit at the FET.

The FET mixer operated in the above mode has been analysed by Pucel et al (1976).

The linear circuit relationships between the various frequency components can be written as

$$[E] = [V] + [Z_T][I] \qquad 6.4$$

and

$$[E] = [Z_m][I] + [Z_t][I] \qquad 6.5$$

where

$$[E] = \begin{bmatrix} E_1^* \\ 0 \\ 0 \\ 0 \\ 0 \\ 0 \end{bmatrix}$$

where E_1^* represents the lower sideband small signal at frequency ω_1

$$\qquad 6.6$$

$$[V] = \begin{bmatrix} V_1^* \\ V_2 \\ V_3 \\ V_4^* \\ V_5 \\ V_6 \end{bmatrix} \qquad 6.7$$

and

$$[I] = \begin{bmatrix} I_1^* \\ I_2 \\ I_3 \\ I_4^* \\ I_5 \\ I_6 \end{bmatrix} \qquad 6.8$$

Z_m and Z_t are the matrices representing the mixer and its terminations. Z_m is given by

$$[Z_m] = \begin{bmatrix} Z_{11}^* & 0 & 0 & Z_{14}^* & 0 & 0 \\ 0 & Z_{22} & 0 & 0 & Z_{25} & 0 \\ 0 & 0 & Z_{33} & 0 & 0 & Z_{36} \\ Z_{41}^* & 0 & Z_{43} & Z_{44}^* & 0 & 0 \\ 0 & Z_{52} & Z_{53} & 0 & Z_{55} & 0 \\ Z_{61}^* & Z_{62} & Z_{63} & 0 & 0 & Z_{66} \end{bmatrix} \qquad 6.9$$

and

$$[z_t] = \begin{bmatrix} Z_1^* & 0 & 0 & \cdot & \cdot & \cdot \\ 0 & Z_2 & & & & \\ 0 & & Z_3 & & & \\ \cdot & & & Z_4^* & & \\ \cdot & & & & Z_5 & \\ \cdot & & & & & Z_6 \end{bmatrix}$$ 6.10

Neglecting harmonics of $g_m(t)$ we obtain

$$Z_{KK}(W_K) = R_G + R_I + R_S + \frac{1}{j\omega_K \bar{C}_{GS}}$$
For $K = 1$ to 3

For $K = 4$ to 6

$$Z_{14} = Z_{25} = Z_{26} = R_S$$

$$Z_{41} = \frac{-g_0 \bar{R}_0}{j\omega_1 \bar{C}_{GS}} + R_S$$

$$Z_{61} = \frac{-g_1 \bar{R}_0}{j\omega_1 \bar{C}_{GS}}$$

$$Z_{52} = \frac{-g_0 \bar{R}_0}{j\omega_2 \bar{C}_{GS}} + R_S$$

$$Z_{62} = \frac{-g_1 \bar{R}_0}{j\omega_2 \bar{C}_{GS}}$$

$$Z_{63} = \frac{-g_0 \bar{R}_0}{j\omega_3 \bar{C}_{GS}} + R_S$$

$$Z_{43} = Z_{53} = \frac{-g_1 R_0}{j\omega_3 \bar{C}_{GS}}$$

where $\bar{C}_{GS}$ represents the time averaged value of the gate to source capacitance C_{GS}.

The transducer conversion gain, G_m, between the RF input and the IF output is given by

$$G_m = \frac{|I_6|^2 \, ReZ_6}{(|E_1|^2/4ReZ_1)}$$ 6.11

and therefore, $\quad G_m = 4R_G R_L \left|\dfrac{I_6}{E_1}\right|^2$ 6.12

where $\quad Z_1 = R_G + jX_G$

and $\quad Z_6 = R_L + jX_L$

where $R_L + jX_L$ is the load on the mixer FET at the IF frequency.

G_m is obviously a complicated function of the terminations on the input and output ports. However, when the intermediate frequency is small compared to the input signal frequency (for example where the signal and IF frequencies are 8 GHz and 30 MHz respectively) some simplifications can be made.

The terminations at the ports other than the signal input and IF output are of secondary importance since $\bar{R}_0 \gg R_S, R_D$ and $g_0 R_S \ll 1$, both being true for well-designed low noise FETs, the solution of equations 6.4, 6.5 and 6.12 gives the mixer transducer gain as

$$G_m = \left\{ \left(\frac{2g_1 \bar{R}_0}{\omega_1 C_{GS}^-} \right)^2 \frac{R_G}{(R_G + R_{IN})^2 + \left(X_G - \dfrac{1}{\omega_1 C_{GS}^-}\right)^2} \right\}$$

$$\cdot \left\{ \frac{R_L}{(\bar{R}_0 + R_L)^2 + X_L^2} \right\}$$ 6.13

where $R_{IN} = R_G + R_I + R_S$ is the input resistance. Considering the case where the source and load are conjugately matched to the FET, i.e. $R_G = R_{IN}$, $Z_G = (\omega_1 C_{GS}^-)^{-1}$ and $R_L = \bar{R}_0$, $X_L = 0$, we obtain the available conversion gain, G_c given by

$$G_c = \frac{g_1^2}{\omega_1^2 C_{GS}^{-2}} \cdot \frac{\bar{R}_0}{R_G + R_I + R_S}$$ 6.14

This can be compared to the simplified expression for the amplifier available gain, G_a, given by

$$G_a = \frac{g_m^2}{\omega_1^2 C_{GS}^{-2}} \cdot \frac{R_0}{R_G + R_I + R_S}$$ 6.15

which can be determined from equations 2.21 and 2.35.

Equation 6.14 indicates that the conversion gain is a strong function of the gate bias and LO drive due to the conversion transconductance, g_1. g_1 is most dependent on gate bias near pinch-off. This may be seen by plotting conversion transconductance versus gate to source bias for the FET used in the experiment of Fig. 6.2. The result is shown in Fig. 6.4 where the instantaneous voltage across the FET's gate and source is given by

$$V_{GS}(t) = V_{GS} + V_0 \cos\omega_0 t$$

where V_{GS} is the d.c. bias and V_0 is the peak r.f. amplitude due to the LO drive. Fig. 6.4(a) and (b) show the relationship between g_m and g_1 as the LO amplitude changes from the positive peak value corresponding to the onset of forward conduction of the Schottky gate to the negative peak value dependent on V_{GS}. g_1 is found by calculating the Fourier component of $g_m(t)$ from the experimental curve of Fig. 6.2 using equation 6.2.

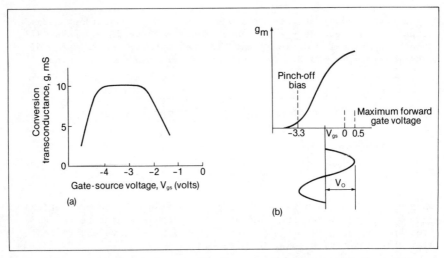

FIG. 6.4.   Calculated Conversion Transconductance at Maximum LO Drive as a Function of Gate-Source Voltage

It is noted from Figs. 6.2 and 6.4 that the maximum value of g_1 is approximately one-third that of the maximum value of g_m. (The ratio obtained for the 'ideal' case when g_m is a step function of the gate bias is $1/\pi$). If the LO power is not at the maximum level then the g_1 value will be lower. This dependence is shown in Fig. 6.5. The steeper the $g_m(t)$ curve is near to pinch-off, the steeper the g_1 curve is near zero LO drive. An illustration of the way in which technology improvements have increased the performance of FET mixers is shown in Fig. 6.6 where the $I_{DS}-V_{GS}$ characteristics of two FETs is shown, one

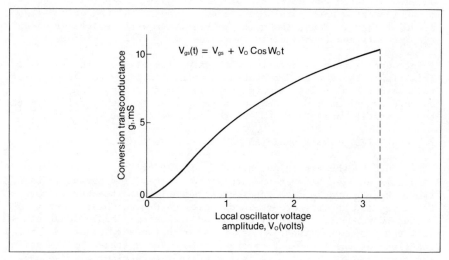

FIG. 6.5.   Conversion Transconductance at Pinch Off Bias as a
Function of LO Drive Level

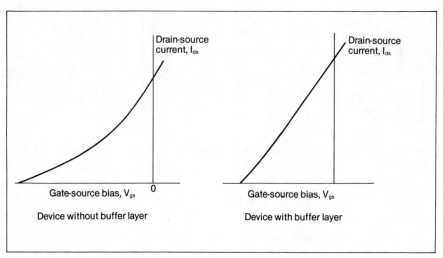

FIG. 6.6.   Effect of VPE Buffer Layer on I_{DS}-V_{GS} Characteristics
of GaAs FETs.

having a buffer layer, the other without. As has been explained in Chapter 4, the buffer layer used with FETs fabricated on epitaxial layers results in the electron mobility being maintained in the active channel up to the interface. This means that combined with a well defined electron concentration profile the value of $g_m(t)$ will be maintained at a higher value close to pinch-off than a device without a buffer layer. Thus, the $g_m(t)$ to V_{GS} curve will be steeper close to pinch-off and consequently the value of g_1 will be higher.

Gate capacitance also varies with gate bias as shown in Fig. 6.7. This figure shows that C_{GS} continues to decrease beyond the pinch-off voltage and can be approximated by a linear function. Thus the average capacitance $\bar{C}$ presented to the signal frequency will be equal to the static capacitance C_{GS} at the gate bias point V_{GS}.

B. LO APPLIED BETWEEN DRAIN AND SOURCE

Fig. 6.8 is the equivalent circuit to that of Fig. 6.3 where the LO signal is applied between the drain and source. The parasitic resistances R_G, R_S and R_D, the gate resistance R_I, gate-to-source capacitance C_{GS} and the gate-to-drain capacitance C_{GD} which are partly voltage dependent as already explained are assumed to be constant. Both the drain resistance R_O and transconductance g_m are time varying functions at the frequency of the LO, ω_O. The voltage amplification factor $\mu(t) = g_m(t)R_O(t)\mu(t)$ is therefore time variant. As before, a conversion matrix can be set up which relates the currents and voltages in the FET.

Assuming $C_{GD} = 0$, we obtain

$$\begin{bmatrix} E_1 \\ 0 \\ 0 \\ 0 \\ 0 \\ 0 \end{bmatrix} = \begin{bmatrix} Z_{11} & 0 & 0 & Z_{41} & 0 & 0 \\ 0 & Z_{22}{}^* & 0 & 0 & Z_{25}{}^* & 0 \\ 0 & 0 & Z_{33} & 0 & 0 & Z_{36} \\ Z_{41} & Z_{42}{}^* & Z_{43} & Z_{44} & Z_{45}{}^* & Z_{46} \\ Z_{51} & Z_{52}{}^* & Z_{53} & Z_{54} & Z_{55}{}^* & Z_{56} \\ Z_{61} & Z_{62}{}^* & Z_{63} & Z_{64} & Z_{65}{}^* & Z_{66} \end{bmatrix} \begin{bmatrix} I_1 \\ I_2{}^* \\ I_3 \\ I_4 \\ I_5{}^* \\ I_6 \end{bmatrix} \qquad 6.16$$

The asterisks indicate complex conjugate values. The matrix elements are given by

$$Z_{KK} = Z_K + R_G + R_I + R_S + \frac{1}{j\omega_K C_{GS}} \qquad \text{for } K = 1 \text{ to } 3$$
$$\text{for } K = 4 \text{ to } 6$$
$$= Z_K + R_D + R_S + R_{OO}$$

$$Z_{14} = Z_{25} = Z_{36} = R_S$$

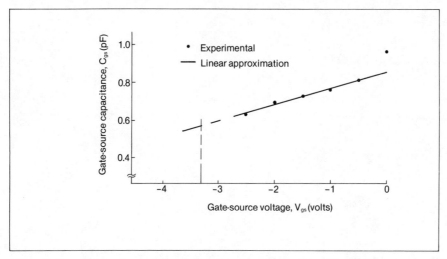

FIG. 6.7.   Gate–Source Capacitance as a Function of Gate–Source
Voltage

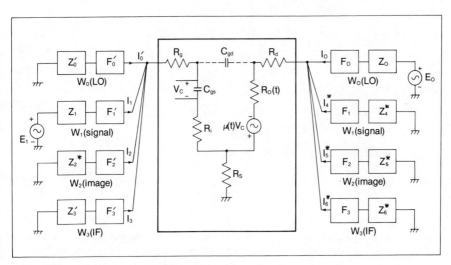

FIG. 6.8.   Diagram of FET Drain Mixer Showing Signal, IF, LO and
Image Circuits Used in Analysis.

$$Z_{41} = R_S - \frac{\mu_o}{j\omega_1 C_{GS}}$$

$$Z_{52} = R_S - \frac{\mu_o}{j\omega_2 C_{GS}}$$

$$Z_{63} = R_S - \frac{\mu_o}{j\omega_3 C_{GS}}$$

$$Z_{43} = - \frac{\mu_1}{j\omega_3 C_{GS}}$$

$$Z_{53} = - \frac{\mu_1{}^*}{j\omega_3 C_{GS}}$$

$$Z_{61} = - \frac{\mu_1{}^*}{j\omega_1 C_{GS}}$$

$$Z_{62} = - \frac{\mu_1{}^*}{j\omega_2 C_{GS}} , \quad Z_{46} = R_{01}$$

$$Z_{56} = Z_{64} = Z_{65} = R_{01}{}^*$$

$$Z_{42} = - \frac{\mu_2{}^*}{j\omega_3 C_{GS}} , \quad Z_{51} = - \frac{\mu_2{}^*}{j\omega_1 C_{GS}}$$

and $Z_{45} = Z_{54} = R_{02}{}^*$

where R_{00}, R_{01} and R_{02} are the Fourier coefficients of $R_0(t)$ and μ_0, μ_1, μ_2 are the Fourier coefficients of $\mu(t)$.

From equation 6.16, the mixer gain is given by

$$G_m = 4R_G R_L \left| \frac{I_6}{E_1} \right|^2 \qquad\qquad 6.17$$

For the simple case, where Z_3, Z_4 and Z_5 are large and conjugate matching is applied at the signal input and IF output ports, equation 6.17 predicts the available conversion gain, G_c to be

$$G_c = \frac{|\mu_1|^2}{4\omega_1^2 C_{GS}^2 (R_G + R_I + R_S)(R_D + R_S + R_{00})} \qquad 6.18$$

The drain-to-source diode can be assumed to be driven by the instantaneous voltage

$$V_{DS}(t) = V_{DS} + V_0 \cos\omega_o t \qquad 6.19$$

where V_0 is the peak local oscillator amplitude and V_{DS} is the static drain bias voltage.

Now, $\qquad \mu(t) = \sum_{K=-\infty}^{\infty} \mu_K e^{jK\omega_o t}$

where $\qquad \mu_K = \frac{1}{2\pi} \int_0^{2\pi} \mu(t) e^{-jK\omega_o t} d(\omega_o t)$

and $\quad R_0(t) = \sum_{K=-\infty}^{\infty} R_{OK} e^{jK\omega_o t}$

where $R_{OK} = \frac{1}{2\pi} \int_0^{2\pi} R_0(t) e^{-jK\omega_o t} d(\omega_o t)$

and $\quad \mu(t) = R_0(t) g_m(t).$

Thus $\mu(t)$ can be found by measuring the transconductance g_m and drain resistance R_0 of the FET as functions of gate and drain bias.

Figure 6.9 (Begemann, et al, 1979) shows, for example, the theoretical conversion gain G_c as a function of the drain bias at the maximum LO drive level for several different gate-to-source bias conditions. Fig. 6.10 shows the influence of the feedback capacitance C_{GD} on the conversion gain of a drain mixer. It may be appreciated that the value of G_c depends strongly on the value of C_{GD}. C_{GD} values are usually in the range of .01 to 0.02 pF for modern FET devices and with such values conversion gain decreases of 4 dB or so are apparent even at a relatively low signal frequency of 4 GHz in the example quoted.

Indeed, Harrop (1979) has compared the conversion gains of FET mixers with the LO injected into the gate and the drain with the signal at 7 GHz and the IF at 1 GHz and shown that there is a 10 dB difference in conversion gain.

226

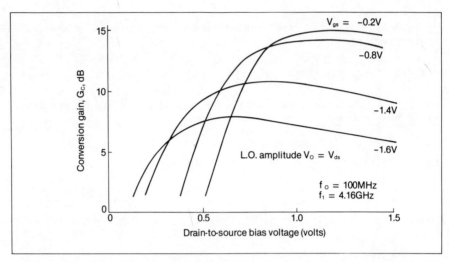

FIG. 6.9.   Conversion Gain of GaAs FET Drain Mixer as a Function of
Drain and Gate Bias Voltages

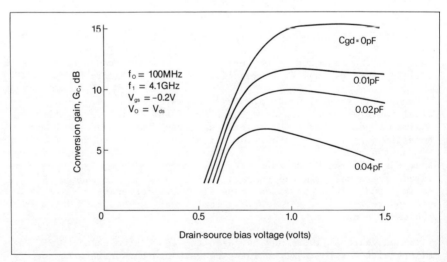

FIG. 6.10.   Conversion Gain of GaAs FET Drain Mixer as a Function of
Feedback Capacitance, C_{GD} and Drain to Source Voltage, V_{DS}

3. EXPERIMENTAL RESULTS ON GATE MIXERS

Fig. 6.11 shows a typical microstrip circuit used to exploit the
single gate FET as a mixer. One of the problems of the single gate
mixer is the requirement to introduce both the LO and signal frequencies into the gate of the FET. This can be achieved using a coupler
but will introduce loss and therefore degrade noise figure and conversion gain unless a high coupling ratio is used which will then
necessarily mean the use of high LO drive. The active signal adding
technique described elsewhere in this chapter is a convenient way of
overcoming this problem.

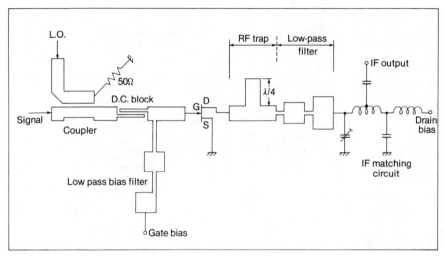

FIG. 6.11.   Typical Circuit Configuration for a Microstrip GaAs
FET Mixer

Cripps et al (1977) have reported 6 dB conversion gains with DSB
(see footnote) noise figures of 8.5 dB including a 3 dB noise figure
IF amplifier at a signal frequency of 10 GHz with an IF of 30 MHz.
Pucel et al (1976) have reported a balanced mixer having a conversion
gain of 3 dB and a noise figure of 7.4 dB DSB having a signal

Footnote:- Noise figures should be measured in a single sideband
(SSB) system where the image noise is rejected. Often, however, noise
figures are measured in a double sideband (DSB) system. Theoretically
if the DSB system were perfect there would be a 3 dB increase in noise
figure in an equivalent SSB measurement. However, since the gain or
insertion loss of the measurement system may not be constant over the
RF band which embraces both sidebands a 3 dB difference should not be
assumed. SSB noise figures are quoted wherever possible.

frequency of 8 GHz with a 30 MHz IF. Harrop (1979) has reported a
FET mixer using 1μm gate length FETs with 10 dB conversion gain and a
8.3 dB noise figure with a 1 GHz IF.

The use of a balanced mixer arrangement with single gate FETs not
only eases the LO and RF injection but also provides advantages in
the form of LO noise suppression, increased dynamic range and good
input match. Fig. 6.12 shows a typical schematic for a balanced FET
mixer approach. The antiphase IF signals produced by the quadrature
shifted r.f. and LO signals have to be brought back into phase. In a
diode mixer this is conveniently achieved by reversing the diodes but
this cannot be done with FETs. Depending on IF frequency, these signals can either be combined using a centre-tapped tuned transformer
with a bifilar wound primary or by the use of leading and lagging
phase shifters as shown in Fig. 6.12. These phase shifters can be
constructed using lumped elements or distributed transmission lines
and can be incorporated into the IF matching networks.

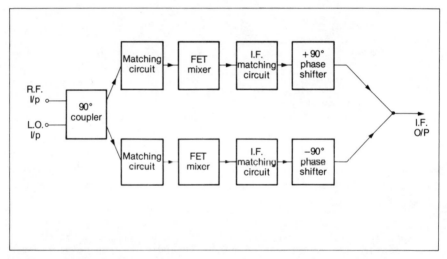

FIG. 6.12.   Diagram of Balanced FET Mixer Showing Phase Shifters in
    IF Circuit for LO Noise Cancellation

4. NOISE FIGURE

The noise figure results in particular that have been quoted above
are disappointing when consideration is given to the fact that the
equivalent noise figures for the same devices operating as amplifiers
are of the order of 3 dB at 12 GHz. The theoretical analysis of noise
in a GaAs MESFET mixer is complicated because the correlation between
the intrinsic drain noise and the induced gate noise cannot be neglected and is also a time varying function.

The relatively high noise figure, compared with that of similar FET amplifier noise figures is attributed to a low frequency (say 30 MHz) component of 1/f baseband noise being amplified where the FETs gain will be very high (Loriou et al, 1976). To overcome this disadvantage IF frequencies must be high enough to work in the region where the thermal noise zone is flat and trapping centres in the GaAs are negligible. The upper frequency for 1/f noise depends on transistor technology and especially on the quality of the n layer to semi-insulating substrate interface. Loriou and Leost (1976) for example, have shown a 3 dB improvement in noise figure for a 1μm FET operated as a mixer at a 1 GHz IF over the noise figure result at a 30 MHz IF. Fig. 6.13(a) shows the variation of gain and noise figure against intermediate frequency. IF's from 0.8 to 1.7 GHz were obtained by altering the signal frequency (curve 1) or by shifting the LO frequency (curve 2). IF output matching impedance and LO power were optimized at each frequency. It may be seen that a noise figure of 7 dB is achieved which includes a 0.7 dB contribution from the IF amplifier at a signal frequency of 8 GHz. Fig. 6.13(b) also indicates the way in which the gate length of the FET influences the noise figure of the mixer. A 5.6 dB noise figure with 9.8 dB associated conversion gain results from the use of a 0.5μm gate length FET compared to a 7 dB noise figure and 5.6 dB associated conversion gain from the 1μm gate length FET.

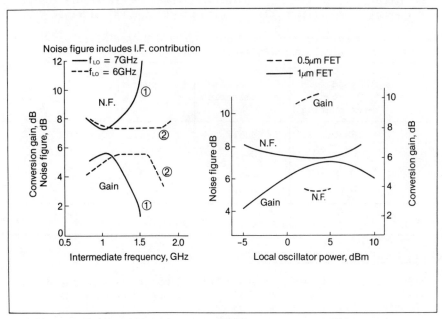

FIG. 6.13(a). Measured Noise Figure and Conversion Gain Against
IF Frequency
(b). Measured Noise Figure and Conversion Gain Versus LO
Power and with FET Gate Length as a Parameter

High IFs in the 1 to 2 GHz region are becoming more popular and are being used, for example, in direct satellite TV receivers for the first IF, 18-20 GHz radio-relay networks as well as certain milli-metre wave systems.

Recently it has been suggested that in order to obtain the lowest noise figures with single gate mixers all the mixing frequencies except the IF at the drain and the r.f. at the gate are short cir-cuited. These conditions also coincide with the requirements for maximum conversion gain. Bengemann et al (1979) has reported noise figures of 4.5 dB at a signal frequency of 3.8 GHz even though the IF was only 100 MHz. Conversion gain was approximately 9 dB. Image frequency gain was some 10 dB less than the r.f. gain.

Harrop et al (1978) have shown that the conversion gain is a maximum where the r.f. and LO frequencies are terminated in short circuits at the FET drain and the i.f. is terminated in a short circuit at the gate. Furthermore, it has been recently shown that it is also neces-sary to terminate the image and sum frequencies in short circuits at the gate and drain of the FET (Tie et al, private communication). Table 6.1 indicates the results for a 0.8μm gate length, 200μm wide FET having a pinch-off voltage similar to that of Fig. 6.2. An ade-quate noise theory for MESFET mixers is not available at the present time although it is already apparent that many of the requirements for low-noise amplifying FETs are also needed for FET mixers.

TABLE 6.1. Conversion Gain of Single-Ended FET GAte Mixer as a Function of RF, LO and IF Terminations

RF and LO output termination	IF input termination	Conversion gain, dB
Short	Short	10.1
Short	Open	5.5
Open	Short	5.4
Open	Open	−0.3
LO Input Level = 6 dBm		

5. SIGNAL HANDLING OF FET MIXERS

Even though the noise performance of FET mixers is somewhat disap-pointing at the present time their other outstanding advantages of high output compression point and conversion gain over diode mixers often make them the obvious choise for certain applications. Also the advent of the double-gate FET followed by a single gate mixer has meant that high conversion gain with noise figures only slightly inferior to that of an equivalent 2 stage FET amplifier can be produ-ced.

The conversion gain of a typical FET mixer versus LO power is shown in Fig. 6.14. The LO power required for maximum conversion gain is dependent on the pinch-off voltage and substantial reductions can be made in LO power by the choice of carrier profile and channel depth to decrease V_p.

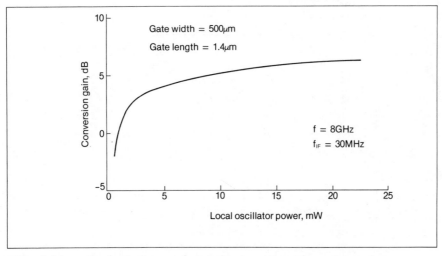

FIG. 6.14.   Typical Conversion Gain as a Function of Local Oscillator Power for a Single Gate FET Mixer

The single gate FET has excellent signal handling properties as evidenced by the results shown in Fig. 6.15. Here is shown the mixer conversion gain as a function of r.f. input power for two different levels of LO power. 1 dB output compression points of the order of 5 dBm are achieved with LO powers of 5 mW or so. The balanced mixer configuration will give a corresponding 3 dB higher output compression point (assuming ideal phase and amplitude matching). The figures for third order output intercept point performance are equally impressive with values of +20 dBm being readily obtainable with LO powers of the order of 10 mW.

6. FURTHER MIXER CONFIGURATIONS USING SINGLE GATE FETs

The recent advent of monolithic microwave circuits has allowed relatively simple single gate FET mixer circuits to be fabricated and evaluated where, perhaps, more than one FET is required. Fig. 6.16 shows three such circuits which have been investigated (Greenhalgh, private communication) and show considerable promise. As with the single gate FET mixer short circuits at r.f. and LO are required at the outputs and i.f. short circuits at the inputs. Fig. 6.16(a) utilizes a common gate input FET TR1 having two separate source

232

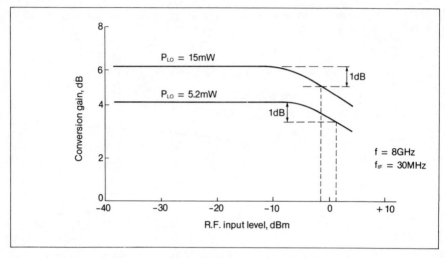

FIG. 6.15. Typical Conversion Gain of a GaAs FET Mixer as a
Function of RF Input Signal Level

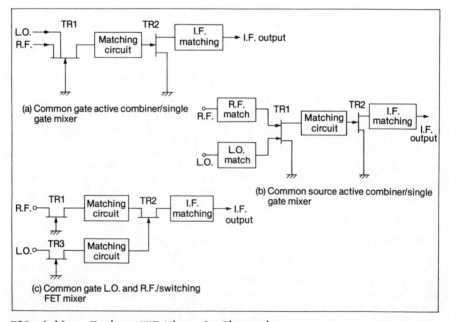

FIG. 6.16.   Various FET Mixer Configurations

electrodes with a common drain. By optimizing the gate width of the device it is possible to bias the device such that average g_m is close to 20 mS.

Thus, at frequencies up to 4 GHz or so, the FET's input impedance will be close to 50 ohms since the device is connected in common gate. The LO to signal isolation provided by this connection is approximately 20 dB at 4 GHz. Since the channel regions corresponding to the two sources are made separate no mixing occurs in the FET. The two signals are therefore added and continue to the single gate mixer TR2. The common gate FET TR1 will provide a small amount of gain at r.f. and LO whilst the mixer TR2 provides considerable conversion gain.

In Fig. 6.16(b) a similar technique is used but here the transistor TR1 provides matched gain to the LO and r.f. frequencies up to a much higher frequency. Again LO to r.f. isolation is of the order of 20 dB up to 8 GHz or so and mixing products in the signal adder are small being primarily determined by the fact that the device is biased as an amplifier rather than as a mixer and that the LO power can be kept low since the FET itself will give LO gains of 10 dB or so up to 10 GHz. Again the signal and LO frequencies are mixed in the single gate FET TR2 having passed through the interstage matching network.

Fig. 6.16(c) shows another common gate connection where the LO is injected into the gate of FET TR2 which acts as a switch. This circuit configuration works well up to 4 GHz or so.

LO drives for all these circuit techniques is optimum around a few milliwatts for 300μm gate width FETs (Greenhalgh, private communication) providing noise figures of the order of 6 dB at 4 GHz for the circuits shown in Fig. 6.16(a) and (c).

The circuit of Fig. 6.16(b) has demonstrated noise figures as low as 3 dB with conversion gains of 15 dB at signal frequencies of 4 GHz with an IF frequency of 200 MHz (Pengelly, private communication).

7. THE DUAL GATE MIXER

The dual gate FET has been shown to be a simple but effective microwave mixer (Cripps et al, 1979). The local oscillator signal is applied to the second gate via a suitable matching network and modulates the transconductance measured at gate 1 at the local oscillator frequency. Mixing action is therefore obtained, analogous to that of the single gate mixer, already analysed with the signal applied to gate 1 and the IF extracted via suitable filtering from the drain, as shown in Fig. 6.17. The way in which the second gate voltage varies the transconductance can be seen from Fig. 6.18 which shows the I_{DS}-V_{G1S} curves as a function of V_{G2S}, the second gate to source voltage. It may also be appreciated from Fig. 6.18 that a considerable change in g_m occurs for relatively small changes in V_{G2S}

especially when V_{G2S} is between zero and the maximum forward diode voltage (approximately 0.5V). This reflects itself in the ability of the dual gate FET mixer to give high conversion gains even when the local oscillator power is very low.

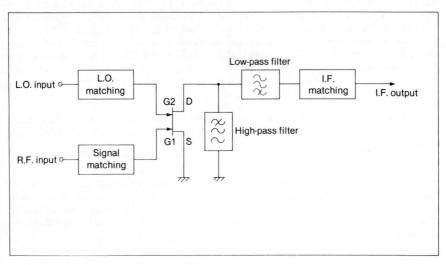

FIG. 6.17.   Dual-Gate FET Mixer

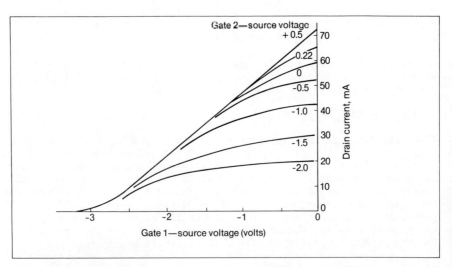

FIG. 6.18.   Transconductance Control of Dual Gate GaAs FET by
Second Gate Bias

Fig. 6.19 shows the conversion gain of a 300μm gate width dual-gate FET mixer as a function of LO drive in a single ended configuration indicating that LO power as low as 100μW gives 10 dB conversion gain.

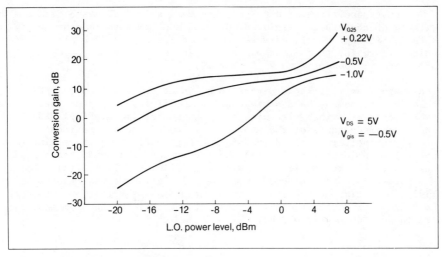

FIG. 6.19.   Conversion Gain of a Dual-gate FET Mixer as a Function of LO Power and Second Gate Bias

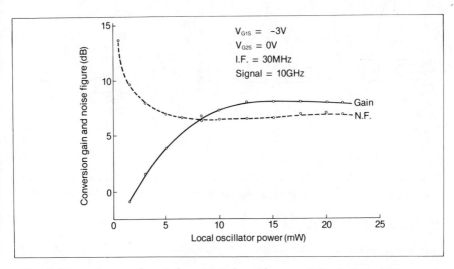

FIG. 6.20.   Conversion Gain and Noise Figure of Dual-Gate FET Mixer

Fig. 6.20 shows a typical gain and noise figure variation with LO power for a simple dual gate mixer where V_{G2S} = 0V. For this case the .optimum LO power drive was approximately 10 dBm in order to obtain the best DSB noise figure of 6.5 dB (Cripps et al, 1977).

The best result achieved to date has been an 8.0 dB noise figure with 14 dB conversion gain at 8 GHz with a 0.5μm gate length dual gate FET (Cripps et al, private communication). The IF amplifier noise figure for this mixer was 3 dB. The mixer circuit used a balanced arrangement with a 150 MHz IF.

Assuming the r.f. input signal to be input at gate 1 of the FET then the device may be described as an amplifying common source MESFET connected in series with a common source (modulating) MESFET. The transconductance of the whole FET is modulated by the local oscillator voltage applied to gate 2. Fig. 6.21 shows the mixing mechanism where in Fig. 6.21(b) the conversion transconductance g_1 is calculated by the Fourier analysis of $g_m(t)$ from Fig. 6.21(a). For the particular FET concerned in these calculations (Stahlmann et al, 1979) the optimum conversion gain occurs at a V_{G2S} of -1.5V with peak LO voltage of 2.35V (the FET having a 1μm gate 1 length, 2μm gate 2 length and 200μm gate width).

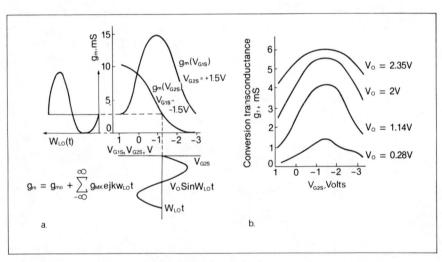

FIG. 6.21.   Dual-Gate FET Mixing Mechanism

8. IMAGE REJECTION MIXERS

A mixer which provides rejection of signals on the image channel has important applications in microwave systems, especially where low intermediate frequencies are used. With the increasing use of GaAs FET preamplifiers to realise low noise front-ends image rejection not only rejects interfering signals but amplifier noise at the image frequency as well.

The development of the single ended mixer into an image rejection configuration is shown in Figure 6.22.

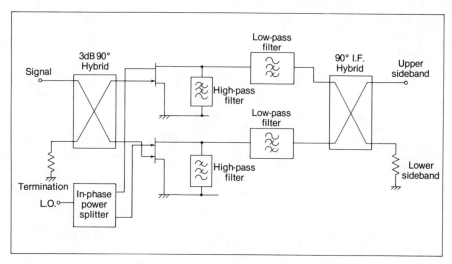

FIG. 6.22.   Image Rejection Mixer Using Two Dual-Gate FETs

The principle of the image rejection mixer circuit is well established (Weaver, 1956) where a balanced arrangement is used in which the two signal gates (the gate 1's of the dual gate FETs) are fed from a broadband 90° hybrid coupler and the two local oscillator gates (the gate 2's) are excited in phase by the signals from a power divider. It can readily be shown that the IF output at the drain of one of the FETs either leads or lags that of the other FET output by 90° depending on whether the signal is in the wanted band or in the image channel. The signal and image channels appear on separate output ports of a 90° IF hybrid circuit, thus adding a further 90° phase shift between the two IF signals from the mixer FETs. In addition to providing image rejection, the circuit of Figure 6.22 also has a low input VSWR due to the balanced arrangement provided that symmetry is maintained at the two FETs. The balanced arrangement also increases the power handling capability of the mixer by 3 dB. Cripps (1979) has reported the results of an X-band image rejection mixer built in microstrip on an alumina substrate, where the input 90° quadrature coupler was of the Lange type (Lange, 1979) and the local oscillator inputs were fed from an in-phase Wilkinson power divider (Wilkinson, 1960). Bias and IF output networks were low-pass filter structures incorporating resistive elements to aid stability. Fig. 6.23(a) shows the conversion gain for the upper and lower sidebands with no attempt to r.f. match the dual-gate FETs.

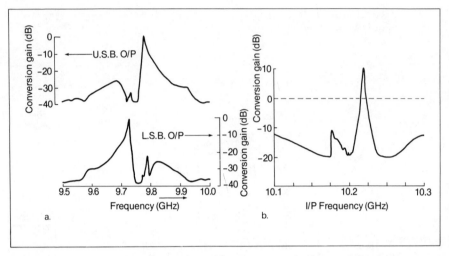

FIG. 6.23(a). Conversion Gain of Dual-Gate IR Mixer without RF
    tuning
        (b). Conversion Gain of Dual-Gate IR Mixer with RF
    tuning.

The 30 MHz IF circuit on the drain of each FET consisted of a tapped
parallel tuned circuit feeding the inputs of a commercial TO-5 packaged 90° IF hybrid. In the tuned state, conversion gains of 10 dB
have been reported with image rejection of 20 dB (Fig. 6.23(b)).
Input VSWR was always better than 2:1 over the 8-12 GHz band being
determined primarily by the Lange coupler. Noise figures of 8.5 dB
have been measured with an IF of 30 MHz and 7.5 dB with an IF of
150 MHz.

9. FREQUENCY UP-CONVERSION USING DUAL-GATE FETs

The dual gate FET can be conveniently used as an up-converter where
the IF signal is applied to gate 2 and the LO to the signal gate 1.

Fig. 6.24 (Tsai et al, 1979) shows the bandwidth and conversion gain
attainable in a single-ended up-converter designed to operate in the
7 to 8 GHz communication band. LO and lower sideband suppression can
be produced by the frequency selectivity of the output matching network. The typical saturated output level in the 7 to 8 GHz band is
8 dBm using a dual gate FET having a gate length of 1μm and width of
500μm. Since the dual-gate up-converter has gain it can eliminate up
to two stages of amplification in a normal heterodyne transmitter
chain at these frequencies.

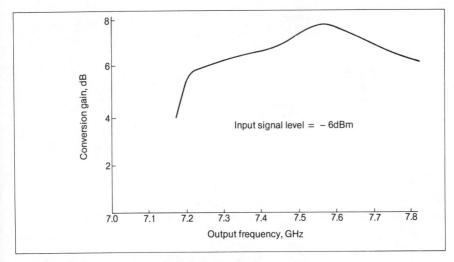

FIG. 6.24. Gain Response of 1.7/7.5 GHz Up-Converter using Dual-
Gate GaAs FET (Tsai et al, 1979)

10. FREQUENCY MULTIPLICATION USING DUAL-GATE FETs

The dual-gate FET is finding increasing applications in the area of
frequency multipliers where the major advantage - greater than 100%
efficiency - results in the replacement of complicated varactor diode
chains with their resultant pre- and post-amplification needs.

The same nonlinearities that produce mixing in a FET, namely the
$I_{DS}-V_{GS}$ and $I_{GS}-V_{GS}$ nonlinearities, are both used in the frequency
multiplier application. The first gate transconductance is modulated
periodically by an input signal and the amplified signal swing at the
second gate is, in turn, modulated by its input nonlinearity, thereby
generating harmonics. The resultant harmonics are further amplified
and extracted from the drain. Fig. 6.25 shows a circuit that has
been used to measure the doubling and tripling efficiency of 1μm gate
length dual gate FETs up to 20 GHz. An input frequency of up to
10 GHz is applied to the first gate through the 10 GHz low pass filter
(Chen et al, 1978) whilst a stub tuner is placed at the drain of the
multiplying FET following a high pass filter which reflects the input
frequency. The sliding short circuit at the second gate is adjusted
to optimize the doubler efficiency. Fig. 6.26(a) and 6.26(b) show the
doubler and tripler conversion efficiencies versus output frequency.
Even though the device has 1μm gate lengths conversion efficiency of
4 dB is possible at 18 GHz in the doubler mode and -2 dB in the tripler mode. Experiments with submicron dual-gate FETs have shown
doubler efficiencies of greater than 6 dB at 20 GHz (Pengelly,

private communication).

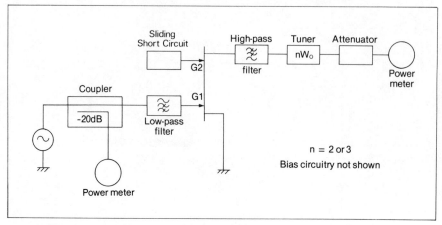

FIG. 6.25.   Dual-Gate FET Multiplier Test Circuit Arrangement

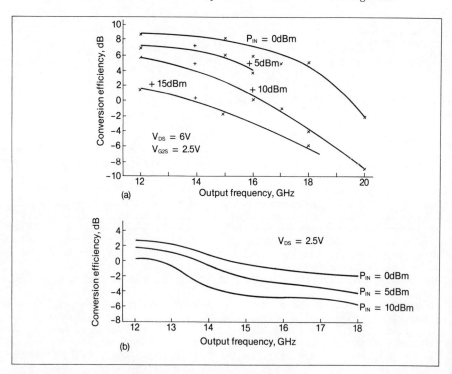

FIG. 6.26(a). Dual-Gate FET Doubler Conversion Efficiency Versus Output Frequency as a Function of Drive Level
          (b). Dual-Gate FET Tripler Conversion Efficiency Versus Out-
put Frequency as a Function of Drive Level

The lack of an idler circuit makes MESFET multipliers reasonably broadband with low AM-PM conversion noise. The GaAs MESFET has low intrinsic and thermal noise although 1/f noise may be a problem as with mixers. However, the use of the dual-gate FET as a frequency multiplier provides design simplicity, high conversion efficiency, good isolation between input and output signals, low noise performance and wide bandwidth. Moreover as may be appreciated from Fig. 6.26 driving power is only approximately 1 mW as compared with 300 mW for a step recovery diode or varactor multiplier and around 100 mW for a bipolar transistor. Performance comparisons are shown in Table 6.2.

TABLE 6.2. Comparison of Microwave Frequency Multipliers

Device Characteristic	Diodes	BJT	FET
Bandwidth	Narrow	Medium	Wide
RF driving power	300 mW	100 mW	1 mW (10 mW)
Output maximum frequency	SRD: 18 GHz Varactor: 120 GHz	11 GHz	30 GHz
Isolation	Poor	Medium	Good
Idlers	Critical	Critical	Less Critical
Power handling (X-band)	1 to 4W	0.5W	1 to 4 W
Stability	Good	Poor	V. good
Higher harmonic distortion	High	Low	Low
RF efficiency	-1.5 to -3 dB	-2 dB	up to +10 dB (up to 3 dB)

Figures in parenthesis are single-gate FET results (Pan, 1978).

11. CONCLUSIONS

This chapter has dealt with the subject of single and dual gate FETs used in frequency conversion applications. It has been shown (Loriou et al, 1976) that both types of MESFET suffer from 1/f noise contributions for IFs much below a few hundred megahertz. The general device requirements for FETs to be used in mixer and other conversion circuits are very similar to those requirements for low noise amplifiers. There is, however, a distinct lack, at the present time, of any solid theoretical foundation for many of the observed noise figure performances achieved. Recent studies into the signal, LO, IF, image and sum frequency terminations have given an insight into the optimum conditions for conversion gain especially with single gate FET mixers. Since the matching of the second gate of a dual gate FET can be difficult over relatively broadbandwidths, as it is in the amplifier case discussed in Chapter 2, the use of dual-gate FETs in mixers, although producing simple injection of LO and signal frequencies, has led to other FET based circuits being realised particularly at the lower frequencies (Van Tuyl, 1980; Suffolk et al, 1980; Ablassmeier, et al, 1980).

Dual-gate FETs are now receiving considerable attention as efficient up-converters and multipliers as well as in self-oscillating mixer circuits (Stahlmann et al, 1979).

12. <u>BIBLIOGRAPHY</u>

Ablassmeier, U., Kellner, W. and Kniepkamp, H. GaAs FET up-converter for TV tuner. IEEE Transactions on Electron Devices, Vol. ED-27, No. 6, June 1980, pp.1156-1159.

Begemann, G. and Hecht, A. The conversion gain and stability of MESFET gate mixers. Conference Proceedings of the 9th European Microwave Conference, Brighton, 1979, pp.316-320.

Begemann, G. and Jacob, A. Conversion gain of MESFET drain mixers. Electronics Letters, 30th August 1979, Vol. 15, No. 18, pp.567-568.

Chen, P.T., Li, C. and Wang, P.H. Dual-gate GaAs FET as a frequency multiplier at Ku-band. 1979 IEEE MTT-S International Microwave Symposium Digest, Ottawa, Canada, June 1978, pp.309-311.

Cripps, S.C., Nielsen, O., Parker, D. and Turner, J.A. An experimental evaluation of X-band GaAs FET mixers using single and dual-gate devices. 1977 IEEE MTT-S, International Microwave Symposium Digest, San Diego, June 1977, pp.285-287.

Cripps, S.C. et al. An experimental evaluation of X-band mixers using dual gate GaAs MESFETs. Proceedings of the 7th European Microwave Conference, Copenhagen, 1977, p.101.

Cripps, S.C., Nielsen, O. and Cockrill, J. An X-band dual-gate MESFET image rejection mixer. 1978 IEEE MTT-S, International Microwave Symposium Digest, Ottawa, Canada, June 1979, pp.300-302.

Cripps, S.C. The All FET front end - a step closer to reality. Microwaves, October 1978, Vol. 17, No. 10, pp.52-58.

Cripps, S.C. and Vudali, T. - private communications, 1979.

Fukui, H. Design of microwave MESFETs for broadband low-noise amplifiers. IEEE Trans. Microwave Theory and Techniques, Vol. MTT-27, No. 7, July 1979, pp.643-650.

Greenhalgh, S.G. - private communication, 1980.

Harrop, P. and Claasen, T.A.C.M. Modelling of a FET mixer. Electronics Letters, 8th June 1978, Vol. 14, No. 12, pp.369-370.

Harrop, P. FET mixers and their use in low noise front ends. Colloquium Digest on low noise microwave front ends - systems needs and component capabilities, IEE Savoy Place, London, March 1979.

Lange, J. Integrated stripline quadrature hybrids. IEEE Trans. MTT, Vol. 17, Dec. 1979, pp.1150-1151.

Loriou, B. and Leost, J.C. GaAs FET mixer operation with high intermediate frequencies. Electronics Letters, 22nd July 1976, Vol. 12, No. 15, pp.373-375.

Pan, J.J. Wideband MESFET microwave frequency multiplier. 1978 IEEE MTT-S International Microwave Symposium Digest, Ottawa, Canada, June 1978, pp.306-308.

Pengelly, R.S. - private communication, 1979.

Pengelly, R.S. GaAs monolithic microwave circuits for phased-array applications. IEE Proc. Vol. 127, Part F, No. 4, August 1980, pp.301-311.

Pengelly, R.S. - private communication, 1980.

Pucel, R.A., Bera, R. and Masse, D. An evaluation of GaAs FET oscillators and mixers for integrated front-end applications. IEEE ISSCC Digest of Technical Papers, p.62, February 1975.

Pucel, R.A., Masse, D. and Bera, R. Performance of GaAs MESFET mixers at X-band. IEEE Transactions on Microwave Theory and Techniques, Vol. MTT-24, No. 6, June 1976, pp.351-360.

Sitch, J.E. and Robson, P.N. The performance of field-effect transistors as microwave mixers. Proceedings of the IEEE, Vol. 61, p.399, 1973.

Stahlmann, R., Tsironis, C., Pouse, F. and Beneking, H. Dual-gate MESFET self-oscillating X-band mixers. Electronics Letters, 16th August 1979, Vol. 13, No. 17, pp.524-526.

Suffolk, J.R., Cockrill, J.R., Pengelly, R.S. and Turner, J.A. An S-band image rejection receiver using monolithic GaAs circuits. Paper 27, Abstracts of the 1980 GaAs IC Symposium, November 1980, Las Vegas, U.S.A.

Tie, G.K. and Aitchison, C.S. - private communication, 1980.

Tsai, W.C., Paik, S.F. and Hewitt, B.S. Switching and frequency conversion using dual-gate FETs. Proceedings of the 9th European Microwave Conference, 1979, Brighton, pp.311-315.

Van Tuyl, R. A monolithic GaAs FET signal generation chip. Digest of Technical Papers, ISSCC, San Francisco, Feb. 1980, pp.118-119.

Weaver, D.K. A third method of generation and detection of SSB signals. Proc. IRE, pp.1703-1705, December 1956.

Wilkinson, E.J. An N-way hybrid power divider. I.R.E. Trans. MTT, Vol. MTT-8, Jan. 1960, pp.116-118.

CHAPTER 7
GaAs FET Oscillators

1. INTRODUCTION

The 1960's saw the development of first practical microwave solid
state sources such as transferred-electron device and IMPATT diode
oscillators and the extension of bipolar transistor oscillators to
microwave frequencies. The development of GaAs MESFET devices in the
early 70's has significantly influenced the choice a system designer
has to make on the selection of a microwave power source.

The transferred electron or the Gunn oscillator, which has been the
mainstay for low power solid-state oscillator applications, suffers
from two main drawbacks. One is the low d.c. to r.f. conversion
efficiency and the other is the threshold current requirement, both of
these resulting in higher d.c. input power consumption. Low operating
efficiency also means that the Gunn device works at high temperatures,
unless the heat can be removed effectively. The GaAs FET oscillator
provides a higher d.c. to r.f. conversion efficiency (>10%) and does
not have any threshold current requirements. Being a three terminal
structure the GaAs FET is an extremely versatile active oscillator
circuit element and by making use of this feature it is possible to
control the behaviour of the oscillator to provide modulation, compensation and stabilisation etc.

The GaAs FET oscillator activity has received much attention in the
recent years. High efficiency fixed frequency oscillators have been
reported in the literature for frequencies up to 25 GHz and beyond.
Electronic tuning of FET oscillators by YIG resonators and varactor
diodes has been extensively researched. As most GaAs FET oscillators
are realised in low Q microstrip circuits, efforts have also been
directed towards improving their close to carrier noise and temperature stability by employing stabilisation techniques. Other applications like monolithic oscillators and pulsed RF oscillators etc. have
also been reported.

In this chapter GaAs FET microwave oscillator design philosophy is
reviewed, steps leading to the design of a free running oscillator are
indicated and their reported performance reviewed. Techniques for

frequency stabilisation of these oscillators using dielectric resonators are reviewed. Design criteria for electronically tunable oscillators are presented and performance capabilities outlined. Finally some novel GaAs FET oscillator circuits are described.

2. INDUCED NEGATIVE RESISTANCE

For solid state devices like the tunnel, Gunn and IMPATT diode, the application of appropriate d.c. bias is generally sufficient to produce negative resistance at the device terminals. This negative resistance condition exists up to microwave frequencies and by suitably coupling an external circuit to the device it is possible to deliver useful r.f. power to the load. These devices are thus inherent negative resistance devices. Broadly speaking, the effort of the circuit designer is mainly directed towards coupling the required power level to the load over the desired frequency range.

Application of d.c. bias to the bipolar or the field-effect transistor, on the other hand, does not generally result in the negative resistance condition. This condition has, therefore, to be induced in these devices so that useful power at microwave frequencies can be obtained. The frequency range over which the inherent negative resistance or conductance devices exhibit the condition is determined by the physical mechanisms in the device. For induced negative resistance devices it is influenced by the chosen circuit topology.

For an active two port device like the GaAs FET, negative resistance condition can be induced at one or both the device ports by suitably coupling the input and output ports of the device. There are two basic feedback arrangements as shown in Fig. 7.1 for a general three terminal device. The device may be in common source, gate or drain arrangement. In the series feedback arrangement the feedback element is the common current carrying element between the device input and output ports. For the parallel feedback arrangement it is the voltage transforming element between the two ports. A combination of series and shunt feedback elements and higher order feedback elements can also be envisaged (Fig. 7.2).

These feedback elements, which are usually reactive, can take the form of lumped or distributed components. In practice, however, an inductive parallel feedback element requires a d.c. block which would introduce problems due to its associated parasitic components at higher frequencies. Also because of d.c. biasing or grounding requirements it is usually difficult to have a capacitive series feedback arrangement.

3. S-PARAMETER MAPPING

The influence of feedback on the behaviour of the transistor can best be evaluated by mapping techniques. Bodway's (1968) general analysis is relevant here. Consider a transistor as a general three port network with its three terminals and a common ground plane forming the three ports.

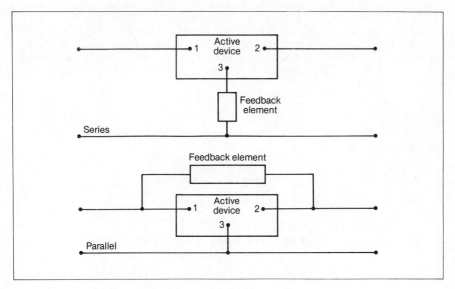

FIG. 7.1.   Basic Feedback Arrangements

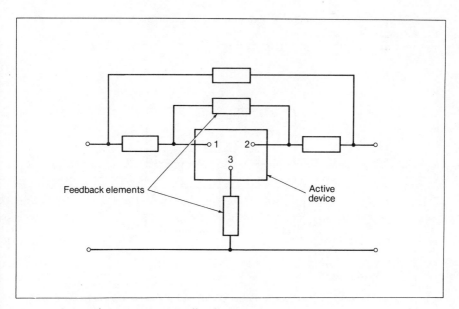

FIG. 7.2.   Higher Order Feedback Arrangement

248

If two device ports are terminated in the characteristic impedance Z_0 and the third port in an arbitrary feedback impedance Z_f, the configuration is equivalent to the series feedback case of Fig. 7.3. The two port S-parameters can be obtained in terms of the three port parameters and the feedback impedance Z_f. In the special case when the common load impedance Z_f is zero one obtains the conventional S-parameters.

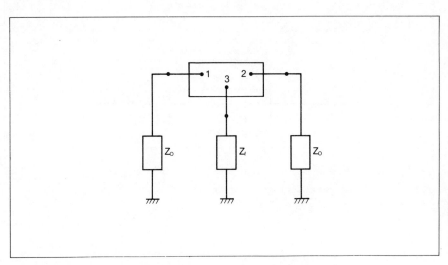

FIG. 7.3.   Series Feedback as a Special Case of a Three Port Network Representation

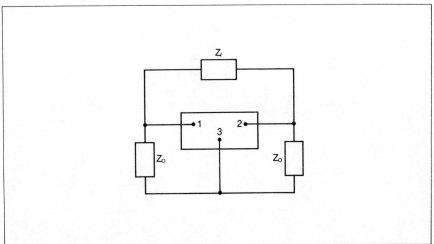

FIG. 7.4.   Parallel Feedback Case Obtained from Transformation of Fig. 7.3.

A further transformation from these S-parameters (referred to a common terminal) would result in a 3 port S-parameter matrix in which the ports are referred to each other. In the special case when two terminals are referred to each other in an impedance Z_f', the configuration is equivalent to a shunt feedback case (Fig. 7.4). Again, the two port S-parameters can be obtained in terms of the three port S-parameters and the impedance Z_f'. The conventional two port S-parameters are obtained when the impedance Z_f' is infinite.

The two port S-parameters with feedback are of the same general form as for terminated two port networks (see Chap. 5).

$$s = a + \frac{b}{\dfrac{1}{r} - c} \qquad\qquad 7.1$$

where a, b and c are related to the three port S-parameters and are complex in nature. The parameter r is the reflection coefficient corresponding to the feedback impedance referred to the characteristic impedance.

Eq. 7.1 is a standard equation in complex variable theory and the relationship between r and s can be written in a bilinear form

$$s = \frac{a + r(b-ac)}{1-rc} \qquad\qquad 7.2$$

This form of equation is similar to Eq. 5.12 for amplifiers which was used to plot constant gain circles as functions of load or source impedance. Similar techniques here would result in constant gain circles as a function of feedback impedance. However, the alternative technique of handling Eq. 7.2 is much more meaningful here. In this the r-plane is mapped onto the s plane. From the bilinear property of Eq. 7.2, circles on the r-plane map onto circles on the s-plane. This means that the Smith Chart for the r-plane can be mapped onto the s-plane giving both the magnitude and phase of s for each value of r.

For a general passive feedback impedance the resistive part may vary from 0 to ∞ and the reactive part from + j0 to + j∞. These values are bounded by the Smith Chart which merely represents magnitude and phase of the reflection coefficient. Thus the transistor S-parameters with feedback form the outline of a distorted Smith Chart due to the bilinear property.

This impedance mapping can be done by converting the transistor two port S-parameters to Y or Z parameters and incorporating the feedback immitance. It can also be done by using the mapping function routine in COMPACT (Users Manual). Figs. 7.5 and 7.6 show the common gate series and common source parallel feedback impedance mapped onto the S-parameter plane at 10 GHz for a Plessey GAT4 transistor with 1 x 300μm gate geometry.

It can be seen that for series inductive feedback the magnitudes of both S_{11} and S_{22} are greater than unity and can thus induce negative resistance conditions at both the input and output ports. The capacitive parallel feedback condition gives rise to negative resistance conditions at the output port only because only $|S_{22}|$ is greater than unity.

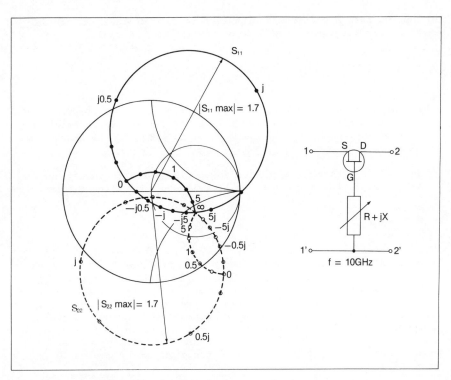

FIG. 7.5.   Grounded Gate with Series Feedback

4. OSCILLATOR DESIGN

A. THEORETICAL ANALYSIS

The GaAs FET can be used in common source, common gate or common drain configuration for oscillator applications. It is first necessary to induce a negative resistance condition by introducing feedback elements. Impedance mapping techniques can be adopted to study the effect of feedback on the input and output reflection coefficients of the circuit. The particular configuration chosen should be capable of providing negative resistance (i.e. $|S_{11}|$ or $|S_{22}| > 1$) conditions at one or both the device ports over the

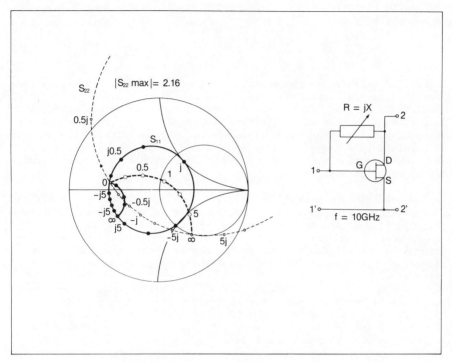

FIG. 7.6.   Grounded Source with Parallel Feedback

desired frequency range. If the input port is now terminated in an impedance so as to maximise the magnitude of the reflection coefficient at the output port, the circuit can be reduced to a one port network with negative resistance $-R_D(\omega)$ and its associated reactance $+jX_d(\omega)$ which are, in general, frequency and amplitude dependent.

If the output port is now terminated in an impedance $Z_L(\omega) = R_L(\omega) + jX_L(\omega)$, which includes the matching network and the load, then for oscillations to occur the following conditions must be satisfied.

The total resistance,

$$R(\omega) = -R_D(\omega) + R_L(\omega) \lesssim 0 \qquad\qquad 7.3$$

and the total reactance,

$$X(\omega) = X_D(\omega) + X_L(\omega) = 0 \qquad\qquad 7.4$$

In general, Eq. 7.3 can be satisfied for a number of frequencies but sustained steady-state oscillations are possible if and only if the stability criteria (Edson, 1953; Kurokawa, 1973) are also satisfied, which states that

$$\frac{\partial R(\omega)}{\partial \omega} > 0 \qquad \text{and} \qquad \frac{\partial X(\omega)}{\partial \omega} > 0 \qquad\qquad 7.5$$

Eq. 7.3 establishes the potential for oscillations. Initially for oscillations to start from noise there should be a net negative resistance in the circuit. As the oscillation amplitude builds up the negative resistance $-R_D$ diminishes and under steady state conditions becomes equal to the load resistance R_L. The oscillation frequency is determined by the zero reactance condition of Eq. 7.4. Eqs. 7.3 to 7.5 thus comprise the necessary and sufficient conditions for steady state oscillations.

The conditions expressed by Eqs. 7.3 to 7.5 apply to both small and large signal operations. However, in order to apply these conditions to large signal operation, it would be necessary to ascertain the dependence of $-R_D + jX_D$ on the oscillation amplitude. This information is not easily obtainable due to nonlinear effects under large signal conditions and therefore experimental measurements are necessary.

Calculations on the basis of small signal S-parameters usually provide good agreement between the theoretical and experimental oscillator frequencies. These are, however, unable to predict the output power level and the conversion efficiency of the oscillator.

The alternative approach in which the input port of the transistor is considered, with the output port terminated in the load and matching network, has also been analyzed. This technique is mainly applied to the design of tunable oscillators in which the tuning elements like the YIG resonator, varactor diode etc. are coupled to the input port. Krowne (1977) has presented an analysis based on the incident and reflected power at the input port. For a unit incident power on port 1, $|S_{11}'|^2$ is reflected back (where prime denotes the modified S-parameters where feedback has been taken into account). If $|S_{11}'|$ is greater than unity then the reflected power is higher than the incident power. If a lossless circuit element or cavity with reflection coefficient magnitude $|\Gamma_r|$ is connected to the input port (Fig. 7.7) then each round trip sees $|\Gamma_r|^2 |S_{11}'|^2$ power returning to the active device port. For oscillations to build up from noise this quantity must exceed unity or

$$|\Gamma_r| \quad |S_{11}'| \ > 1 \qquad\qquad 7.6$$

Also the first round trip reflected power $|1 + \Gamma_r S_{11}'|^2$ is maximised when

$$\theta_r + \theta_{11}' = 2n\pi, \quad n = 0,1,2,3 \ldots \qquad\qquad 7.7$$

where θ_r and θ_{11}' are the phase angles associated with Γ_r and S_{11}'.

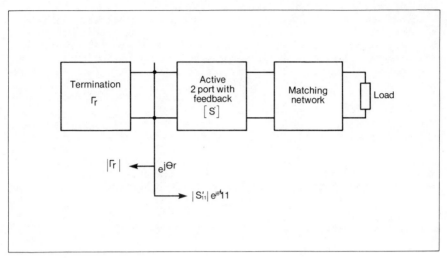

FIG. 7.7.   Oscillator Schematic

In general, $|\Gamma_r|$ which corresponds to a resonator or a reactive termination with negligibly small losses will be nearly equal to unity and Eq. 7.6 can be reduced to

$$|S_{11}'| > 1 \qquad\qquad\qquad 7.8$$

The small signal S-parameters of the device may be used to establish this condition but once oscillations build up, due to large signal operation the steady state condition becomes

$$\left.|S_{11}'|\right._{LS} = 1 \qquad\qquad\qquad 7.9$$

For broadband tunable oscillators, the feedback arrangement is selected such that Eq. 7.8 is satisfied over the desired frequency range. For fixed frequency oscillator applications the topology is selected to maximise $|S_{11}'|$ at the desired frequency. This technique is also applicable when the output port is terminated in impedances other than the characteristic impedance. This merely modifies S_{11}' in accordance with the standard equation.

$$S_{11}' = S_{11} + \frac{S_{12}\,S_{21}}{\dfrac{1}{\Gamma_m} - S_{22}} \qquad\qquad\qquad 7.10$$

where Γ_m represents the reflection coefficient corresponding to the output matching network.

It has been seen that having incorporated the necessary feedback elements to the basic transistor, the input or the output port has to be terminated to reduce it to an equivalent two-terminal negative resistance device. The input port is generally terminated for fixed frequency oscillator applications and the output port termination is generally used for tunable oscillators. The aim in both cases is to ensure that $|S_{11}'|$ or $|S_{22}'|$ is greater than unity. Golio and Krowne (1978) analysed the condition when the output port of the transistor with feedback is terminated and the terminating reflection coefficient values which ensure $|S_{11}'| > 1$ were investigated.

A general analysis applicable to either the input or the output port with the other terminated can be enunciated. The terminated reflection coefficient at one port with the other port terminated in reflection coefficient Γ_K can be written as

$$S_{ii}^{\ T} = S_{ii}' + \frac{S_{ij}' \ S_{ji}'}{\frac{1}{\Gamma_K} - S_{jj}'} \qquad\qquad 7.11$$

where $i = 1$, $j = 2$ and $K = \ell$ corresponds to the input port with the output port terminated in Γ_ℓ and $i = 2$, $j = 1$ and $K = s$ corresponds to the output port with the input port terminated in Γ_s. Eq. 7.11 can be handled by computer analysis techniques but it would be useful to get some physical insight. Rewriting

$$S_{ii}^{\ T} = S_{ii}' + A_{jK} \qquad\qquad 7.12$$

The magnitude of $S_{ii}^{\ T}$ is determined by the magnitude and phase of S_{ii}' and A_{jK}. S_{ii}' depends upon the two port network characteristics while A_{jK} is dependent upon the terminating reflection coefficient Γ_K which is the controllable parameter. Thus for a given magnitude of A_{jK}, the magnitude of $S_{ii}T$ is greatest when the phase angles of A_{jK} and S_{ii}' are equal. This is the condition of phase addition. Also from Eq. 7.11 $A_{jK} \rightarrow \infty$ when $\Gamma_K S_{jj}' = 1$ resulting in the largest possible value of $|S_{ii}T|$. Additionally, $|A_{jK}| = 0$ when $|\Gamma_K| = 0$ which corresponds to the condition when the jj port is terminated in the characteristic impedance. These two points thus represent the end points of the phase addition curve and the point $\Gamma_K = \frac{1}{S_{jj}'}$ is called the limiting match for a given frequency.

For most practical applications the reflection coefficient Γ_K corresponds to passive impedance elements only. If $|S_{ii}'|$ is less than unity, then there is a minimum value of $|A_{jK}|$ corresponding to some Γ_K beyond which $|S_{ii}T|$ is greater than unity. This satisfies the relationship $|A_{jK}|_{min} = 1 - |S_{ii}'|$. On the other hand if $|S_{ii}'| > 1$

this value is zero corresponding to $\Gamma_\ell = 0$ and the device feedback itself is sufficient to produce a negative resistance condition.

Now considering the terminated port jj, if $|S_{jj}'| > 1$ then the limiting condition in which $A_{jK} \to \infty$ corresponds to a Γ_K value realisable by passive elements. If, on the other hand, $|S_{jj}'| < 1$ then the $|\Gamma_K|$ is greater than unity for the limiting condition of $|A_{jK}| \to \infty$. Under this condition the point at which $|\Gamma_K| = 1$ on the inphase addition curve gives the upper limit of $|A_{jK}|$ realisable by passive elements.

Eq. 7.11 can also be used to determine Γ_K contours for constant $S_{ii}T$ magnitude and phase (Golio and Krowne, 1978). This information provides the designer with the overall picture of the effect of the terminating impedance.

The form of Eq. 7.11 is similar to Eq. 7.1. Thus it is also possible to use mapping techniques to study the effect of terminating reflection coefficient Γ_K on the value of $S_{ii}{}^T$ in manners similar to those considered in Section 3.

B. SMALL SIGNAL AND LARGE SIGNAL ANALYSIS

Large signal device characteristics should be used for the analysis and design of GaAs FET oscillators. The approach taken by some workers has been to measure small signal S-parameters of the device over the desired frequency range and to extend the applicability of the measured values to large signal operation (Maeda et al, 1975a; Maeda et al, 1975b; Finlay et al, 1978; Joshi et al, 1979). This procedure is capable of a reasonable prediction of the frequency of operation as the change in S-parameters is largely resistive under large signal conditions. Because of this the oscillation frequency is not affected since it is determined by the reactive part of the circuit impedance. This technique would however, not be able to predict performance characteristics like the output power efficiency.

Large signal S-parameter measurements can be used to characterise the GaAs FET at a particular frequency and bias arrangement (Mitsui et al, 1977a; Mitsui et al, 1977b). This procedure provides enough information to predict oscillator performance but involves extensive measurements and therefore has limited application.

Johnson (1979) did extensive small-signal S-parameter measurements for the FET covering a range of bias conditions and frequencies together with a limited number of large signal S-parameters at several power levels for the desired frequency. An equivalent circuit model for the FET which includes non-linear circuit elements for large signal behaviour was developed from this data. The nonlinearity was assigned to up to 7 elements of the equivalent circuit and was expressed in terms of the nonlinearity of the transconductance. This provided a reasonably complete large signal FET model and gave good agreement between the theoretical and experimental results for an FET oscillator.

Rauscher and Willing (1978, 1979) characterised the performance of the FET under small signal conditions using extensive small-signal S-parameter measurements over a wide range of bias conditions and frequencies. This data was used to develop an equivalent circuit model capable of large signal description. With this information the form of nonlinearity of the equivalent circuit elements was derived to predict the large signal behaviour. This was achieved through the relationships between the "small signal incremental values" of non-linear circuit elements (predicted from small-signal S-parameter measurements) and the "instantaneous values" (applicable to large signal oscillations in bias conditions) in the form of sets of differential equations. This method has the advantage that only small signal S-parameters are required but needs many computationally intensive steps to fit the equivalent circuit model to the measured S-parameters.

Rauscher used this to establish design techniques for fixed frequency and varactor tuned GaAs FET oscillators for optimum large signal performance. The technique requires the synthesis of a lossless three port coupling network which simultaneously provides the optimum terminating impedances for the transistor ports and the optimum feedback between the drain and gate terminals to maximise the r.f. power delivered to the load. The coupling network must yield only one stable state of oscillation, thus suppressing spurious oscillations. In his design example the external load to which the power is to be coupled was considered an integral part of the coupling network. Maximum net output power from the oscillator was obtained by maximising the difference between the power available at the drain port and the power necessary to be fed back into the gate port to sustain oscillations.

5. FREE-RUNNING OSCILLATOR-PERFORMANCE REVIEW

The first detailed work on microwave GaAs FET oscillators in which a packaged device with inductive impedance at the gate and capacitive impedance at the source was utilised was reported by Maeda et al, (1975a, 1975b). These impedances were optimised to yield the maximum negative resistance at the drain-source output port of the FET. The output was coupled to a 50 ohm load through a matching network. The imaginary part of the device impedance was resonated with the imaginary part of the load impedance at 10 GHz (the desired oscillation frequency) (Eq. 7.4). The circuit was realised on a 0.6 mm thick alumina substrate using thin film technology. The oscillator provided a peak power of 40 mW and a maximum efficiency of 17% (Fig. 7.8). The oscillation frequency was in the 10 to 10.6 GHz range for a number of oscillator circuits. The output power and frequency variation with temperature from -20 to $+60^{\circ}$C was also measured (Fig. 7.9). The frequency variation was dependent upon the gate bias value. The frequency variation with temperature was reduced as the gate was biased. The best value obtained was -0.45 MHz/$^{\circ}$C at -3.0V gate bias but was -0.9 MHz/$^{\circ}$C at the maximum output power gate bias level of -2.0V.

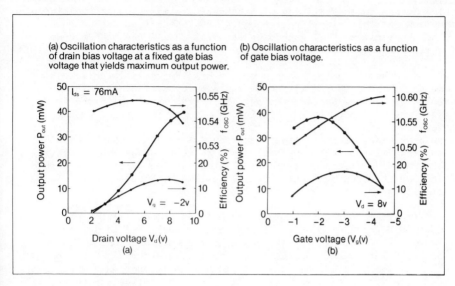

FIG. 7.8.   Bias Dependence of a GaAs FET Oscillator

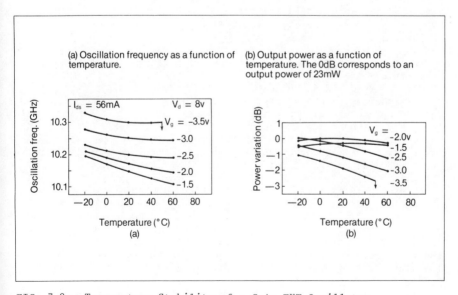

FIG. 7.9.   Temperature Stability of a GaAs FET Oscillator

The output power variation with temperature was found to be minimal at the gate bias level corresponding to the maximum output power. For the minimum dF/dT condition of -0.45 MHz/$^{\circ}$C the output power variation was -0.02 dB/$^{\circ}$C.

The AM and FM noise spectra of the GaAs FET oscillator were also measured and compared with Gunn and Impatt oscillators. The AM noise could be reduced by reducing the drain bias level while the FM noise was insensitive to this variation. The measured AM and FM noise results are shown in Figs. 7.10 and 7.11. The AM noise of these free running GaAs FET oscillators was seen to be higher than that of the Gunn oscillators but comparable to that of the GaAs Impatt oscillator. The FM noise of the GaAs FET oscillator was found to be worse than that of the Gunn oscillator but better than the GaAs Impatt diode oscillator.

A figure of merit for comparison of GaAs FET oscillator output power was introduced by Joshi and Turner (1979). A term called the Peripheral Power Density (PPD) was introduced which normalises the output power of the oscillator to the gate width of the FET used. The maximum value reported so far is 0.33 mW/micron gate width and was realised with an oscillator circuit using series and shunt feedback. The gate width of the FET was 300μm and the oscillator efficiency was 24% (Joshi et al, 1979).

Low power oscillators using smaller gate width FETs (e.g. 300μm) have been realised with both series and parallel feedback elements. The design is generally based on small-signal S-parameters and provides 'ball-park' results on oscillator frequency. The maximum reported conversion efficiency from FET oscillators is 45% and was obtained by Tserng and Macksey (1977) for a 0.5 x 300μm device at 10 GHz.

High power free running oscillators using large gate width multi-cell devices are usually designed using large signal S-parameters. Rector and Vendelin (1978) used large signal S_{11} and S_{22} data with small signal S_{21} and S_{12} data. The large signal S_{11} and S_{22} were derived by conjugately matching the transistor at the desired signal level and frequency and measuring the tuner impedances. Rector and Vendelin (1978) used these parameters to design a common gate oscillator which gave 1.2 watt output power at 8 GHz. The device geometry was 1 x 3000μm and an efficiency of 27.2% was achieved.

Abe et al (1976) used a similar technique to measure large signal S_{11} and S_{22} parameters of a 'subnetwork' on a chip carrier which included the 1.5 x 2500μm FET with parallel feedback elements. Such a technique for oscillator design was first demonstrated by Gonda (1972). On the basis of this information the gate port was terminated in an open circuited line and the output was coupled to the load. The oscillator provided 250 mW output power with 22% efficiency at 6 GHz.

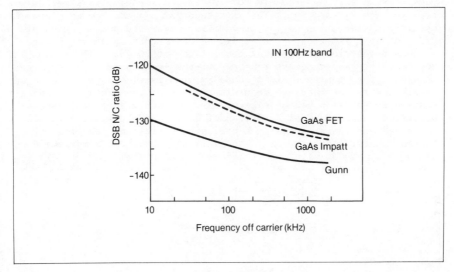

FIG. 7.10. AM Noise of GaAs FET, Gunn and Impatt Oscillators as a
Function of Frequency off the Carrier

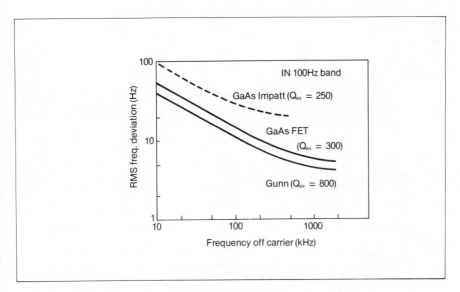

FIG. 7.11. FM Noise of GaAs FET, Gunn and Impatt Oscillators as a
Function of Frequency off the Carrier

Mitsui et al conducted large signal S-parameter measurements for power levels varying from −10 to +20 dBm. A standing wave method was employed for S_{11} and S_{22}. The magnitudes of S_{12} and S_{21} were measured by direct observation of the transmitted power and phase angles by using a magic T method. The amplitude dependence is shown in Fig. 7.12 as a function of incident power level at 10 GHz. The phase angles of the S-parameters remained virtually unchanged. The marked change in the magnitude of S_{22} and S_{12} with increased power level was attributed to the increase in drain conductance and feedback capacitance between drain and gate terminals. These large signal S-parameters were expressed as functions of drain and gate current amplitudes and were used to design a composite series and parallel feedback oscillator network. The oscillator provided a maximum power of 45 mW at 10 GHz with 19% efficiency.

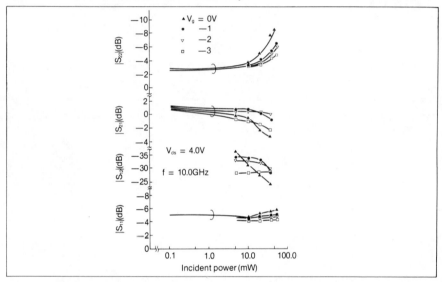

FIG. 7.12.   Dependence of the Amplitude of the S-parameters on the Incident Power Level

The grounded gate configuration does not provide adequate heatsinking when used for power FET oscillators. A grounded source configuration can be utilised but the parallel feedback element coupling the gate and drain terminals has to be small enough to minimise parasitic elements at the desired frequency. The source terminal is either flip-chip mounted or via hole techniques are used to obtain low thermal resistance (see Chapter 3). Common drain configurations possess a natural tendency (due to $|S_{11}| \simeq 1$) for oscillations. The drain terminal, however, does not have low thermal resistance. Wade and Camisa (1978a, 1978b) and Sechi (1979a, 1979b) have used the so called "reverse-channel" or common-drain oscillator in which the roles of source and drain terminals are interchanged. The new drain terminal (formerly source) is flip chip mounted for good electrical

and thermal grounding. This voltage reversal is possible for some
FETs in which the gate is symmetrically placed in the channel and thus
exploits the symmetry of the self-aligned gate processing. Oscillator
applications with a single power supply are possible with this topology.

These reverse channel oscillators have given output powers up to
1 watt with 24% maximum efficiency at 7 GHz (Wade). Camisa and
Sechi obtained 370 mW output power at 8.5 GHz with 24% efficiency.
A maximum efficiency of 26% was obtained.

A. OUTPUT POWER

When small signal S-parameters are used for the oscillator design
good agreement between measured and predicted oscillation frequencies
can be obtained. It is, however, not possible to predict the output
power level. The large-signal characterisation work undertaken by
Mitsui et al (1977) was able to give good agreement between the
measured and calculated power level and frequency of the oscillator.

Pucel et al (1975) and Johnson (1979) have used the power gain
saturation characteristics of a FET amplifier to predict the oscillator output power and to derive an expression for the maximum oscillator power obtainable from a particular device. Typical characteristics are shown in Fig. 7.13.

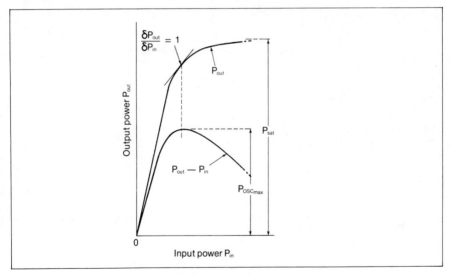

FIG. 7.13.   Power-Gain Saturation Characteristics

The output power asymptotically reaches a value P_{sat}, the saturated
power which is device and bias point dependent. The difference
between the output power, P_{OUT}, and the input power P_{IN}, which would
indicate the oscillator power is also known. This in principle

represents the power available from the device as an oscillator.

Pucel et al (1975) indicated that this characteristic can be approximated as

$$P_{out} \simeq \frac{G_0\, P_{in}}{1 + \dfrac{G_0\, P_{in}}{P_{sat}}} \qquad\qquad 7.13$$

where G_0 is the small signal gain. From Eq. 7.13 the maximum oscillator power is given by

$$P_{osc\ max} \simeq P_{sat}\ (1 = \frac{1}{\sqrt{G_0}})^2 \qquad\qquad 7.14$$

This value is lower than the saturated power output of the corresponding amplifier and reaches it when the small signal gain is very large.

Johnson's approximation for this is given by Eq. 7.15 and was derived from experimental observations,

$$P_{out} = P_{sat}\ (1 - \exp\frac{-G_0\, P_{in}}{P_{sat}}) \qquad\qquad 7.15$$

and the point of maximum oscillator power was computed to be

$$P_{osc(max)} = P_{sat}\ (1 - \frac{1}{G_0} - \frac{\ln G_0}{G_0}) \qquad\qquad 7.16$$

and the maximum efficient gain G_{ME} defined as the power gain which maximises the two port added power, is given by

$$G_{ME} = \frac{G_0 - 1}{\ln G_0} \qquad\qquad 7.17$$

Using these equations an estimate of the output power obtainable from a particular device can be obtained.

B. NOISE

Although GaAs FET oscillators have the attraction of high efficiency and convenient biasing requirements their noise performance is poorer than transferred-electron devices. This has made them less attractive for applications where noise performance is critical. Although the noise behaviour of GaAs FETs used as small-signal amplifiers has been quite extensively investigated and theories adequately describing their performance presented there has been no similar investigation of the noise in a GaAs FET when it is used in oscillator applications.

Experimental results on the close to carrier AM and FM noise of FET oscillators have been presented by several authors for free-running oscillators (Omori et al, 1975; Mitsui et al, 1977b; Abe et al, 1976; Finlay et al, 1978). FM noise for a number of oscillators reported in the literature and their Q values are shown in Fig. 7.14.

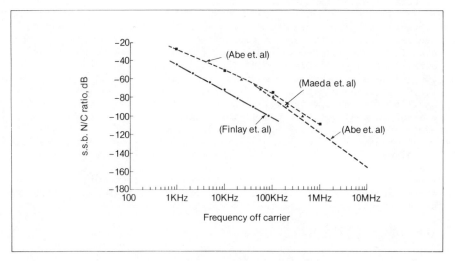

FIG. 7.14.   GaAs FET Oscillator FM Noise Results

It has been generally observed that FM noise is very device dependent. Ruttan (1977) reported FM noise data at X-band for a number of FETs in the same circuit and observed a variation of up to 20 dBs.

Noise in FET oscillators is thus one important area which needs investigation. Any understanding of the origins of noise in GaAs FET oscillators and device improvement may go a long way towards achieving practical application of these oscillators.

6. STABILISED OSCILLATORS

Microwave radio relay and communication systems require a highly stable source with extremely low noise performance. A conventional method of obtaining such high stability is to employ a quartz crystal oscillator followed by frequency multiplier chain. With this the frequency change with temperature is increased by the order of multiplication n while the close to carrier FM noise power is increased by a factor of n^2. The conversion efficiency also worsens as n increases.

It is attractive to use fundamental frequency solid state sources for these applications but it puts severe restraint on their frequency-temperature stability and FM noise performance. Such solid-state sources would therefore require frequency stabilisation before they could be used for such applications. This can be achieved by several techniques like frequency or phase locking to a stable reference signal, discriminator stabilisation or high-Q cavity stabilisation.

A. STABILISATION TECHNIQUES

Stabilisation by frequency or phase locking is a fairly standard technique adopted at low microwave frequencies. Post multipliers are used if higher frequency is desired. In discriminator stabilised oscillators a small fraction of the oscillator output signal is coupled to a discriminator and its output signal which is proportional to the difference between the discriminator centre frequency and the oscillator frequency is amplified and fed back to a frequency control element in the microwave oscillator. Glance and Snell (1976) reported this technique with a microstrip Gunn oscillator with a varactor diode as the control element. Cantle et al (1978) used this technique with coaxial GaAs FET oscillators and MIC discriminator. The gate terminal was used as a frequency control element. The performance of such an oscillator depends upon the discriminator characteristics and the frequency-temperature variation is controlled by the characteristics of the discriminator itself.

(i) Cavity Stabilisation

In the cavity stabilisation method, the oscillator is coupled to a high-Q resonator and a trade-off between the oscillator power and stability can be established. The advantages of this technique are low cost, simplicity and small size. With the use of high Q ceramic resonators the technique is now compatible with MIC circuits also.

Fig. 7.15 shows the basic cavity stabilisation techniques. In the absorption (or reaction) mode a high Q cavity is coupled to the transmission line between the oscillator and the load. The cavity acts as a band reject filter and at the resonant frequency a small amount of r.f. power is reflected by the cavity which is fed back and injected into the oscillator and improves stability. The reflected power is dependent upon the mismatch introduced by the cavity. This technique can also be termed "passive injection locking" as all the band reject filter spectral characteristics are transferred to the free running oscillator. In the reflection stabilisation mode the cavity is coupled to one oscillator port while the other port is coupled to the load. In the transmission cavity stabilisation method, the resonator is treated as a two port network acting as a band pass filter connected between the oscillator and the load. This arrangement has its drawbacks in that coupling losses at two ports have to be accounted for.

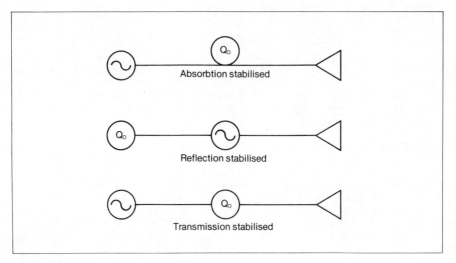

FIG. 7.15.   Basic Oscillator Stabilization Techniques

When the physical size of the cavity is small, as in the case of
compact ceramic resonators, it is possible to use it as a feedback
element coupling the input and output port of the active device.
This is essentially a special case of the transmission stabilisation
method.

A cavity stabilised oscillator is basically composed of two resonant
circuits: the oscillator resonant circuit and the stabilising cavity.
An oscillator may show many possible modes which manifest themselves
in mode jumping, frequency hysteresis etc.

James et al (1975) used Invar and titanium silicate band pass
filter cavities to stabilise GaAs FET oscillators. The cavity size
used was large and was not compatible with the MIC techniques used
for the oscillator itself. With the use of recently developed dielectric material it is possible to obtain thermally stable low noise
oscillators. Improved performance is possible because the ceramic
resonators provide high Q in extremely small outline and the ease of
coupling to the MIC circuits render them very attractive for use with
GaAs FET oscillators.

7. DIELECTRIC RESONATORS

Material scientists have been engaged in the pursuit of a suitable
ceramic material for cavity resonator applications for about a decade
or so (Day, 1970; Masse, 1971; Plourde et al, 1977). The requirements for such a material are low loss (or high Q), relatively large
dielectric constant and low frequency-temperature coefficient.

A large dielectric constant is required in order to realise compact resonators at the required operating frequency and also to ensure that most of the electromagnetic field be confined in them.

Rutile or titanium oxide (TiO_2) has a high dielectric constant ($\varepsilon_r = 100$) and a high-Q (10,000 at 4 GHz). The material would thus seem ideal for microwave applications as the high dielectric constant ensures that the fields are contained within a compact resonator. The frequency-temperature coefficient of rutile is, however, poor (~ 100 ppm/^{o}C) thus rendering the material unusable where high temperature stability is required.

Rutile has, however, been used as a constituent part of a new family of ceramics for this application. $BaO-TiO_2$ system ceramics and a series of zirconate based compositions have been investigated. For $BaO-TiO_2$ system ceramics, as the relative percentage of TiO_2 is varied two stable phase conditions corresponding to $BaTi_4O_9$ and $Ba_2Ti_9O_{20}$ are possible. The barium nanotitanate provides the best combination of dielectric properties with $\varepsilon_r = 39$, $Q = 8000$ and $\tau_f = 1$ to 3 ppm/^{o}C (Masse, 1971; Plourde et al, 1977; O'Bryan et al, 1974).

The resonant frequency temperature coefficient $1/f_r \cdot \Delta f_r/\Delta T$ of a $BaO-TiO_2$ system ceramic can be approximated as

$$-\frac{1}{2}\frac{1}{\varepsilon_r}\frac{\Delta\varepsilon_r}{\Delta T} - \frac{1}{\ell}\frac{\Delta\ell}{\Delta T} \qquad\qquad 7.18$$

In the room temperature range the thermal expansion coefficient $\frac{1}{\ell}\frac{\Delta\ell}{\Delta T}$ is approximately 9 ppm/^{o}C and the dielectric constant temperature coefficient is nearly -24 ppm/^{o}C, giving the frequency temperature coefficient of 3 ppm/^{o}C. Thus the temperature effects on expansion and dielectric constant offset each other to provide extremely small frequency change with temperature.

It is useful to characterise the material properties at microwave frequencies because measurements at lower frequencies may provide different results due to the dielectric relaxation phenomena. Different types of test cavities or dielectrometers can be used to evaluate the dielectric constant and frequency-temperature coefficient of the material (Hakki et al, 1960; Courtney, 1970). Extreme care is required for the measurement of loss tangent (and hence Q) as its value is very small for a good resonator material.

A. RESONANT FREQUENCIES OF DIELECTRIC RESONATORS

Electromagnetic theory analyses of dielectric resonators have been conducted by a number of authors (Cohn, 1968; Okaya et al, 1962; Yee, 1965; Chow, 1966; Itoh et al, 1977; Konishi et al, 1968; Guillon et al, 1977; Pospieszalski, 1977; Pospieszalski, 1979). The analyses usually considered the circular cylindrical resonator in free space and calculated the resonant frequencies for various TE, TM and HE modes. Pospieszalski's analyses of cylindrical resonators have direct practical application for microstrip, stripline and TEM circuit configurations. The first analysis considered the case of a cylindrical sample of a low-loss high dielectric constant material placed between two parallel conducting planes. Such a configuration, known as the dielectric post resonator has been used for the measurement of dielectric constant and loss tangent of insulating materials (Courtney 1970). The analysis derived resonance frequency expressions for the $HE_{n1\ell}$, $TM_{0m\ell}$ and $TE_{0m\ell}$ modes as functions of dielectric constant and the dimensions of the resonator. The properties of the TE_{011} mode were discussed in detail and applied to the measurement of complex permittivity of microwave dielectrics.

Pospieszalski's second work considered a microwave cavity formed by a cylindrical dielectric material of low-loss and high ε_r when placed a finite distance away between two parallel conducting planes perpendicular to the sample axis. The resonant frequency expressions were derived for symmetrical and unsymmetrical cases (Fig. 7.16).

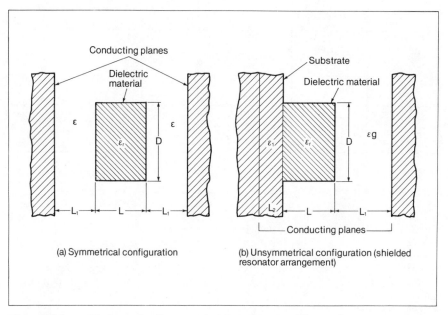

(a) Symmetrical configuration

(b) Unsymmetrical configuration (shielded resonator arrangement)

FIG. 7.16. Dielectric Resonator Mounting Configurations

The unsymmetrical case represents a typical configuration used in microwave integrated circuits and is also known as the shielded resonator arrangement. The dominant TE_{010} mode indicating constant field in the axial direction in this configuration is commonly designated as $TE_{01\delta}$ mode (Itoh et al, 1977). The calculation of the resonant frequency by Pospieszalski's work involves finding the solutions of a Bessel and modified Hankel function equation which also satisfies an expression arrived at by imposing boundary conditions at the dielectric interface. These calculations are best undertaken by a numerical technique.

Table 7.1 shows resonance frequencies for $TE_{mn\ell}$, $TM_{mn\ell}$ and $HE_{mn\ell}$ modes for a cylindrical dielectric material contained between infinite parallel plates as calculated from Pospieszalski's work.

TABLE 7.1.   Calculated Resonance Frequencies (in GHz) for $TE_{mn\ell}$, $TM_{mn\ell}$ and $HE_{mn\ell}$ modes for a Cylindrical Dielectric Resonator Contained Between Two Infinite Conducting Parallel Plates

D = 7.2 mm   L = 2.3 mm ε_r = 39.1

	$TE_{01\ell}$	$TE_{02\ell}$	$TE_{03\ell}$
ℓ = 1	12.464	22.119	32.198
2	14.014	23.094	32.908
3	15.786	24.255	33.764
	$TM_{01\ell}$	$TM_{02\ell}$	$TM_{03\ell}$
ℓ = 1	13.223	22.387	32.328
2	15.080	23.532	33.131
3	17.083	24.863	34.089
	$HE_{11\ell}$	$HE_{21\ell}$	
ℓ = 1	11.435	12.989	
2	21.380	22.250	
3	31.639	32.233	

FIG. 7.17 shows the $TE_{01\delta}$ mode resonant frequency as a function of the air gap dimension for the same sample when placed on a 0.635 mm thick alumina substrate (dielectric constant = 9.8). The theoretical values were obtained from Pospieszalski's analysis. The "tuning characteristics" of the resonator indicate that for an air gap larger than a critical value (3 mm in this case) the resonant frequency remains virtually invariant. The available "tuning range" is a function of dielectric constant of the material and its dimensions. Unloaded Q of these resonators is typically greater than 5000 at X band frequencies.

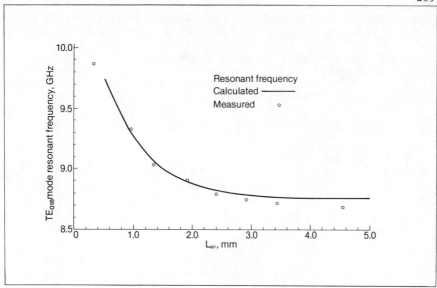

FIG. 7.17.   Tuning Characteristics and Quality Factor of the
   Resonator of Table 7.1 in a Shielded Arrangement

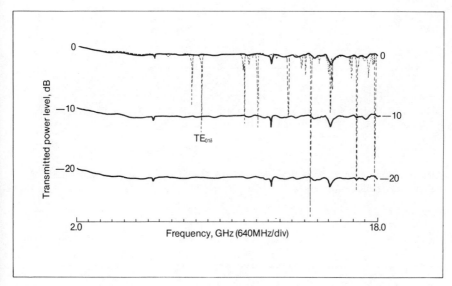

FIG. 7.18.   Resonance Frequencies for the Case of Fig. 7.17
   (L$_{air}$ = 4.55 mm)

The microwave cavity formed by the dielectric resonator has multiple resonant frequencies corresponding to various TE, TM and HE modes. The circuit designer has to overcome this by suppressing or isolating the unwanted modes which might interfere with the proper circuit performance. Fig. 7.18 shows the transmission characteristics of a 50 ohm microstrip line when coupled to a shielded ceramic resonator. The desired $TE_{01\delta}$ mode is identified.

(i) Dielectric Resonator as a Microwave Circuit Element

When a dielectric resonator is placed in the vicinity of a microstrip line, on an alumina substrate, magnetic coupling between the resonator and the line is caused by the $TE_{01\delta}$ mode in the resonator. For a match-terminated line of characteristics impedance Z_0, an equivalent circuit of Fig. 7.19. can be established.

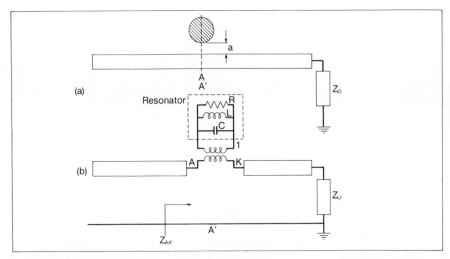

FIG. 7.19. (a) Schematic Diagram of a Shielded Dielectric Resonator Coupled to a MIC Line of Characteristic Impedance Z_0
(b) Equivalent Circuit

The impedance at reference plane A-A' can be expressed near the resonance frequency as

$$Z = Z_0 \left[1 + \frac{K^2 R/Z_0}{1 + j2Q_r\delta_r} \right]$$ 7.19

where Q_r : unloaded Q factor of the resonator

$$\delta_r : \frac{f-f_r}{f_r}$$

f_r : resonant frequency

$K^2 R/Z_0 = \beta$: coupling factor.

Abe et al obtained a family of impedance curves for the configuration shown in Fig. 7.19 as a function of the distance between the
resonator edge and the microstrip edge. These are shown in Fig.7.20.
By fitting the impedance locus of Fig. 7.20 to Eq. 7.19 the values
for β and Q_r were found to be 2.2 and 4000, respectively for ℓ_y =
2.5 mm.

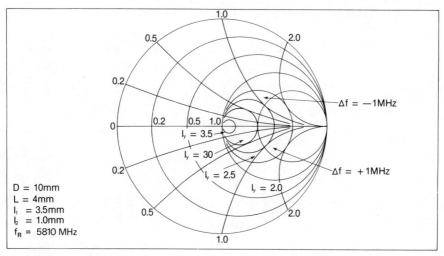

FIG. 20.   Reflection Coefficients of the Resonator Coupled to a
   Terminated 50 ohm Microstripline

8. DIELECTRIC RESONATOR STABILISED FET OSCILLATORS

James et al stabilised GaAs FET oscillators with Invar and titanium
silicate resonant cavities used as band pass filters. At lower
frequencies the required cavity size becomes prohibitively large.
Also the need for coupling a resonant cavity directly to a FET oscillator fabricated using MIC techniques presents difficult problems.

Abe et al used the dielectric resonator in the absorption or reaction mode to stabilise a free-running GaAs FET oscillator at 6 GHz.
A GaAs FET chip for medium power applications with parallel feedback
was first fabricated on a chip carrier. A free running oscillator
was obtained by connecting an open circuited microstrip line at the
gate terminal and an output matching network at the drain terminal
(Abe et al, 1976). The free running oscillator gave 400 mW output
power at 6 GHz with a maximum d.c. to r.f. conversion efficiency of 38%.

A triple-layered or shielded dielectric resonator configuration was employed at the drain terminal to stabilise the oscillator. The dielectric resonator of dimensions D = 10 mm and L = 4 mm was characterised by reflection coefficient measurements indicated in the previous section.

The large signal reflection coefficient of the oscillator and hence the oscillator impedance appearing at the drain terminal was measured. A variable tuner was connected at the output terminal for adjustment of power output and frequency. The reflection coefficients presented to the oscillator by the tuner were measured for constant power levels and frequencies. The results (Abe et al, 1978) are shown on the so called Rieke diagram of Fig. 7.21. The impedance contour of the stabilising band rejection filter is also shown.

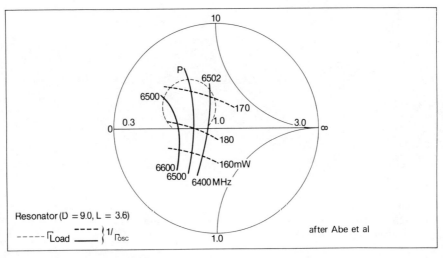

FIG. 7.21.   Rieke Diagram

Theoretical analysis of oscillator stabilisation was presented by considering a parallel resonant circuit coupled to a transmission line terminated by a load with reflection coefficient $|\Gamma|e^{j\delta}$ and applying the condition that the oscillation frequency is determined by the zero susceptance condition together with $\partial B_T/\partial f > 0$.

Unstabilised oscillator oscillation frequency f_o and the external Q value Q_o and the resonant circuit frequency f_r and Q value were defined. Taking β as the coupling constant between the resonant circuit and the transmission line, expressions for stabilisation range, temperature coefficient p and pushing figure q were derived. The stabilisation range which includes hysteresis was given by

$$\Delta_a \simeq \frac{1}{2Q_o} \frac{\beta}{1 + \beta} f_r \qquad\qquad 7.20$$

and the one which does not include hysteresis was given by

$$\Delta_b \simeq 2\sqrt{\frac{\beta}{Q_o Q_r}} \cdot f_r \qquad\qquad 7.21$$

The temperature coefficient and pushing figure were given by

$$p = \frac{1}{f} \frac{\Delta f}{\Delta T} = \frac{1}{f_r} \frac{\Delta f_r}{\Delta T} + \frac{1}{1 + \dfrac{\beta}{(1+\beta)^2} \cdot \dfrac{Q_r}{Q_o}} \cdot \frac{1}{f_o} \frac{\Delta f_o}{\Delta T} \qquad 7.22$$

and

$$q = \frac{1}{1 + \dfrac{\beta}{(1+\beta)^2} \dfrac{Q_r}{Q_o}} \cdot \frac{\Delta f_r}{\Delta V_{GS}} \qquad\qquad 7.23$$

The output power delivered to the load is reduced by the insertion of the band reject filter by a factor given by

$$K = \frac{1}{1 + \beta} \frac{P_o(Y_o/(1 + \beta))}{P_o(Y_o)} \qquad\qquad 7.24$$

where P_o is the unstabilised oscillator output power level depending upon the load conductance G_L.

The performance of the oscillator with GaAs FET voltages, temperature and as a function of resonator air gap was measured. Good agreement between the theoretical and the experimental values was obtained. Performance as bias voltages are varied is shown in Fig. 7.22. Mode jumping and hysteresis were observed. A stabilised output power of 100 mW at 5810 MHz with 17% efficiency was obtained. Frequency and output power variations with ambient temperature are shown in Fig. 7.23. A frequency temperature coefficient of +2.3 ppm/^{o}C was obtained with output power variation of $\pm$ 0.4 dBm from 0 to 50°C temperature change. The mechanical tuning characteristics of the oscillator as the resonator air gap was varied is shown in Fig. 7.24. For L_{air} = 2.1 to 2.6 mm the oscillator frequency followed the resonator frequency without any hysteresis or mode jumping. The FM noise improvement obtained by stabilisation was better than 30 dB.

Sone and Takayama (1978) have presented similar results on an absorption stabilised FET oscillator. The design procedure adopted was similar to Abe et al (1978) but a grounded drain configuration as opposed to grounded source was utilised at 7 GHz.

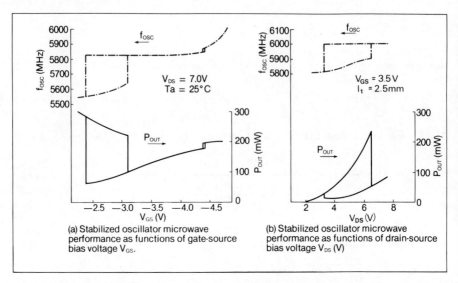

FIG. 7.22. Stabilized Oscillator Microwave Performance as Functions
of Gate-Source and Drain-Source Bias Voltages

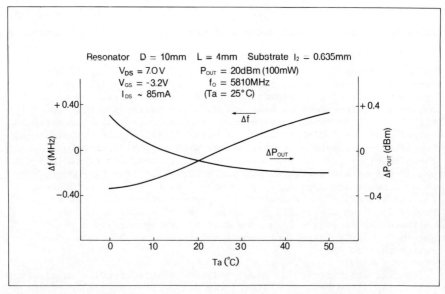

FIG. 7.23. Deviations in P_{out} and f_{osc} for Ambient Temperatures
Between 0 and 50°C.

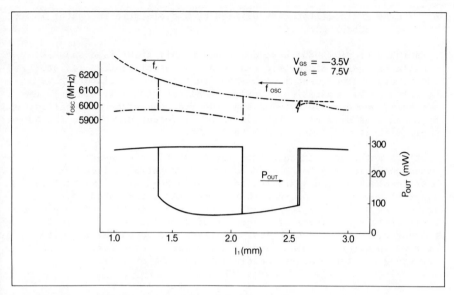

FIG. 7.24. Mechanical Tuning Characteristics of Stabilized
Oscillator

A reflection type dielectric resonator stabilised FET oscillator
has been published by Podcameni and Bermudez (1981). The circuit
topology with feedback was selected to maximise the reflection coefficient magnitude at the gate-source port with the drain-source port
terminated in a 50 ohm load. The reflection coefficient at the input
port was measured under large signal conditions and a dielectric
resonator was coupled to the match-terminated input microstrip line
to obtain stabilisation.

Transmission-type stabilisation, in which the dielectric resonator
is used as a band-pass filter, has been presented by Alley and Wang
(1979) at 1 GHz with bipolar transistor oscillators and by Shinokazi
et al (1978) with FET oscillators at 12 GHz. In the latter the
resonator is placed between two parallel microstrip lines and oscillation frequency between 10 and 12 GHz is obtained by using different
dielectric resonators. Frequency tuning over a limited range was
obtained by varying the associated air gap.

Lesarte et al (1978) reported the first feedback stabilised FET
oscillator using a dielectric resonator. The resonator ($Ba_2Ti_9O_{20}$)
was mounted on a quartz spacer on an alumina circuit and was used as
part of a feedback circuit coupling the drain and gate of the FET.
The oscillator provided a maximum output power of 22 mW at 18%
efficiency at 11 GHz. The output power variation for an ambient
temperature change of 0 to 60°C was less than 1 dB and the frequency
variation was less than 1 ppm/°C. Saito et al (1979) reported a

6 GHz GaAs FET oscillator stabilised by a dielectric resonator in an external feedback circuit.

Ishihara et al (1980) have described highly stabilised GaAs FET oscillators using dielectric resonator and stabilisation resistors. Their mainly experimental work shows that these oscillators have hysteresis-free operation, provide excellent stability against temperature, possess low noise and a wide frequency range capability. By using five dielectric resonators of different dimensions the same basic circuit was tuned from 9 to 14 GHz.

It was noted in Section 2 that for GaAs FETs the negative resistance condition has to be induced by incorporating a feedback element. Most feedback elements are formed by using bond wires, microstrip lines etc. which can be in either lumped and distributed form. These components have low Q and can not be easily altered to produce a different power level or frequency of oscillation. A dielectric resonator, however, due to its small size and compatibility with microstrip circuits, when used as a feedback element offers high Q, low frequency-temperature coefficient and ease of tuning.

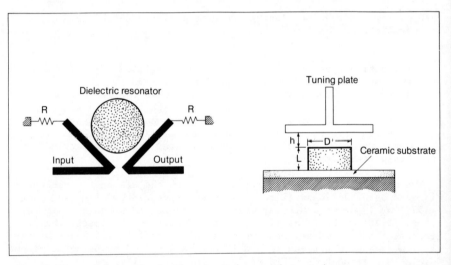

FIG. 7.25.   (a) Feedback Circuit Using a Dielectric Resonator and
    Stabilization Resistors
            (b) Cross Section of a Dielectric Resonator Feedback
    Circuit

Fig. 7.25 shows a feedback circuit in which a dielectric resonator is used for coupling two microstrip lines terminated with resistors equal to the line characteristic impedance. The transmission characteristic is shown in Fig. 7.26 in which the transmitted power and phase are indicated.

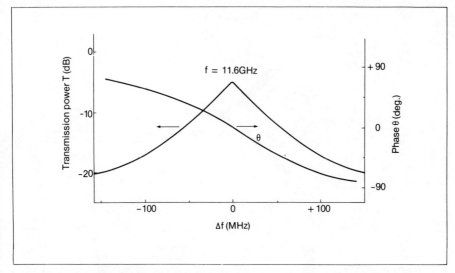

FIG. 7.26.   Transmission Characteristics of a Feedback Circuit

The resonator is in a shielded arrangement (Fig. 7.16b) and the
resonance frequency can be altered by varying the air gap. When this
filter configuration is applied to the drain-to-gate feedback circuit
of a FET amplifier, a highly stabilised GaAs FET oscillator can be
obtained, provided the power gain of the amplifier is higher than the
filter transmission power loss. Because such an oscillator has a
dielectric resonator feedback circuit it was termed the DRF GaAs FET
oscillator by Ishihara et al (1980). The transmission power and phase
of the filter can be varied by changing the distance between the
dielectric resonator and microstrip line, by varying the characteristic impedance of the lines and by varying the angle between the
microstrip lines.

Schematic MIC patterns of these DRF oscillator types investigated
by Ishihara et al (1980) are shown in Fig. 7.27. In these oscillators microwave power is incident on the gate terminal and the amplified output power from the drain terminal is fed back through the
dielectric resonator to the gate. The output power was taken either
from the source terminal (Fig. 7.27(a) and (c)) or from the drain
terminal (Fig. 7.27(b)). The former arrangement allows operation
from a single power supply. Due to heat dissipation considerations
the drain output arrangement is preferable for high power oscillators.
Comparison of Fig. 7.27(a) and (c) shows that the drain resistance R_D
can be eliminated by having an open circuited line of appropriate
length (Ishihara 1980). The dielectric resonator made of SnO_2-TiO_2-

ZrO system with a dielectric constant of 37.5 was used. The oscillator configuration of Fig. 7.27(c) is preferable for low power
oscillators due to its lower component count and operation from a
single power supply.

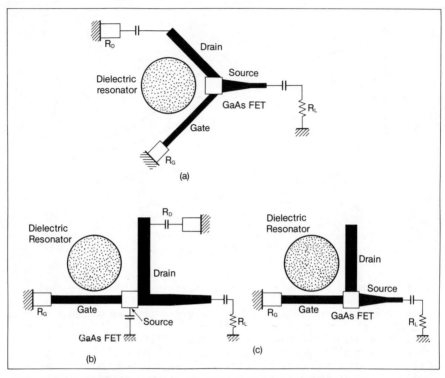

FIG. 7.27. MIC Patterns of the DRF GaAs FET Oscillators.
(a) Source Output Type, (b) Drain Output Type, (c) Source Output
Type with only a Gate Resistor

The mechanical tuning characteristic as a function of air gap for
such a DRF stabilised GaAs FET oscillator is shown in Fig. 7.28.
Fig. 7.29 shows that an oscillator with the same MIC pattern and
GaAs FET can be tuned from 9 to 14 GHz by using dielectric resonators
of different thicknesses.

Oscillator performance with ambient temperature indicated that from
-20 to $+60^{\circ}C$ change the output power variation was ± 0.01 dB/$^{\circ}C$ and
frequency variation was ± 0.16 ppm/$^{\circ}C$ (± 150 KHz total). The FM
noise of the oscillator was found to be comparable with Gunn oscillators.

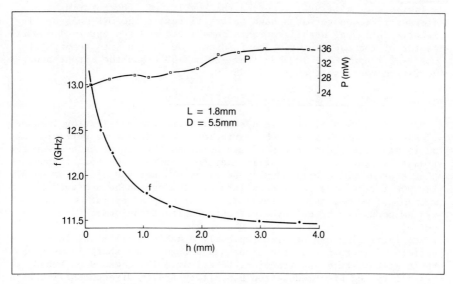

FIG. 7.28. Mechanical Tuning Characteristics as a Function of
Air-Gap Thickness (h)

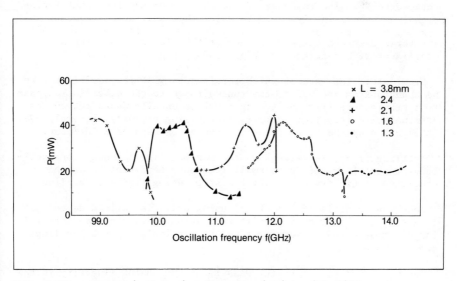

FIG. 7.29. Mechanical Tuning Characteristics with Different
Resonator Thickness (L) as a Parameter

Ishihara et al (1980) reported the results of incorporating a dielectric resonator as a band reject filter to the DRF GaAs oscillator. This dual stabilisation scheme increased the external quality factor of the oscillator. The frequency deviation with temperature was $(\pm 0.1 \text{ ppm/}^{\circ}\text{C}) \pm 100$ KHz from -20 to $+60^{\circ}\text{C}$ and the output power deviation was less than $+0.022 \text{ dB/}^{\circ}\text{C}$.

9. ELECTRONIC TUNING OF GaAs FET OSCILLATORS

Many of today's electronic systems require the oscillating frequency of the microwave source to be variable. This frequency variation may be in discrete finite steps or may be a smooth transition between two frequency states. This facility may be required from both transmitting sources which will usually be of high power and also from mainly low power receiving local oscillators. Tuning can be achieved by mechanical means, but for ease of, and high speed operation most system needs can be met by electronic tuning techniques.

The increasing use of electronically tunable oscillators in a variety of microwave applications has brought into sharp focus both their steady-state and transient behaviour. The demands on broad tunability and high modulation capability are in direct conflict with high thermal stability and low noise requirements.

A. ELECTRONICALLY TUNABLE OSCILLATOR REQUIREMENTS

Consider an electronically tunable oscillator in a steady-state condition oscillating at a frequency f_1 with a tuning input V_1 applied at the control terminals. At an instant $t = 0$ the tuning input is changed to V_2. The response of the oscillator and the input tuning signal are shown in Fig. 7.30 in which the oscillator frequency reaches a steady state frequency f_2 after a finite time. The following terms describe the performance of the electronically tunable oscillator with reference to Fig. 7.30.

Slew rate: The rate at which the frequency of the oscillator can be changed from one end of the tuning range to the other in response to a step change from the tuning input is usually dependent on the external circuitry (for example, linearising circuits, if employed). It is given by the value $\Delta f/\Delta t$ and can be as high as 1 GHz/μsec.

Settling time: is defined as the maximum time taken by the oscillator to settle within some small frequency band centred on the predicted static frequency after a step tuning signal is applied.

Post tuning drift (PTD): Post tuning drift is defined as the largest value of frequency drift when measured over a time interval large compared to the setting time but not including the settling time frequency drift.

Tuning linearity: A useful measure of the tuning linearity is the maximum frequency deviation of the experimental tuning curve from the best linear fit. This is an important requirement for some applications.

These parameters along with low a.m. and f.m. noise and spurious and harmonic free response define the broadband electronic tuning oscillator requirements. Varactor diodes and YIG devices are the two main components used for electronic tuning applications and the performance obtained from these two are complimentary rather than competitive in nature.

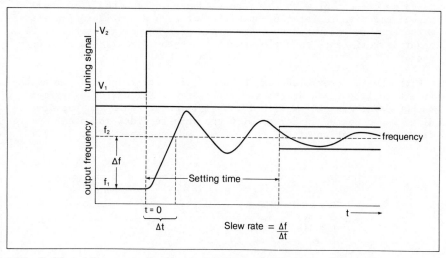

FIG. 7.30. Characteristic Parameters of Electronically Tuned Oscillators

The oscillation conditions for a negative resistance oscillator are given by Eqs. 7.3 and 7.4. The stability criteria of Eq. 7.5 must also be satisfied for sustained steady state oscillations. It is thus evident that, if the circuit reactance is varied by some means, the oscillator frequency will alter to satisfy the zero reactance condition. If the oscillator circuit incorporates a varactor diode or a YIG resonator the circuit reactance can be changed by either varying the d.c. bias level on the varactor diode or by varying the magnetic field strength applied to the YIG resonator.

The gate tuning of free running GaAs FET oscillators can also be considered as electronic tuning. The variation of the gate bias alters the depletion capacitance of the FET and consequently alters the oscillation frequency.

10. VARACTOR TUNED FET OSCILLATORS

Varactor diodes as tuning elements in conjunction with GaAs FET oscillators are quite attractive for high speed and moderate tuning range applications. The circuit designer has to ensure that negative resistance conditions are maintained over the desired frequency range of the oscillator. There are mainly two topological arrangements in which a varactor diode can be incorporated in a FET oscillator circuit - as a part of the feedback element or as a terminating element to one of the device ports.

Tserng and Macksey (1977) used small signal and power FETs in grounded gate topology with series feedback to obtain varactor tuning. The varactor diode was either placed in series with the output matching circuit or with the gate feedback inductance. The small signal negative resistance capability of such a configuration was measured. It indicated that for drain bias values greater than 5V negative resistance conditions exist over the 7 to 12 GHz frequency range.

With the varactor diode in the drain circuit, frequency variation from 8 to 11.5 GHz was achieved with a power FET device. Maximum output power was 210 mW with 17.5% efficiency. When the varactor diode is connected in series with the gate feedback inductance frequency variation of 8.2 to 13.2 GHz was obtained (Fig. 7.31).

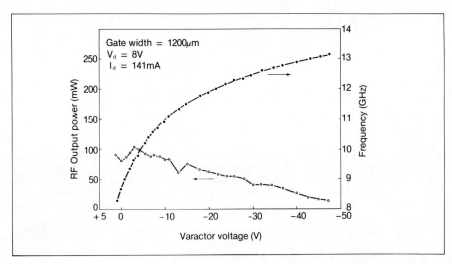

FIG. 7.31.   Performance of an X-band GaAs MESFET VCO with Varactor Tuning in the Gate Circuit

For X-band operation a varactor swing of 30 volts was sufficient. It provided less than 3 dB power variation over the tuning range with 50 mW minimum power. A similar configuration was adopted for a small-signal 0.5μm gate length FET Ku band oscillator. A tunable frequency range from 12.8 to 16.8 GHz was obtained with a nominal output power of 20 mW.

When the varactor diode is used as a termination at the source port for a grounded gate arrangement with series feedback design techniques similar to those outlined in Section 4 can be utilised. The change in varactor bias gives rise to a variation in the phase angle θ_r. With such a circuit a tuning range of 1.7 GHz with less than 1 dB power variation and 16% minimum efficiency was achieved. The performance is indicated in Fig. 7.32.

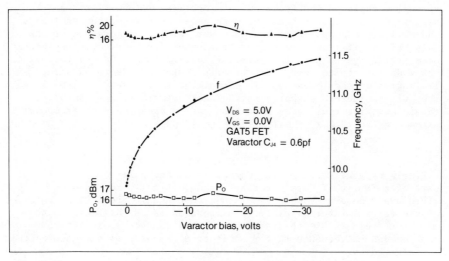

FIG. 7.32.   Varactor Tuned FET Oscillator Performance

Camisa and Sechi (1979a and b) used a varactor diode in the gate circuit of a common-drain oscillator to realise a tuning range from 10.5 to 12.5 GHz. The minimum output power and efficiency was 25 mW and 1.6 percent respectively. Similar techniques were used by Wade.

Rauscher (1980) used the large signal design approach (Section 4.2) for varactor tuned FET oscillators. Varactor tuning was provided in both the gate and source leads of the device while the drain was connected to the ground. A tunable range of 7.4 to 13.1 GHz was achieved and good agreement between measured and predicted output power level was obtained.

Scott et al (1981) used a circuit arrangement with varactor diodes located in series with the gate feedback inductor and across the source to ground port. A common power supply was used to bias the varactors. As the tuning voltage was varied from 0 to 16 volts the oscillator tuned from 7.3 to 15.6 GHz.

It can thus be seen that high power wideband varactor tuned microwave oscillators can be obtained using GaAs FETs. The efficiency of
these oscillators is generally higher than Gunn or IMPATT diode
tunable sources. The noise performance of GaAs FET tunable oscillators is also better than IMPATT oscillators. These tunable FET
oscillators are now beginning to influence the design of microwave
systems.

11. <u>YIG TUNED GaAs FET OSCILLATORS</u>

Over the last several years, the YIG tuned Gunn oscillator has
virtually replaced the Backward Wave Oscillator (BWO) for X and Ku
band signal sources such as in sweep generators, spectrum analysers
and military ECM applications. However, with the emergence of the
GaAs FET and its advantages over the Gunn-effect device it has become
more attractive to use it in similar applications. One factor which
has made the choice more attractive is that there is no physical
limitation on the lowest possible oscillation frequency of the FETs.
Thus with a judicious choice of feedback network it is possible to
obtain negative resistance condition over a broad band of frequencies
extending from 3 GHz or below to up to 18 GHz.

A. YIG RESONATORS

Single crystal Yttrium Iron Garnet (YIG) and gallium doped YIG are
part of a family of ferrites that gyromagnetically resonate at
microwave frequencies when immersed in a magnetic field. The rate
of the ferrite internal electron precession is 28 GHz/Tesla of the
magnetic field within the ferrite and thus can be simply altered.
The resonance frequency f_o for an isotropic sphere in a magnetic field
H_o is given by

$$f_o = \nu H_o \qquad\qquad 7.25$$

where ν is the gyromagnetic ratio (28 GHz/Tesla).

Microwave energy can be coupled to the YIG sphere through a loop in
the plane of applied magnetic field H_o encircling the sphere. The
equivalent circuit is shown in Fig. 7.33 where L_1 is the self inductance of the loop and the parallel tuned circuit elements are related
to the unloaded Q (Q_u) of the YIG resonator (Carter 1961, Olliver
1972).

$$R_o = \mu_o V k^2 \omega_m Q_u$$

$$L_o = R_o/Q_u \omega_o \qquad\qquad 7.26$$

$$C_o = 1/L_o \omega_o^2$$

where μ_o = permeability of free space

V = volume of the YIG sphere

$k = {}^1/d_1$ = coupling factor

d_1 = loop diameter

$\omega_m = 2\pi f_m = 2\pi \nu (4\pi M_s)$

$4\pi M_s$ = saturation magnetisation.

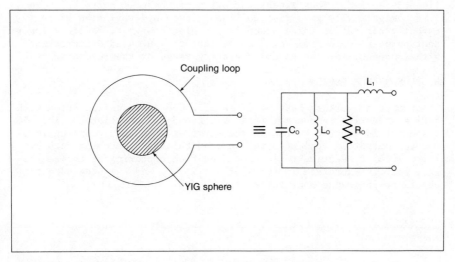

FIG. 7.33.   YIG Resonator and Its Equivalent Circuit

The input impedance of the combination is given by:

$$Z_{in} = j\omega L_1 + \frac{(j\omega\mu_o/Q_u)R_o}{\omega_o^2 - \omega^2 + j(\omega\mu_o/Q_u)} \qquad 7.27$$

The unloaded Q, Q_u of the resonator is related to the saturation magnetization and is given by

$$Q_u = \frac{H_o - \frac{1}{3}(4\pi M_s)}{\Delta H} \qquad 7.28$$

$$= \frac{f_o - f_m/3}{\nu \Delta H}$$

where ΔH is the resonance line width.

The premature decline phenomena or low level limiting (Olliver 1972) inside the YIG sphere limits its use as a high Q resonator to

286

frequencies above f_{min}, where f_{min} is given by

$$f_{min} = \frac{2}{3} f_m = \frac{2}{3} \nu (4\pi M_s) \qquad\qquad 7.29$$

From Eq. 7.25 it can be seen that by linearly varying the magnetic field applied to the YIG sphere the oscillation frequency can be linearly varied. Thus linear variation of the coil current in the electromagnet will give rise to linear frequency variation if no magnetic saturation takes place in the electromagnet. Another important point to note from Eq. 7.28 is that the unloaded Q increases with frequency for the useful operating range of the resonator.

B. PERFORMANCE REVIEW

A number of workers have used the grounded gate configuration with series inductive feedback for the YIG tuned FET oscillator. The YIG resonator is coupled to the source terminal through a coupling loop and the output is connected to the load through a matching network (Fig. 7.34). Negative resistance condition appearing at the source-gate port is expressed in terms of $|S_{11}T| > 1$ and has been used to obtain tuning range over the entire X-band.

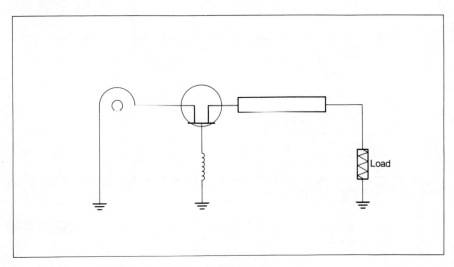

FIG. 7.34. YIG Tuned Oscillator Schematic

For a circuit configuration in which the drain port is match terminated the gate feedback inductance which results in $|S_{11}| > 1$ can be evaluated and the inherent trade-offs studied. In most cases the feedback inductance itself is not sufficient to provide $|S_{11}| > 1$ over an octave or larger bandwidth.

Trew (1980) used additional parallel feedback between the source
and drain terminals which caused an effective increase in the total
bandwidth. With this compound feedback scheme and without any matching at the output port it was possible to obtain oscillations from
5.9 to 13.25 GHz. A minimum output power of 3 mW was obtained from
5.9 to 12.5 GHz with a peak power of 22 mW at 6 GHz and a peak
efficiency of 8%.

The alternative technique of presenting a negative resistance condition is to mismatch the output or drain port in such a way so as to
have the same phase as the S_{11}' vector at all frequencies of interest
and consequently to result in $|S_{11}T|$ being greater than for the
desired frequency range. (Section 4.A). When the drain port is
terminated in a reflection coefficient Γ_L, $S_{11}T$ is given by

$$S_{11}^{T} = S_{11}' + \frac{S_{12} \, S_{21}}{\frac{1}{\Gamma_L} - S_{22}}$$

7.30

The objective is to choose a Γ_L at all frequencies of interest so
that $|S_{11}T| > 1$ for a particular feedback inductance. The values of
Γ_L which satisfy this condition define a circle in the Γ_L plane.

This Γ_L circle can be constructed by solving for

$$\Gamma_L = \frac{S_{11}^{T} - S_{11}'}{S_{12}' \, S_{21}' + S_{11}^{T} \, S_{22}' - S_{11}' \, S_{22}'}$$

7.31

where $S_{11}^{T} = 1 \, e^{j\theta}$ and $\theta = -180$ to $+180^{\circ}$.

Papp and Koyano (1980) plotted stability circles for $|S_{11}T| = 1$.
The centre of the circle and its radius are given (Carson 1975) in
Eqs. 7.32 and 7.33 respectively.

$$C = \frac{S_{22}'^* - S_{11}' \, (S_{11}'^* \, S_{22}'^* - S_{12}'^* \, S_{21}'^*)}{|S_{22}'|^2 - |S_{11}' \, S_{22}' - S_{12}'S_{21}'|^2}$$

7.32

$$R = \frac{S_{12}' \, S_{21}'}{|S_{22}'|^2 - |S_{11}' \, S_{22}' - S_{12}' \, S_{21}'|^2}$$

7.33

The region inside or outside the circle represents the unstable
region for which $|S_{11}T| > 1$ and can be determined in a similar manner
as for amplifiers.

Once the stability circles are plotted the load trajectory of Γ_L
with frequency that would be necessary to produce oscillations in the
required frequency range can be determined. The design problem then

reduces to the construction of a matching circuit which simulates the behaviour of Γ_L.

Fig. 7.35 indicates the stability circles for a 1 nH feedback inductance for a grounded gate NEC 388 FET as reported by Papp and Koyana (1980). The arrows point to the desired unstable region and the trajectory for output reflection coefficient from 8 to 18 GHz is also indicated. Realisation of a circuit which behaves in the desired manner is done by trial and error - direct synthesis techniques have not been devised.

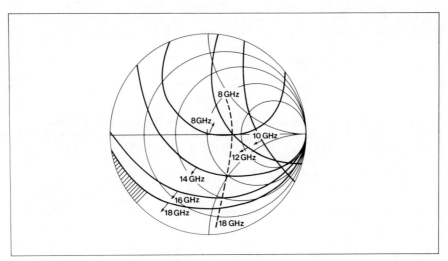

FIG. 7.35. Stability Circles Together with Load Trajectory of Γ_L Across the 8-18 GHz Range

Trew (1980), Basawapatna and Stancliff (1979) and Papp and Koyano (1980) used empirical techniques to design the drain matching network which provided the necessary performance. Basawapatna and Stancliff (1979) did load pulling measurements to verify the required trajectory. Their YIG tuned FET oscillator followed by a single stage amplifier provided tuning from 5.5 to 14.1 GHz with 15 dBm minimum power from 5.9 to 12.4 GHz. The design technique was applied to a bipolar oscillator and provided tuning from 1.8 to 9.3 GHz. The circuit realised by Trew had a tunability of 7.9 to 14.4 GHz with 10 mW minimum power from 7.9 to 14 GHz and a conversion efficiency better than 6%. A post amplifier stage was not used in this case. Papp and Koyano (1980) achieved a tunable range of 7.9 to 18.5 GHz with the basic oscillator and with the oscillator and a buffer amplifier. The results are indicated in Fig. 7.36 and represent the highest power achieved so far with an unbuffered oscillator in this tuning range. Oyafuso (1979) presented the results of the first YIG tuned FET oscillator which provided a tuning range of 8-18 GHz.

Buffer amplifiers were used to obtain a power level of +10 dBm.
Heyboer and Emery (1976) and Ruttan (1977) reported initial performance results on YIG tuned FET oscillators.

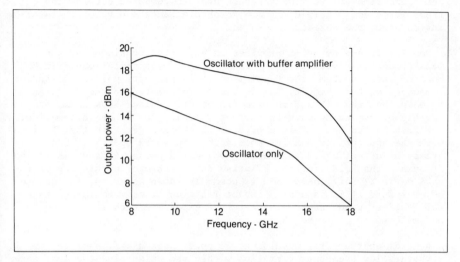

FIG. 7.36.   Power Output Versus Frequency Across the 8-18 GHz Band
    for the Oscillator with and without Buffer Amplifier

Careful circuit layout is necessary in order to prevent spurious
oscillations.   These can occur when the oscillation conditions and
the stability criteria are satisfied even with no YIG bias.   These
oscillations are termed spurious because they do not tune with the
YIG resonator.   The active circuit and the resonator need to be
carefully designed to eliminate the possibility of spurious oscillations.   The coupling loop inductance should be minimised so that
the spurious oscillation frequency occurs out of the negative resistance condition.   Parasitic source to ground capacitance should also
be minimised.   Trew (1980) and Papp and Koyano (1980) have discussed
the spurious oscillation problem in some detail.

A rather unique approach has been taken by Le Tron et al (1979) for
the design of YIG tuned FET oscillators.   They used a common YIG
resonator coupled to both the source and gate port of the FET.   In
this configuration the negative resistance condition is obtained only
for a narrow range of 50 MHz.   However, as the magnetic bias is
changed both the source and gate ports track together providing a
tuning range of 3.5 to 14 GHz.

12. PULSED RF OSCILLATORS

During the last several years solid state microwave devices have
made possible many novel applications in radar techniques for

measurement of range and velocity and have enabled utilisation of
these systems for many commercial non-military applications. The
radar problem consists in the unambiguous determination of the range
and/or velocity of the target remote from the sensor. The properties
which may be used to characterise target behaviour are frequency,
phase and time delay. These parameters are compared between the
transmitted and received signals to obtain range and velocity information about the target.

Doppler frequency shift provides an accurate technique for the
measurement of radial target velocity for many applications. In CW
Doppler radar systems the Doppler shift is proportional to the radial
component of the velocity and is therefore related to the difference
between the transmit and receive frequency. Pulsed RF systems are
capable of providing both the target velocity and range information.
The transmitted and received signals when mixed generate the IF signal
with the appropriate Doppler modulation impressed upon it. One
transient effect which can cause problems in pulsed RF systems is the
frequency change during the RF pulse due to changes within the microwave device. This phenomenon is commonly known as chirp, and has to
be minimised if it is a significant fraction of or much larger than
the receiver bandwidth. Elaborate and expensive techniques have to
be employed to minimise the chirp.

A novel approach for obtaining low power pulsed r.f. oscillators
with very low frequency variation within the pulse uses the dual gate
GaAs field effect transistor and indicates yet another application of
this flexible device.

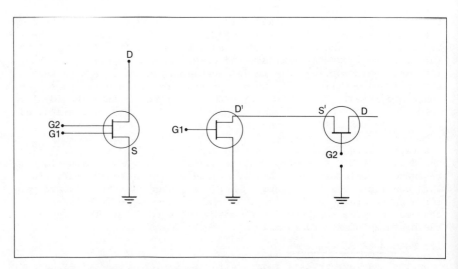

FIG. 7.37. Dual Gate GaAs FET and its Equivalent

GaAs dual gate FETs have an inherent property which make them extremely suitable for low chirp pulsed r.f. oscillator applications. This property can be best illustrated by considering a simple physical model of the dual gate FET. Fig. 7.37 shows the equivalent of a dual gate FET. If the first FET containing gate 1 is used as an active element for obtaining steady state oscillations, the second FET containing gate 2 can be operated as a high speed switch to obtain pulsed r.f. output. This can be easily done by applying a negative pulse to the second gate.

TABLE 7.2. Input Reflection Coefficient S_{11} and Forward Gain S_{31} (between Drain and Gate 1) for the Dual Gate GaAs FET as a Function of V_{G2}. $V_{G1} = -1.03V$.

V_{G2}, Volts	S_{11}		S_{31}	
	Mag	Phase	Mag	Phase
0	0.56	−42.0	1.15	−58.5
−0.4	0.58	−42.6	0.93	104.8
−0.9	0.61	−46.4	0.57	116.7
−3.2	0.62	−95.3	0.03	171.0

Table 7.2 indicates the magnitude and phase of the gate 1-to-source and gate 1-to-drain scattering parameters S_{11} and S_{31} for different gate 2 bias levels. It can be seen that as the second gate voltage is increased from 0 to −0.9V the phase of the input reflection coefficient remains virtually unchanged while the forward gain of the device decreases by as much as 6 dB. This minimal phase change in S_{11} maintains the operating frequency. However, the phase of S_{11} changes very rapidly beyond this level as V_{G2} is biased towards pinch off. But this phase change does not give rise to any change in oscillation frequency because the gain of the device has fallen to the extent that it is not able to sustain oscillations. Thus these two factors − minimal phase change in S_{11} and reduction of gain $|S_{31}|$ − act in unison to result in minimal frequency change during the pulse.

On the basis of the above discussion low chirp GaAs dual gate FET pulsed r.f. oscillators can be realised by incorporating feedback and matching elements between gate 1 and source. The drain/source port is coupled to the load while modulation is applied to gate 2. An important factor to note here is that drain current is virtually zero when the device is off. This results in less heat dissipation in the device and thus improves device reliability because of low operating junction temperature (Joshi et al, 1980).

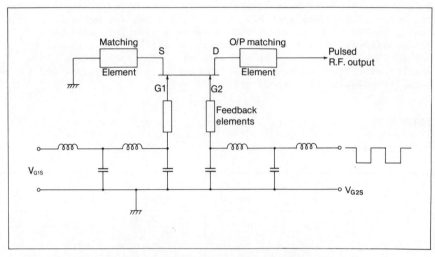

FIG. 7.38.   Schematic Diagram of Dual Gate GaAs FET Pulsed
Oscillator

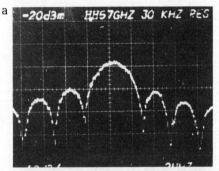

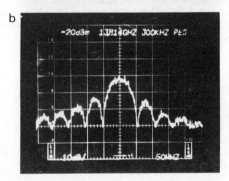

FIG. 7.39.   Spectra of X and J Band Pulsed FET Oscillators

Fig. 7.38 shows a schematic diagram of the dual gate GaAs FET pulsed oscillator. A series feedback element is introduced at the gate 1 terminal while the source terminal is capacitively terminated by an appropriate 50 ohm short-circuited line length. The design procedure for this oscillator was along the lines indicated in Section 4.

FIG. 7.39 shows the spectra of the r.f. output for the X and J band pulsed FET oscillators. From the side lobes of the output spectra of Fig. 7.39 it can be calculated that the frequency variation for the duration of the pulse is approximately 0.3 MHz. The output spectrum was virtually symmetrical in both cases indicating linear frequency variation during the pulse duration (M. Brookbanks, 1980).

13. CONCLUSIONS

In this chapter the design and application of the GaAs FET as microwave oscillators has been covered. It was seen that the introduction of external elements is usually necessary to induce negative resistance conditions. Apart from the relatively poor close to carrier noise performance of free running GaAs FET oscillators, their other features appear quite attractive. The compatibility of FET oscillator circuits and ceramic resonators have resulted in improved performance of stabilised sources. Electronic tuning capabilities of GaAs FET oscillators using varactor diodes and YIG resonators are impressive and will form important building blocks for microwave systems.

14. BIBLIOGRAPHY

Abe, H. et al. A high power microwave GaAs FET oscillator. ISSCC, Digest, 1976, pp.164-165.

Abe, H. et al. A stabilised low noise GaAs FET integrated oscillator with a dielectric resonator at the C-band. 1977 IEEE-ISSCC Digest, pp.168-169.

Abe, K. et al. A highly stabilised low noise GaAs FET integrated oscillator with a dielectric resonator in the C-band. IEEE Trans. MTT Vol. MTT-26, No. 3, March 1978, pp.156-162.

Alley, G. and Wang, H. An ultra low noise microwave synthesiser. IEEE Trans. MTT, Vol. MTT-27, No. 12, December 1979.

Basanpatna, G.R. and Stancliff, R. A unified approach to the design of wideband microwave solid state oscillators. IEEE Trans. MTT, Vol. MTT-27, No. 5, May 1979, pp.379-385.

Bodway, G. Circuit design and characterisation of transistors by means of three port scattering parameters. Microwave Journal Vol. 11, No. 5, May 1968.

Brookbanks, M. - private communication, 1980.

Camisa, R. and Sechi, F. ISSCC 1979a Digest, pp.160-161.

Camisa, R. and Sechi, F. Common drain flip chip GaAs FET oscillators.
  IEEE MTT-27, No. 5, May 1979b, pp.391-394.

Cantle, et al. A coaxial GaAs FET local oscillator and its
  stabilisation by a temperature-compensated MIC discriminator.
  Proc. Eu MC, Paris 1978, pp.259-263.

Carson, R. High frequency amplifiers. New York, Wiley 1975.

Carter, P.S. Magnetically tunable microwave filters using single
  crystal yttrium-iron-garnet resonators. IRE Trans. MTT, May 1961,
  pp.252-260.

Chow, K.K. On the solution and field pattern of the cylindrical
  dielectric resonator. IEEE Trans. MTT, Vol. MTT-14, p.439,
  September 1966.

Cohn, S.B. Microwave bandpass filters containing high Q dielectric
  resonators. IEEE Trans. MTT, Vol. MTT-16, pp.210-217, April 1968.

COMPACT User's Manual, Compact Engineering Inc., U.S.A.

Courtney, W. Analysis and evaluation of a method of measuring the
  complex permittivity and permeability of microwave insulators.
  IEEE Trans. MTT Vol. MTT-18, pp.476-489, Aug. 1970.

Day, W.R. Dielectric resonators as microwave circuit elements.
  IEEE Trans. MTT, Vol. MTT-18, pp.1175-1176, December 1970.

Edson, W. Vacuum pulse oscillators. New York, Wiley, 1953.

Finlay, H., Joshi, J. and Cripps, S. An X-band FET oscillator with
  low FM noise. Elec. Letts. Vol. 14, No. 6, March 1978.

Glance, B. and Snell, W. A discriminator stabilised microstrip
  oscillator. IEEE Trans. MTT, Vol. MTT-24, Oct. 1976, pp.648-650.

Golio, J.A. and Krowne, C.M. Microwave Journal, October 1978,
  pp.59-61.

Gonda, J. Large signal transistor oscillator design. IEEE MTT-S
  Digest, p.110-112, May 1972.

Guillon, P. and Garault, Y. Accurate resonant frequencies of dielectric resonators. IEEE Trans. MTT, Vol. MTT-25, pp.916-922,
  November 1977.

Hakki, B.W. and Coleman, P.D. A dielectric resonator method of
  measuring inductive capacities in the millimeter range. IRE Trans.
  MTT, Vol. MTT-8, pp.402-410, July 1960.

Heyboer, T. and Emergy, F. YIG tuned GaAs FET oscillators. IEEE MTT-S Int. Microwave Symposium Dig., pp.48-50, 1976.

Ishihara, O. et al. A highly stabilised GaAs FET oscillator using a dielectric resonator feedback circuit in 9-14 GHz. IEEE Trans. MTT, Vol. MTT-28, No. 8, April 1980.

Itoh, T. and Rudokas, R. New method for computing the resonant frequencies of dielectric resonator. IEEE Trans. MTT, Vol. MTT-25, pp.52- , January 1977.

James, D. et al. Stabilised 12 GHz MIC oscillators using GaAs FETs. Proc. 5th Eu MC, pp.296-300, Sept. 1975.

Johnson, M.M. Large signal GaAs MESFET oscillator design. IEEE Trans. MTT-27, No. 3, March 1979, pp.217-227.

Joshi, J.S. and Turner, J.A. High peripheral power density GaAs FET oscillator. El. Letters, Vol. 15, No. 5, March 1979, pp.163-164.

Joshi, J. and Pengelly, R. Ultra low chirp GaAs dual gate FET microwave oscillator. 1980, MTT-S Symposium Digest.

Joshi, J.S. - unpublished.

Konishi, Y. et al. Resonant frequency of TE_{01} dielectric resonators. IEEE Trans. MTT, Vol. MTT-24, pp.112-114, February 1968.

Krowne, C.M. Network analysis of microwave oscillators using microstrip transmission lines. Elec. Letts. Vol. 13, No. 4, pp.115-117, Feb. 1977.

Kurokawa, K. Injection locking of microwave solid state oscillators. Proc. IEEE, Vol. 61, No. 10, October 1973, pp.1386-1410.

Lesartre, P. et al. Stable FET loscal oscillator at 11 GHz with electronic amplitude control. Eu MC Paris Digest, pp.264-268, 1978.

Le Tron et al. Multioctave FET oscillators double tuned by a single YIG. ISSCC, February 1979, pp.162-163.

Maeda, M., Takahashi, S. and Kodera, H. CW oscillation characteristics of GaAs Schottky barrier gate FETs. Proc. IEEE, February 1975a, pp.320-321.

Maeda, M., Kimura, K. and Kodera, H. Design and performance of X-band oscillator with GaAs Schottky gate FETs. IEEE Trans. MTT-23, No. 8, pp.661-667, August 1975b.

Masse, D. A new low-loss high $_r$ temperature compensated dielectric for microwave applications. Proc. IEEE Vol. 59, pp.1628-1629, November, 1971.

Mitsui, Y., Nakatani, M. and Mitsui, S. Design of GaAs MESFET
oscillator using large signal S-parameters. IEEE MTT-S, Symposium
Digest, San Diego, 1977a, pp.270-272.

Mitsui, Y., Nakatani, M. and Mitsui, S. Design of GaAs MESFET
oscillator using large signal S-parameters. IEEE Trans. MTT-25,
No. 12, December 1977b, pp.981-984.

Mori, T., Ishihara, O. et al. A highly stabilised GaAs FET oscillator using a dielectric resonator feedback circuit in 9-14 GHz band.
IEEE MTT-S, Symposium Digest, Washington DC, 1980.

O'Bryan, H., Thomson, J. and Plourde J. A new BaO-TiO$_2$ compound
with temperature stable high permittivity and low microwave loss.
J. Am. Ceramic Society, 57(10), 1974, pp.450-452.

Okaya, A. and Barash, L.F. The dielectric microwave resonator.
Proc. IRE Vol. 50, pp.208--2092, October 1962.

Olliver, P.M. Microwave YIG-tuned transistor oscillator amplifier
design: Application to C-band. IEEE Jnl. Solid State Circuits,
Vol. SC-7, No. 1, February 1972.

Omori, M. and Nishimoto, C. Common gate GaAs FET oscillator. Elec.
Letts, Vol. 11, No. 16, pp.369-371, August 1975.

Oyafuso, R. An 8-18 GHz FET YIG tuned oscillator. IEEE MTT-S Sym.
Digest, pp.183-184, 1979.

Papp, J. and Koyano, Y. An 8-18 GHz YIG-tuned FET oscillator.
IEEE Trans. MTT, Vol. MTT-28, No. 7, July 1980, pp.8-14.

Plourde, J. et al. A dielectric resonator oscillator with 5 ppm long
term frequency stability at 4 GHz. 1977 IEEE MTT-S Symp. pp.273-
276.

Plourde, J. et al. Ba$_2$Ta$_9$O$_{20}$ as a microwave dielectric resonator.
Journal of American Ceramic Society, Vol. 58, No. 9-10, pp.418-420.

Podcameri, A. and Bermudez, L.A. Stabilised FET oscillator with
input dielectric resonator: large signal design. El. Letts.
Vol. 17, No. 1, pp.44-45, January 1981.

Pospieszalski, M. On the theory and application of the dielectric
post resonator. IEEE Trans. MTT, Vol. MTT-25, No. 3, pp.228-231,
March 1977.

Pospieszalski, M. Cylindrical dielectric resonators and their
applications in TEM line microwave circuits. IEEE Trans. MTT,
Vol. MTT-27, No. 3, March 1979, pp.233-238.

Pucel, R., Bera, R. and Masse, D. Experiments on integrated GaAs FET
oscillators at X-band. Elec. Letts. Vol.11, pp.219-20, May 1975.

Rauscher, C. and Willing, H.A. 1978 MTT, Vol. 26, p.1017.

Rauscher, C. and Willing, H.A. Simulation of nonlinear microwave FET performance using a quasi-static model. IEEE Trans. MTT, Vol. MTT-27, pp.834-840, 1979.

Rauscher, C. Optimum large signal design of fixed frequency and varactor tuned GaAs FET oscillators. Proc. 1980, IEEE MTT-S Int. Microwave Symp. pp.373-375.

Rauscher, C. Broadband varactor tuned GaAs FET oscillator. Electronics Letts. Vol. 16, No. 14, July 1980, pp.534-535.

Rector, R.M. and Vendelin, G.D. A 1.0W GaAs MESFET oscillator at X-band. IEEE MTT-S, Digest, 1978, Orlando, Florida.

Ruttan, T. X-band GaAs FET YIG tuned oscillator. IEEE MTT-S, Symposium Digest, 1977a, San Diego, pp.264-266.

Ruttan, T. GaAs FETs rival Gunns in YIG tuned oscillators. Microwaves, July 1977b.

Saito, T. et al. A 6 GHz highly stabilised GaAs FET oscillator using a dielectric resonator. 1979, IEEE MTT-S Symposium Digest, pp.197-199.

Scott, B. et al. Octave band varactor tuned GaAs FET oscillators. 1981 ISSCC Symposium Digest, pp.138-139.

Shinokazi, S. et al. 6-12 GHz transmission type dielectric resonator transistor oscillators. IEEE MTT-S Symposium Digest, Ottawa, 1978, pp.294-296.

Sone, J. and Takayama, Y. A 7 GHz common-drain GaAs FET oscillator stabilised with a dielectric resonator. NEC R & D No. 49, pp.1-8, April 1978.

Trew, R.J. Octave band GaAs FET YIG tuned oscillators. Electron. Letts. Vol. 13, pp.625-630, October 1977.

Trew, R.J. Design theory of broad band YIG tuned FET oscillators. IEEE Trans. MTT, Vol. MTT-27, No. 1, January 1980, pp.8-14.

Tserng, H., Macksey, H. and Sokolov, . Performance of GaAs MESFET oscillators in the frequency range 8-25 GHz. Elec. Letts. Vol. 13, No. 3, February 1977, pp.85-86.

Tserng and Macksey. Wide band varactor tuned GaAs MESFET oscillators at X and Ku-bands. 1977, IEEE MTT-S, Digest, San Diego, pp.267-269.

Wade, P.C. X-band reverse channel GaAs FET power VCO. Microwave Journal, April 1978a, pp.92.

298

Wade, P.C.   Novel FET power oscillators.   El. Letts. Vol. 14, No. 20,
    pp.672-674, August 1978b.

Yee, H.Y.   Natural resonant frequencies of microwave dielectric
    resonators.   IEEE Trans. MTT, Vol. MTT-13, pp.256, March 1965.

CHAPTER 8
Microwave FET Packaging

1. INTRODUCTION

As has been seen in previous chapters the gallium arsenide field
effect transistor has revolutionised the design of low-noise and
power microwave frequency amplifiers as well as providing an excellent device for oscillators, modulators and mixers. The GaAs FET is
often used in its unencapsulated or 'bare' chip form in microwave
circuits but for many applications it is desirable to package the
device or devices in well characterised hermetic enclosures. For
high frequency applications these packages need to be as small as
possible to minimize the effect of reactances associated with the
encapsulation. It is desirable to make the electrical length between
the packaged device and the circuit as small as possible to avoid
large phase angle changes in the input and output reflection coefficients as the frequency is changed. Such large phase angle differences lead to an inability to match the packaged FET over wide bandwidths. Small packages, however, tend to increase the effect of
feedback paths thus increasing the $|S_{12}|$ of the FET and decreasing
the frequency range over which the device is stable.

Microwave packaging techniques are only just catching up with the
rapid advances of the GaAs FET in its frequency performance and its
applications and it is only recently that some fundamental problems
encountered with certain packaging techniques have been understood
and overcome at frequencies above 15 GHz.

"Pre-matching" the FETs, i.e. including matching circuits within the
package so that the influence of the package on device performance is
minimised, is one effective method particularly suitable for power
FETs where device impedances tend to be somewhat lower than 50 ohms.
However, the pre-matching technique is to a certain extent a compromise solution since it invariably limits bandwidth.

Fig. 8.1 shows a photograph of a number of popular microwave packages used for field effect transistors. Some of these packages, such
as the HPAC 100, are more readily available than others since they
are high volume industry standard packages used for bipolar transis-

tors. Packages which have four leads such as the HPAC 100 result in device performance compromises since such leads and their connections to the FET inside the package are electrically long at the higher frequencies.

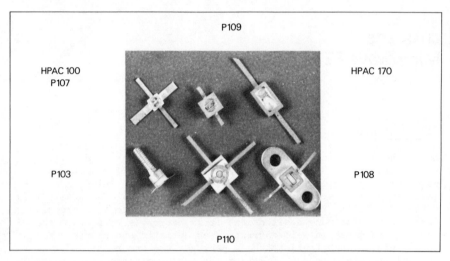

FIG. 8.1. Some GaAs FET Packages (courtesy of Plessey Co. Ltd.)

2. PACKAGES AND SEALING

Package styles, materials, properties and manufacturing processes all influence their design. Such designs in turn influence device performance whilst assembly and sealing techniques effect performance, yield, cost and reliability.

Packages can be divided into three main constructional types:-

1. Cofired ceramic

2. Hard fired ceramic

3. Glass-metal.

A cofired multilayer ceramic package involves the lamination and metallisation of two or more layers of ceramic in the unfired or 'green' state and subsequent firing of the structure to form a simple, homogeneous package. Fig. 8.2 shows an example of such a package, the P109, which enables GaAs FETs to operate with good performance up to 18 GHz. The ceramic walls of the package are brazed to an oxygen-free copper base enabling the FET source pads to be bonded to an effective microwave ground. The base to ceramic interface is important and care has to be taken to avoid cracking the ceramic during brazing caused by the different coefficients of thermal expansion between the alumina ceramic and the copper. Packages of the cofired type are usually sealed with a gold-plated thin metal lid using a gold-tin solder preform.

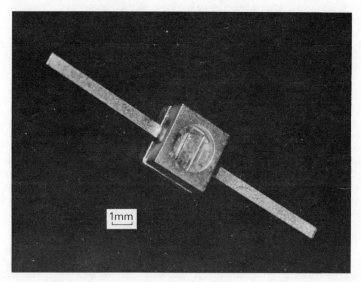

FIG. 8.2.   P109 GaAs FET Package (Courtesy of Plessey Co. Ltd.)

FIG. 8.3.   Glass to Metal Seal Package

Fig. 8.3 shows an example of a glass-metal package which although popular for such microwave components as PIN diode limiters, attenuators and switches has not found popularity as a FET device package because of its relatively high cost in comparison to the cofired ceramic package. The glass-metal package is, however, finding increasing application for compact sub-system integration (Galli et al, 1980) as well as being a suitable packaging technique for GaAs integrated circuits (ICs) (covered in Chapter 10). Hard-fired ceramic packages usually consist of separate ceramic and metal parts which are held together by the use of brazes, epoxies or glasses.

Packages are sealed with either a ceramic or metal lid usually using a gold-tin solder preform which has a melting point of 280°C. This temperature is ideal for the GaAs FET as it does not lead to degradation in the device's performance. Temperatures in excess of approximately 350°C can affect the gallium arsenide due to arsenic outdiffusion and for this reason sealing techniques are somewhat limited. Glass sealing using a ceramic lid can be a very low-cost alternative to metal seals and the elimination of the metal sealing ring on the top of a ceramic package (Fig. 8.1) can lead to reduced package capacitance.

The use of epoxies for lid sealing can be applied to certain high frequency microwave packages such as the dual gate FET package of Fig. 8.4. where hermetic sealing can be achieved. However, there is no epoxy available to date which can provide an adequate moisture barrier and indeed many epoxies will not withstand temperature cycling over the usual military temperature range of -40°C to $+80^{\circ}$C.

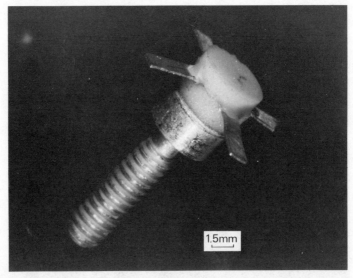

1.5mm

FIG. 8.4. P106 Epoxy Sealed GaAs FET Package (Courtesy of Plessey Co. Ltd.)

Glass sealing using a ceramic lid can be a very low cost alternative to metal seals. The development of a low melting-point glass which does not absorb moisture and is hermetically inert (at least, to the types of reagents used in the manufacture of FETs and packages) is mandatory to the success of this approach. Glasses approaching these desirable conditions are now available including materials which melt at 385°C. By using heated package holders and pulsed sealing techniques or using lasers it is possible to seal effectively small packages without raising the FET chip temperature higher than 300°C.

The design of a small, high frequency package is often constrained by the need to allow ease of wire bonding of the chip. This results in severe constraints on metal-to-metal sealed ceramic packages.

The metal sealing ring of many multilayer ceramic packages itself forms a resonant structure when coupled by the ceramic walls to the input and output leads. This produces a low Q resonance in the package input to output isolation characteristic. This resonance is usually only modified by the presence of the metal sealing lid, careful package design and lid grounding being required to move the resonance to a higher frequency where the FET is not to be operated or to remove it altogether. Even though the lid may be 'grounded' by package metallization the resonance may only be moved in frequency.

3. PACKAGE MODELLING

Microwave modelling of packages is most important especially before committing considerable effort and money to tooling for a new design. An equivalent circuit for a package is shown in Fig. 8.5(a) together with the origins of the package parasitics in Fig. 8.5(b). Wire bonds to the FET are modelled as the gate inductance L_G, the source inductance L_S and the drain inductance L_D. There is also an inductance L_M created by the source lead metallization of the package and the package mounting. In Fig. 8.1 for example some of the packages have a larger value of L_M than others due to the length of metallisation between the source pads and the 'through wall' metal to the outside leads. To aid in decreasing this inductance the two source pads are connected both internally and externally as shown in Fig. 8.5(b).

There are a variety of capacitances which include input and output shunt values C_{IN} and C_{OUT} created by the package geometry and mounting. Feedback capacitances C_{F1} and C_{F2} are extremely important and need to be minimised. They are caused by the internal package metallisation geometry and the external lead sizes respectively. The way in which these capacitances affect performance is demonstrated by calculating and measuring the gains and stability factors of FETs in different package styles (Barrera et al, 1979). Fig. 8.6 shows for example, the relative gains of two packaged 1μm gate length, 300μm gate width FETs one in a cofired ceramic package (HPAC 100) and the other a low parasitic hard-fired package, the P103 (Pengelly, 1979). Grounding the lid and reducing lead pad size and proximity can be used to minimise C_{F1} and C_{F2}. The need for small, yet repeatable parasitics

304

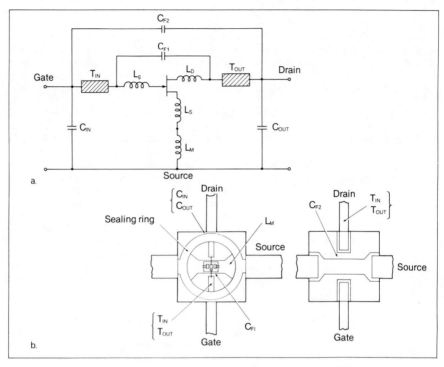

FIG. 8.5(a). Typical Equivalent Circuit for a GaAs FET Package
(b). Origins of Package Parasitics

often forces the package designer to impose tight dimensional tolerances on the already difficult to manufacture package.

Finally there are transmission line segments T_{IN} and T_{OUT}. Such
packages as the HPAC 100 have the gate and drain connections brought
into the package sandwiched between two ceramic layers which are
bounded by the metal lid and the backside metallisation. Such a
configuration sets a limit on the characteristic impedance of the
lines leading to and from the chip.

Providing C_{F1} and C_{F2} can be minimised, the one parameter in a
common source bonded FET that needs to be minimized is the source
lead inductance. This is made up of the wire bonds, package metallisation and package to ground path defined by the chip mounting.
Fig. 8.7 shows the effect of common lead inductance on a packaged
0.5μm length small signal FET where it may be seen that well over
6 dB change in gain results from introducing only up to 0.2 nH inductance in the source lead at 8 GHz. More importantly, the device is

potentially unstable with inductances greater than 0.05 nH.

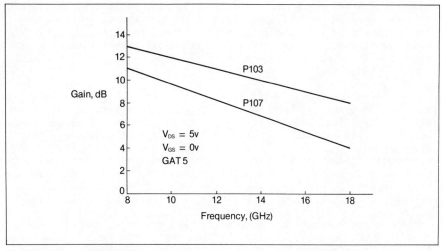

FIG. 8.6.   Comparison of Gains as a Function of Frequency for 1μm
Gate Length FET in Low-Parasitic P103 Package and Co-fired P107
(HPAC 100) Package

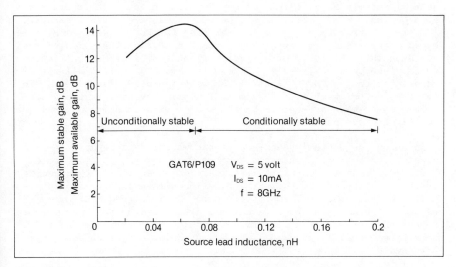

FIG. 8.7.   Effect of Source Lead Inductance on Gain of Packaged
GaAs FET

4. PREMATCHED GaAs FETs

The microwave design engineer often requires a packaged FET which will enable him to produce broadband amplifiers – typically of octave bandwidths – at frequencies up to 20 GHz. The scattering parameters of such packaged devices (particularly S_{11}) rotate in phase on a Smith Chart much too quickly to allow optimum matching over the required frequency range. This is particularly true of power FETs where the value of gate-to-source capacitance is much larger than in a small signal device.

Conventional microwave amplifiers are usually designed using a distributed approach which necessitates using a relatively large area of dielectric for the microstrip circuitry. The only 'lumped' components used are usually the wire-bonds connecting the device either to the package or, in the case of chip devices, to the circuit. The pre-matched technique in which the matching circuit is incorporated in the device package is a low cost approach to avoiding the problems of package parasitics or, indeed, incorporating them as part of the matching circuit. For many chip FETs, $S_{11}*$ or optimum noise impedance are conveniently matched over wide bandwidths using simple inductive and capacitive components. This takes the form of wire bond lengths and chip capacitors (using MOS or MIS structures). The FETs can be self-biased by the use of bypass capacitors on the source together with chip resistor arrays whose value is selected on test. test.

For example Fig. 8.8(b) shows the response of a single stage GaAs FET which has been prematched over the 10 to 20 GHz frequency range using bond wires as inductors and chip capacitors as matching components (Pengelly et al 1979), as may be seen in the circuit diagram of Fig. 8.8(a). Fig. 8.8(c) shows the general form of such a circuit. A resistor array is used for self-biasing the FET as well as a source bypass capacitor. It may be seen from Fig. 8.8(b) that greater than 5.5 dB gain is achieved with a gain flatness of $\pm$ 0.2 dB. A medium power prematched FET result is shown in Fig. 8.9 where small signal gains of 6 to 7 dB over 12 to 15 GHz were achieved with a 1 dB gain compression point of over 20 dBm at 14 GHz. The placement of wire bonds can be made accurately by the use of an automatic bonder leading to reproducible performance.

An extension of this technique is the matching and combining of power GaAs FETs within a package to produce powers, at present, up to 15 to 20W (Takayama et al, 1979) with good power added efficiencies and reasonable bandwidths. Fig. 8.10 shows an example of an internally matched power FET amplifier where the inductors are formed using wire bonds and shunt capacitors are formed by the parallel plate capacitance of the metallized areas on high dielectric constant substrate material. Fig. 8.11 shows an equivalent circuit of this arrangement.

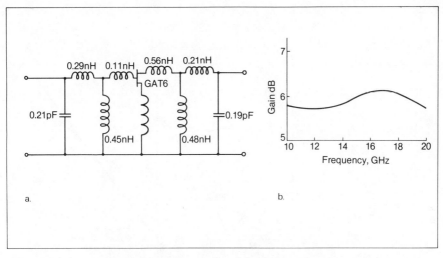

FIG. 8.8(a)   Lumped-Element Broadband Matching Circuit for GAT6
  in P105 Package
          (b)   Measured Response of Broadband Pre-matched GAT6

FIG. 8.8(c)   Typical Prematched Transistor in Package (J-band
  Medium Power Amplifier)

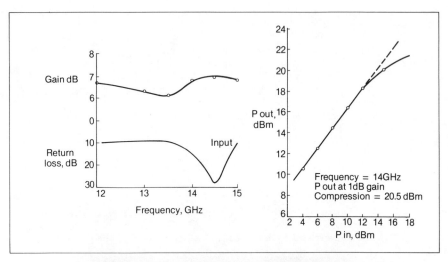

FIG. 8.9.   Response of Pre-matched J-band Medium Power Transistor

FIG. 8.10.   Compact Power FET Module Showing Wire Bond Inductances
and Parallel Plate Capacitances

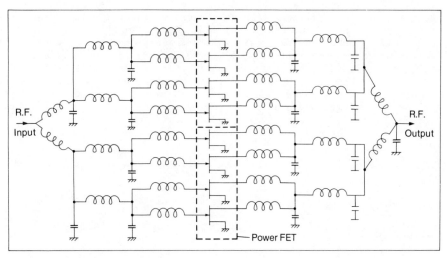

FIG. 8.11.   Equivalent Circuit of 'Internally Matched' Power FET
shown in Fig. 8.10.

5. PACKAGING AND THERMAL RESISTANCE

One of the most important properties of a package relating to
device reliability and its derating of maximum power dissipation above
ambient temperature is the thermal resistance of the structure.   The
total thermal resistance of a packaged device will depend on the chip
thickness, the way the chip is secured to the package and the package
itself.   In the latter case, for example, the P108 package of Fig.8.1
has better thermal properties than the HPAC 100 package since in the
first case the chip is mounted directly onto the grounded metal base
of the package whilst in the second the chip is mounted on a thin
metal strip on the ceramic.   This thin strip is connected to the heat
sink outside the package by the same leads that act as the source
connection.

Fig. 8.12 indicates the way in which the channel temperature of a
FET is composed of

$$T_{CH} = T_A + P_D (\theta_{chip} + \theta_{solder} + \theta_{amount})$$

where T_A is the ambient temperature, P_D is the dissipated power and
θ_{chip}, θ_{solder} and θ_{mount} are the various thermal resistances where
we have assumed the chip is secured to the package using a eutectic
preform.   There are two general cases for thermal resistance

(Pritchard, 1967), depending on the mode of heat flow. If the thickness of the chip material is small compared to the lateral dimensions of the device and chip, heat will tend to flow in a vertical 'column'. If the chip material is thick compared to the device size, and the device dimensions are less than 20 percent of the chip side dimension a 'spreading' heat flow can be assumed. Generally FET dimensions, particularly power devices, fall into a region somewhere between spreading and columnar flow. Unlike the calculations performed for silicon transistors, the calculations performed for GaAs FETs (Cooke, 1978) have taken into account the fact that the heat source is made up of long, thin lines – the channels. Thermal resistance for a FET can be approximated by drawing an analogy between fringing capacitance for an electrical conductor and thermal heat-flow spreading.

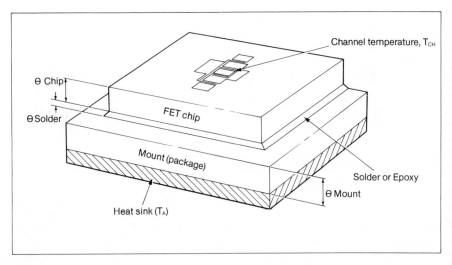

FIG. 8.12.   Origins of the Thermal Resistance of Mounted GaAs FET
      Chip

Since the capacitance per unit length of a transmission line is $(120\pi\varepsilon)/Z_0$ where Z_0 is the characteristic impedance, formulas for transmission line characteristic impedance may be used to calculate thermal resistance. Using a formula for stripline characteristic impedance (Cohn, 1954) and the equivalent ideal line (Oliver, 1955), we can derive the following equation for a FET consisting of a single gate.

$$\theta W_G = \frac{K(k)}{2K_{TH}} \frac{K(k)}{K(k')} \qquad\qquad 8.1$$

where $k = \text{sech}\left[\pi L_G/4F\right]$

and $k' = \tanh\left[\pi L_G/4F\right]$

where L_G is the gate length

$\quad\quad$ F is the chip thickness

$\quad\quad W_G$ is the gate width

$\quad\quad K_{TH}$ is the thermal conductivity (0.44 for GaAs)

and $\quad$ K is the complete elliptical integral of the first kind.

Equation 8.1 will be modified for multigate FETs since there is thermal coupling and heat transfer between gates (see Chapter 3).

Equation 8.1 has been evaluated for GaAs FETs on a 100μm thick chip for gate lengths of 0.1 to 4μm (Fig. 8.13) and Table 8.1 compares the thermal resistance of FETs with various gate length and gate-width combinations.

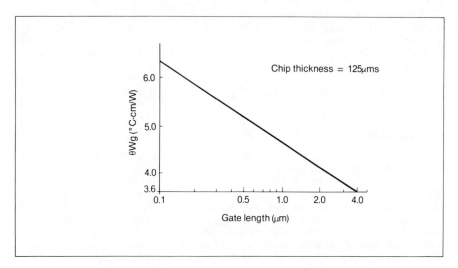

FIG. 8.13.   Thermal Resistance of Single Gate GaAs FET

From reliability data (Drukier, et al, 1979) it is generally concluded that a GaAs FET channel temperature should not exceed 125°C.

Consider, for example, the GaAs FET in Table 8.1 having a thermal resistance of 19.5°C/watt. To this must be added the solder thermal resistance and package thermal resistance giving a typical total of 40°C/W. Under CW conditions with a heat sink temperature of 25°C:

$\quad\quad$ 125 = P_D (70) + 25°C

and therefore, P_D = 2.45 watts.

TABLE 8.1. Thermal Resistance of Various Gate Length/Gate Width FETs

Gate Length (μm)	Gate Width (μm)	θ_{WG} ($^{\circ}C\ W^{-1}\ cm^{-1}$)	θ ($^{\circ}C\ W^{-1}$)
0.5	150	5.17	345
0.5	300	5.17	172
1.0	500	4.68	93.6
1.0	2400	4.68	19.5

Thus the device must not dissipate more than 2.45 watts. A FET with a gate width of 2.4 mm will give an r.f. output power of approximately 1.5 watts with a power-added efficiency of 35% and 10 dB power gain at 4 GHz, for example. Dissipated power will, therefore, be 2.35 watts, i.e. within the figure calculated. However, for higher ambient temperatures the power output and/or gain of this power FET would have to be derated. It may thus be appreciated that, particularly for power FETs, the thermal resistance of a package plays an important role since without due attention particularly to chip mounting channel temperature limits can be easily exceeded. Thus many power and small signal FET packages are designed to exploit the good thermal and electrical properties of a copper base at the bottom of the structure.

6. CONCLUSIONS

GaAs FETs in small signal form are, at present, available in hermetic packages up to 20 GHz or so with power FET packages containing some internal matching being available up to 15 GHz. Such packaged devices offer considerable advantages to the design engineer. The devices can be pretested and screened to the quality levels demanded by commercial, military and space applications much more easily than bare-chip devices. The user is provided with a much easier to handle component especially if he does not have wire bonding and chip handling facilities. With the advent of better performance FETs, and packaging techniques, packaged devices are now achieving good performance up to 18 GHz that was, until a few years ago, only achievable with bare-chip FETs. The thermal resistance of GaAs FETs, particularly power devices, is receiving much attention with the advent of flip chip mounting etc. (Fukui, 1980).

7. BIBLIOGRAPHY

Barrera, J.S. and Huang, C.L. Why use a packaged FET? Microwave Systems News, Aug. 1979, Vol. 9, No. 8, pp.144-152.

Cohn, S.B. IRE Transactions on Microwave Theory and Techniques, July 1954, Vol. MTT-2, No. 2, pp.52.57.

Cooke, H.F.   FETs and bipolars differ when the going gets hot. Microwaves, Feb. 1978, Vol. 17, No. 2, pp.55-61.

Drukier, I. and Silcox, J.F.   A reliability study of power GaAs FETs. Proceedings of the 9th European Microwave Conference, Brighton, Sept. 1979, pp.277-281.

Fukui, H.   Thermal resistance of GaAs field effect transistors. Technical Digest of the International Electron Devices Meeting, Washington, USA 1980, pp. 118-121.

Galli, J.G., Gilchrist, B.E. et al.   Integration shrinks microwave front ends.   Microwave Systems News, Sept. 1980, Vol. 10, No. 9, pp.119-130.

Oliver, A.A.   IRE Transactions on Microwave Theory and Techniques, March 1955, Vol. MTT-3, pp.134-143.

Pengelly, R.S., Arnold, J., Cockrill, J. and Stubbs, M.G.   Prematched and monolithic amplifiers covering 8 to 18 GHz.   Conference Proceedings of the 9th European Microwave Conference, Brighton, Sept. 1979, pp.293-297.

Pengelly, R.S.   Packaging of Miniature, high frequency microwave amplifiers using GaAs FETs.   Proceedings of the 1979 Internepcon Semiconductor Symposium 1979, October 1979, pp.274-277.

Pritchard, P.L.   Electrical characteristics of transistors, McGraw-Hill, Chapter 9.4, New York (1967).

Takayama, Y., Ogawa, T. and Aono, Y.   11 GHz and 12 GHz multiwatt internal matching for power GaAs FETs.   Electronics Letters, Vol.15 No. 11, 24 May 1979, pp.326-328.

CHAPTER 9
Novel FET Circuits

1. INTRODUCTION

This chapter is intended as an introduction to the use of GaAs FETs
in circuit roles other than the usual ones of amplification, oscillation and frequency conversion. Both single and dual gate FETs,
whether they be low noise or power devices have been used successfully
in a variety of microwave applications such as switches, attenuators,
phase shifters and modulators. The introduction of monolithic circuits, dealt with in more detail in Chapter 10, has enabled many of
these circuits to be fabricated on single chips of GaAs. However,
many of the circuits have also been implemented using bare chip FETs
and hybrid microstrip techniques (Pengelly et al, 1980).

2. SWITCHES

For many years the PIN diode as a control element has dominated
microwave circuits such as switches and phase shifters (Garver, 1972).
However, recently the use of both single and dual-gate FETs has
received considerable attention in the design of fast switches.

Several configurations are possible to exploit the low-noise small
signal and the power FET as switches (Fig. 9.1). The series configuration (Fig. 9.1(a)) makes use of the saturation and pinch-off
conditions of the device and provides a broadband (untuned) response
with zero d.c. bias power when using small signal devices.

The use of the series configuration to switch medium levels of r.f.
power (i.e. a few watts) can be achieved using a power FET structure.
However, such a configuration produces poor frequency response due to
the capacitance between the source and drain metallizations and
ground. Fig. 9.2(a) shows the simple equivalent circuit of a FET
series switch in its ON and OFF states where in the OFF state the
capacitance C3 will dominate, the channel resistance R being large.
In the ON state, capacitances C_1 and C_2 limit the insertion loss but
can be tuned out, using a low-pass filter synthesis technique, by
inductors L_1 and L_2.

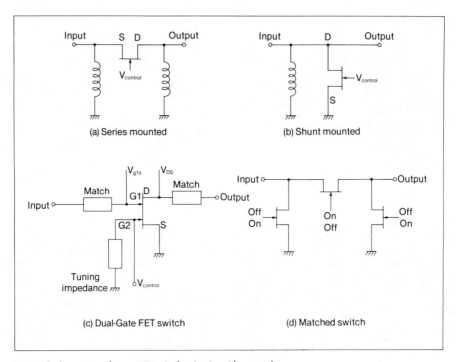

FIG. 9.1.   Various FET Switch Configurations

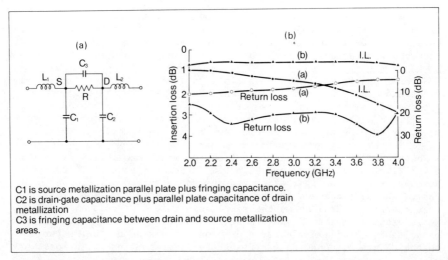

FIG. 9.2.(a).   Simple Equivalent Circuit of Power FET Switch
      (b). Frequency Response of Tuned and Untuned Series Power FET Switch

Such a technique is demonstrated by the performance of a 1 watt
series switch shown in Fig. 9.2(b) in the 'ON' state where the effect
of tuning is shown on both the insertion loss and input VSWR obtainable over the octave band 2 to 4 GHz.

However, in order to maximise the 'OFF' isolation it is also necessary to resonate the capacitance C_3 with an inductor L_3 as shown in
Fig. 9.3(a). Such a simple resonant circuit results in the useful
bandwidth in the 'OFF' condition being decreased.

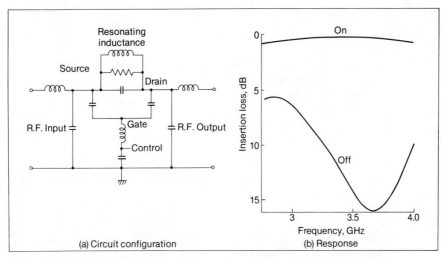

(a) Circuit configuration   (b) Response

FIG. 9.3.   Resonated Series Power GaAs FET Switch Performance

Fig. 9.3(b) demonstrates the result of a 1W switch which has been
optimized in its ON and OFF state. McLevige et al (1980) have demonstrated such a circuit technique in X band whilst Gaspari and Yee
(1978) have reported an 8 way switch utilizing tuned series connected
FETs using small signal devices.

The shunt mounted FET (Fig. 9.1(b)) is somewhat easier to operate
as a broadband switch. Fig. 9.4(a), for example, shows the small and
large signal operation of a 1 watt power FET (Pengelly et al, 1980)
whilst Fig. 9.4(b) shows simple equivalent circuits for such a device
in its ON and OFF states. Again tuning can be used to reduce the
insertion loss and broaden the bandwidth using low pass filter techniques similar to those adopted in PIN diode switch design.

Figs. 9.5(a) and (b) show the way in which the insertion loss and
attenuation of FET switches depends on the equivalent circuit parameters of the devices used without the addition of tuned circuits.
The RC networks shown in Figs. 9.5(a) and (b) are simple equivalent

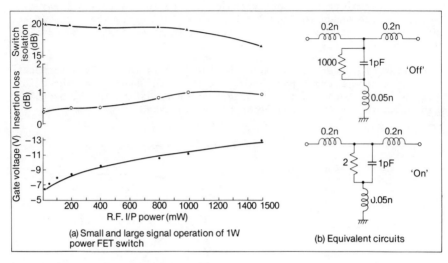

FIG. 9.4. Power FET Shunt Switch

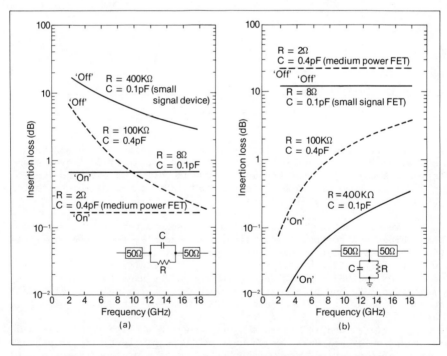

FIG. 9.5.(a) Insertion Loss and Attenuation of Series Switch
      (b) Insertion Loss and Attenuation of Shunt Switch

circuits for the FET switch where R is the channel resistance between 'source' and 'drain' and C is the depletion capacitance.

The shunt mounted FET can be incorporated easily into a single pole, double throw switch (SPDT) as shown in Fig. 9.6(a). When one FET is 'ON' and the other is 'OFF', where the 'OFF' FET is in its 'short-circuit' state, the OFF FETs impedance is transformed through the quarter wavelength line to an open-circuit at the switch input port. Thus the input impedance of the switch is dicated by the impedance of the 'ON' FET.

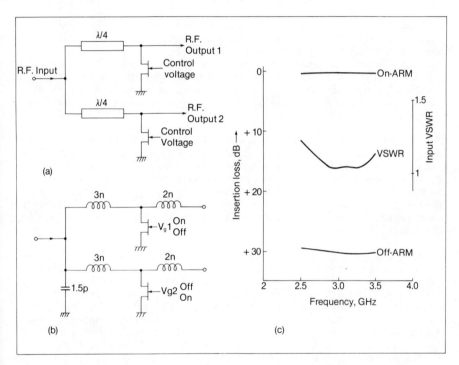

FIG. 9.6. (a) SPDT Switch Using Shunt-Mounted GaAs FETs, (b) SPDT Switch Incorporating Lumped Components and GaAs FETs, (c) Performance of SPDT Switch

Such a circuit can be realised in either transmission line form or in lumped component form. In the distributed form (Fig. 9.6(a)) the FET can, in fact, be incorporated into the transmission lines themselves when the circuit is fabricated in monolithic form (Ayasli et al 1980) whilst in the lumped form (Fig. 9.6(b)) the FETs' capacitances again become part of the circuit when converting the quarter wavelength lines to equivalent inductances and capacitances (Pengelly et al, to be published). Figure 9.6(c) shows the response of a 1 watt SPDT switch designed in lumped form showing the excellent isolation and input VSWR that can be obtained.

The FET in shunt connection, is limited in its power handling capability by two factors. By monitoring the gate current under negative gate bias conditions, i.e. when the switch is in the low insertion loss state it is possible to measure an input power level threshold at which significant gate current flows. This is due to avalanche breakdown on the negative peaks of the r.f. cycle between the gate and drain. This breakdown is often accompanied by light emission from the channel region (Furutsaka, et al, 1978; Yamamoto et al, 1978).

To maximise power handling of the switches several improvements can be made to the device. These include increasing the carrier level in the channel region and decreasing channel depth to reduce pinch-off voltages. Optimum pinch-off appears to be around -6 volts (McLevige, private communication). Also gate to drain spacing can be increased (although this will increase the drain to source resistance) as well as changing the channel recess from an abrupt form to a graded form (Higashisaka et al, 1979).

Gate current is also observed when the FET switch is in the grounded gate state (i.e. the isolation condition) which is due to rectification of the r.f. power by the gate-source diode on negative half cycles of the r.f. drive. On positive half-cycles the gate goes negative with respect to the drain tending to pinch-off the device thus reducing the isolation. By applying a small positive bias to the gate in this state the isolation can be restored to a value close to the small signal case.

The shunt mounted FET is a reflective structure in its 'OFF' state but may be combined with a series mounted device in the π configuration of Fig. 9.1(d) to produce a broadband high isolation switch with low VSWRs. The action of this switch or attenuator is similar to that of a PIN π attenuator (White, 1974) and relies only on having known voltage to attenuation laws for the series and shunt mounted devices. π attenuators having close to 40 dB switching range with insertion losses of less than 1 dB in S band have been built giving VSWRs in all states of greater than 15 dB (Pengelly et al, 1980).

The major advantage of the FET over the PIN diode switch is that it requires no control current and is inherently a very fast device (typically 1 nsec or less rise times can be achieved with small signal devices).

The dual gate FET can also be used as a high isolation switch using a circuit as shown in Fig. 9.1(c). Unlike the previous examples the dual gate FET is operated in the normal manner with drain to source voltage. The disadvantages of the dual-gate FET approach are:

1. The switch is non-reciprocal;

2. The switch takes d.c. power in its 'ON'-state, and

3. The device requires considerably more complicated matching circuits
   to operate over the same bandwidths as single-gate switches.

However, the advantages of such a configuration are that gain is
available and the phase change between the 'ON' and 'OFF' states can
be maintained close to zero. Fig. 9.7 shows the phase response for a
300μm gate width dual-gate FET as the attenuation is increased from
the 'ON' state to the 'OFF' state. It may be seen that for a 30 to
35 dB switching range it is possible to maintain the phase change to
within 1 or 2 degrees.

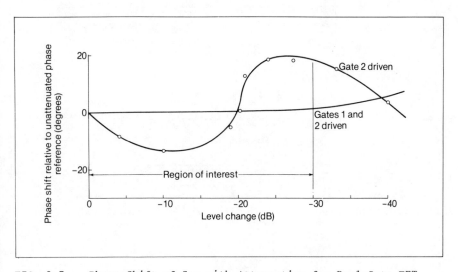

FIG. 9.7.   Phase Shift of S_{21} with Attenuation for Dual-Gate FET

Vorhaus (1979) has reported a multithrow dual-gate FET switch having
isolations in excess of 25 dB at X band using a novel four sided
structure with a common source connection to ground. The ground is
supplied by introducing a via through the GaAs substrate.

Figures 9.8(a), (b) and (c) illustrate the inherent bandwidth of the
dual-gate FET as a switch as well as the on-off ratio attainable with
second gate voltage (Tsai et al, 1979).

For example, the Raytheon dual gate FET (LND-841) having nominal
gate dimensions of 1μm by 500μm has a gain of 6.5 dB over the frequency range 2 to 8 GHz without any matching networks. The RF attenuation
with the second gate pinched off is over 30 dB whilst input and output

return losses with the second gate biased 'ON' and 'OFF' are plotted in Figs. 9.8(b) and (c). One significant characteristic of the device is the relative insensitivity of input/output VSWRs to the second gate bias.

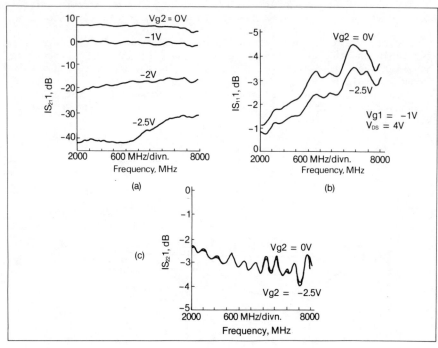

FIG. 9.8.   Magnitude of S-Parameters with FET Biased 'ON' and 'OFF'

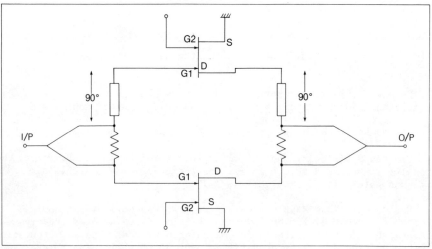

FIG. 9.9.   Schematic Diagram of Biphase Modulator

The dual-gate FET has become a popular device for switching in PSK modulators where the dual-gate FETs can supply high ON to OFF isolation over relatively wide bandwidths (Tsai et al, 1979) whilst maintaining good input and output return losses in both the 'ON' and 'OFF' states. Fig. 9.9 shows the way in which two SPDT switches using dual-gate FETs can be combined with unequal line lengths (over narrow bandwidths) or Schiffman type meander line sections for broadband applications (Schiffman, 1958; Schiek et al, 1977) to form a BPSK (bi-phase shift keyed) modulator.

3. PHASE SHIFTERS

Recently dual-gate GaAs FETs have been used to provide relatively wide band phase shifts which can be controlled with the voltages applied to the gates of the device. The input signal can be applied to the first gate, the output taken from the drain whilst the second gate is terminated in a 50 ohm load. Matching is applied to all three ports to produce maximum gain. For example Pengelly et al (1981) have shown that, if some of this gain is sacrificed by adjusting the bias voltages on gates 1 and 2, a multitude of voltage settings on the first and second gates are available for a specific gain. The transmission phase is found to vary however with the voltage settings as illustrated in Fig. 9.10.

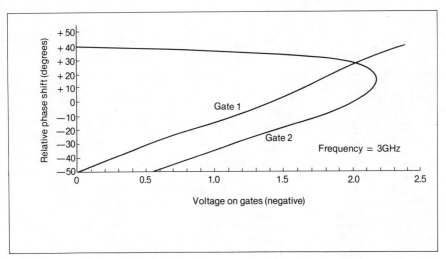

FIG. 9.10.   Transmission Phase Shift as a Function of Gate 1 and
   Gate 2 Voltages for a Dual-Gate GaAs FET

For example, -30° phase shift at 0 dB gain was measured with V_{G1} = -0.5V, V_{G2} = -1.2V, and $+10^{\circ}$ phase shift, at 0 dB gain, with V_{G1} = -1.7V and V_{G2} = -2.2V. Obviously the phase shift will depend on frequency.

To determine whether the FET itself or the interaction between any circuit and the FET is the major contribution to this phase shift, various circuit configurations have been tested with the results being summarised in Table 9.1.

TABLE 9.1.   Variation in Phase Shift with Different Amplifier Tuning Conditions

Device Type	Tuning Conditions	Gain	Set Gain	Max. Phase Shift
1	All 3 ports tuned for maximum gain at 3 GHz	14 dB	0 dB	90°
1	Both gates tuned	12 dB	0 dB −2 dB	80° 85°
1	Gate 2 tuned	6 dB	0 dB −8 dB	30° 36°
1	No tuning	6 dB	0 dB −8 dB	13° 15°
2	All 3 ports tuned for maximum gain at 3 GHz	14 dB	0 dB	58°
2	Both gates tuned	10 dB	0 dB −4 dB	53° 61°
2	Gate 1 tuned	10 dB	0 dB −4 dB	55° 67°
2	No tuning	4 dB	0 dB −10 dB	10° 23°

There will be a transmission phase difference between gate 1 and drain and gate 2 and drain due to the different physical spacings of the gate 1 and gate 2 electrodes to the drain.   Two dual gate FETs were used, one having a gate 1 to gate 2 spacing of 1µm, the other a spacing of 1.5µm.   The results show that for both FET types the tuning conditions affect the amount of phase shift achievable to a considerable extent.   The most critical part of the circuit is the input matching network since high phase shifts could be obtained with tuning on gate 1 only.   The inherent device phase shift, as appreciated from Table 9.1 is small.   It is concluded that in such a dual-gate FET phase shifter, most of the phase shift which occurs at the gate voltages are varied is due to interactions between changing device parameters and the input matching network.

FIG. 6.15.
Measurements made with a sliding short circuit as the termination

on gate 2 result in phase shifts of up to 140° at 4 GHz.

As the S_{11} of the dual-gate FET changes with the first gate voltage, amplitude changes are incurred at the band-edges as shown in Fig.9.11.

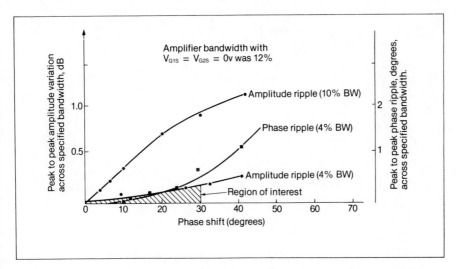

FIG. 9.11.   Amplitude and Phase Variations with Bandwidth for a Dual-Gate FET Phase Shifter

However, by matching the device over a bandwidth typically 100% greater than is needed phase and amplitude ripples can be substantially reduced. For example (Fig. 9.11) if a phase shifter element giving 22.5 degrees needs to operate to an amplitude ripple of ± 0.1 dB over the 2.7 to 3.2 GHz bandwidth then the matching circuits need to produce flat gain over the 2.2 to 3.7 GHz band in the reference phase shift state.

Tsironis et al (1980) have also used the dual gate FET as a phase shifter at X-band where the input signal is applied to the second gate and the first gate is made parallel resonant with an inductor. The dual-gate FET can be modelled as the connections in Fig. 9.12(a) where the parallel resonant circuit consists of L + L_{G1}, C_{GS1}, R_{GS1} + R_{S1} and L_{S1}. The bias voltage is varied on the first gate to change the value of C_{GS1} resulting in a phase shift in S_{21} as shown in Fig. 9.12(b). By optimizing the matching on gate 2 and drain it is possible to realize phase shifts of around 70° over 11.9 to 12.2 GHz, for example whilst maintaining a gain of 3 dB.

Single gate FETs are used as switching elements in circuits such as the switched line and high pass/low pass configurations (Pengelly, 1980). The switched line and its derivative the loaded line phase shifter, operate over relatively narrow bandwidths particularly as the phase shift required increases. The high pass/low pass

326

characteristic is produced using lumped elements as shown in Fig. 9.13(a).

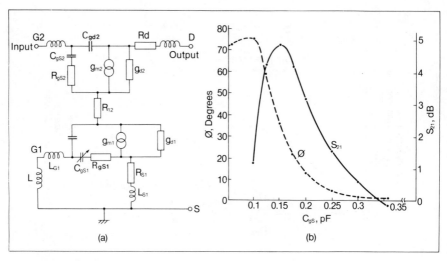

FIG. 9.12(a). Equivalent Circuit. (b) Transmission Phase and Gain Behaviour of Dual-Gate FET at 11.5 GHz.

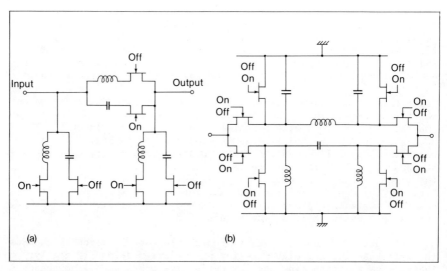

FIG. 9.13(a). High/Low Pass Phase Shifter using FET Switches
(b) High/Low Pass Phase Shifter using SPDT FET Switches

Since the insertion loss and phase difference of the circuit depends not only on the series resistance of the series FETs but also on the shunt capacitance of the 'OFF' FETs the circuit of Fig. 9.13(b) produces broader band performance since the lead/lag networks are separated from the switching elements. Such circuits can provide up to 180° phase shift over 20% bandwidths. The flexibility of GaAs MESFETs as control devices is exemplified in the vector modulator circuit (Brandwood, 1978) which provides both phase and amplitude control of a microwave signal and finds application in phased array radars and interference cancellation systems (Hicks et al, 1978).

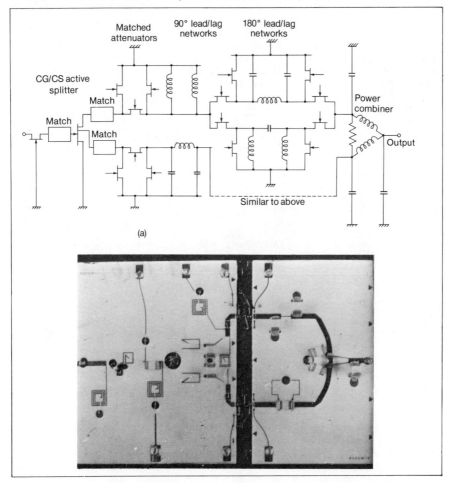

(a)

FIG. 9.14(a).   Schematic Diagram of Vector Modulator
     (b).   Photograph of Hybrid Vector Modulator

Two circuit examples are shown in Fig. 9.14(a) and Fig. 9.15. In Fig. 9.14(a) the input r.f. signal is equally divided with a single

328

gate/double drain FET to produce 14 dB gain over the 2.5 to 3.5 GHz band. This signal divider uses a common gate connected FET at the input to provide a 50 ohm match by employing a 300µm FET operating at a drain current such that its transconductance is 20 mS. The common source connected FET is matched passively on the drain to provide a low output VSWR. The π attenuators use small signal FETs where the devices are chosen for low on-resistance and small 'OFF' capacitance as explained earlier. Phase lead/lag networks supplying $+45^{\circ}$ and -45° phasing follow the attenuators. The phase shifted vectors are then recombined in a lumped element combiner which provides low insertion loss and signal isolation between its input ports. Fig. 9.14(b) shows a photograph of a hybrid microstrip realisation of such a circuit which provides 0° to 90° phase shift between 2.5 to 3.5 GHz with an overall loss of -15 dB in the 'OFF' state. Fig. 9.14(a) also shows the manner in which the other three quadrants of phase shift 90° to 180°, 180° to 270° and 270° to 360° can be produced using FET switched $+90^{\circ}$ and -90° phasing networks.

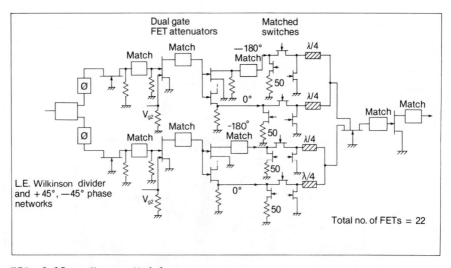

FIG. 9.15.   Vector Modulator

Fig. 9.15 shows an alternative circuit configuration for a vector modulator where the vector amplitudes are produced using dual-gate FET attenuators, followed by common source and source follower FETs for the 180 and 0° vectors. Matched broadband switches are produced by modifying the π attenuators of Fig. 9.14 such that the shunt FET closest to the common source and source follower FETs, has a 50 ohm r.f. resistor in its source to ground lead. Together with the $\lambda/4$ sections the switch provides a match at all its ports in both is 'ON' and 'OFF' states. The insertion loss of the switches and phasing networks is compensated by the active combiner at the output of the

vector modulator. The active combiner consists of a FET in common
gate configuration with two separate source electrodes and one common
drain electrode. If two independent channels in the common gate FET
are present then the signals are added and no mixing takes place.

The circuit of Fig. 9.15 provides full 0° to 360° phase shift in a
continuously variable manner whilst providing an overall gain of
14 dB with input and output VSWRs of 1.5 to 1 or better operating over
2.5 to 3.5 GHz (Pengelly et al, private communication).

4. DISCRIMINATORS

Frequency discriminators or IFMs (instantaneous frequency measurement) usually use two diode mixers following unequal phase length
circuits from a power divider. Since the phase difference in the 2
circuits prior to the mixers will increase with frequency, frequency
can be measured as a phase change by monitoring the baseband voltage
output of the two diode mixers. This output will have a cosine relationship to the phase difference. Over a large range of frequencies
(depending on the phase length differences used) the difference
between $\cos \theta$ and θ (radians) is small. Thus frequency is approximately equal to the baseband output voltage.

The dual-gate FET has recently been shown to exhibit intrinsic phase
and therefore frequency discrimination properties when the r.f. signal
is applied to both the 1st and 2nd gates (Pengelly, 1979).

To be able to use successfully a dual-gate GaAs FET as a frequency
discriminator several factors have to be considered in circuit
design. Firstly, the mixing action of the device must be considered.
Thus, the device needs to operate in a bias region where the transconductance of the device can be varied periodically as strongly as
possible by one or both of the two applied gate voltages.

As may be seen from Fig. 9.16 the actions on the drain current
characteristics of a dual-gate device by the first or second gate
voltages, V_{G1S} or V_{G2S}, are similar when the gate voltages exceed a
certain value. In a discriminator circuit the r.f. applied to gates
1 and 2 periodically modulate the g_m of each FET. The relative phase
difference between the first gate-to-drain and the second gate-to-
drain transmission coefficients will determine the action of the
device as a microwave discriminator.

Considering the simplified equivalent circuit of Fig. 9.17 it may
be appreciated that the conversion gains of FETs 1 and 2 of the dual-
gate connection depend on the time-averaged gate-to-source capacitances C_{GS1} and C_{GS2} as well as the time-averaged output resistances
R_{O1} and R_{O2}. An analogous equation to Eq. 6.13 for the 'mixer' gains
of each of the FETs in the dual-gate FET connection can be formed
which can be reduced in a simpler form to, for example

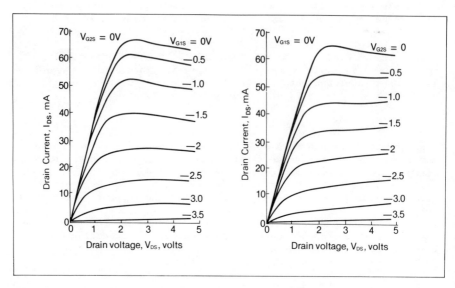

FIG. 9.16.   Dual-Gate FET d.c. Characteristics

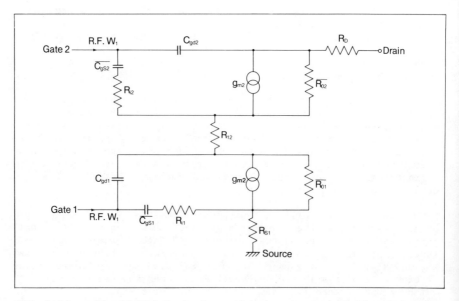

FIG. 9.17.   Simplified Equivalent Circuit of GaAs FET Discriminator

$$g_{C2} \simeq \frac{g_{12}^2}{4\omega_1^2 \, C_{GS2}^-} \cdot \left\{ R_{G2} + R_{12} + R_{i2} + \left[\left(\frac{1}{j\omega_1 C_{GD1}} + R_{i1} + \frac{1}{j\omega_1 C_{GS1}^-} \right)^{-1} \right. \right.$$

$$\left. \left. + \left(R_{01}^- + {}^1\!/g_{m1} \right)^{-1} \right]^{-1} + R_{S1} \right\}$$

A similar expression exists for g_{C1}, the conversion gain of the other FET in the long-tail connection of Fig. 9.17.

Since C_{GS1}^- and C_{GS2}^- are directly related to the S_{11}, S_{21} and S_{33}, S_{23} of the dual-gate FET S-parameters where port 1 is gate 1, port 2 is the drain and port 3 is gate 2, it may be expected that the addition of the two 'IF' outputs from the action of FET 1 on FET 2 and vice versa will be related to the relative phases of S_{11}, S_{21} and S_{33}, S_{23}. The transmission coefficients' angle difference $\Delta\phi$ $(S_{23} - S_{21})$ is found to be linear over a large range of frequencies as shown in Fig. 9.18. The linear portion of the relationship is variable in slope and frequency by the applied second gate voltage since $\angle S_{23}$ is effected by second gate bias much more than $\angle S_{21}$.

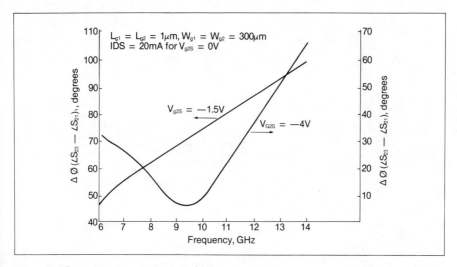

FIG. 9.18. Measured Phase Difference Between Gate 1-Drain and Gate 2-Drain Transmission Coefficient Angles

As the second gate voltage is increased until the device is near pinch-off the linear portion of this phase difference increases in slope and frequency. The change in the frequency coverage of the linear portion of this phase difference is due to the rotation of the angles of S_{23} and S_{21} with second gate bias. This phase shift can, therefore, be brought into a particular frequency range by changing the matching on gates 1 and 2. Frequency ranges of 2 to 6 GHz, 6 to 12 GHz and 12 to 18 GHz have been reported. In order to achieve reasonable conversion efficiency in the discriminator (i.e. to maintain efficient mixing) and also to allow operation in the linear region of the phase difference graph, the first and second gate bias voltages are adjusted such that the drain current is approximately 10% of the saturated value I_{DSS}.

By dividing the r.f. signal into two equal amplitude and phase components and applying these signals to the dual-gate FET's two gates it is possible to produce a discriminator output voltage whose magnitude is proportional to $\cos\Delta\phi$, where $\Delta\phi$ is the phase difference between the gate 1 to drain and gate 2 to drain transmission coefficient phase angles. In order to maintain the linearity of the discriminator, broadband matching circuits are needed on the gate 1 and 2 ports to maintain flat gain between gate 1 and drain and gate 2 and drain over the band of interest.

Fig. 9.19 shows a schematic diagram of the dual-gate GaAs FET discriminator. A signal limiter is needed before the discriminator to ensure that a constant signal level is available whatever the input signal level over a certain dynamic range.

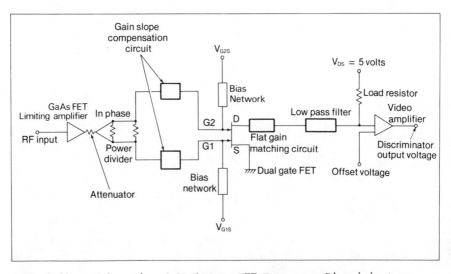

FIG. 9.19.   Schematic of Dual-Gate FET Frequency Discriminator

The response time of the discriminator to pulsed r.f. inputs depends on the cut-off frequency of the low-pass filter following the drain of the FET.

5. GaAs FET OSCIPLIER

Chapter 6 has indicated that the dual gate FET can be used as a frequency multiplier because of its transfer nonlinearities controlled by either the first or the second gate bias. Since the dual-gate FET has also been used as a self oscillating mixer (Tsironis et al, 1979) it becomes clear that the device should be capable of acting as an oscillator and multiplier simultaneously. Fig. 9.20 shows a schematic diagram of the so-called osciplier.

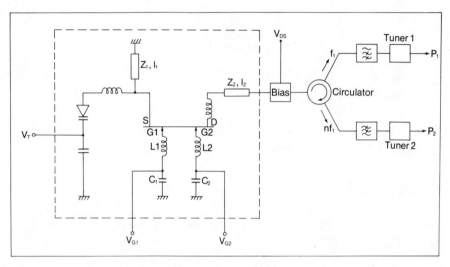

FIG. 9.20.   Osciplier Using Dual-Gate FET

The circuit within the dashed lines is a varactor tuned oscillator designed using the feedback element L_1 at the first gate and a GaAs varactor at the source. Chu (1980) has reported oscillators capable of operating from 6 to 11.5 GHz using 1μm long, 400μm wide gate MESFETs. A capacitance load C_2 at the second gate improves the conversion efficiency of the multiplier. The optimum load impedances for maximum oscillation efficiency at frequency f_1 and for maximum multiplication conversion efficiency at frequency nf_1 (where n is the multiplication factor) will be different. Thus, frequency selective tuners can be used, as shown in Fig. 9.20, to provide at least two distinct impedances at different frequencies. Thus, once the oscillation builds up at the first gate, the harmonic of this oscillating signal is generated by two kinds of nonlinearities, namely, the second gate input nonlinearity and the drain output nonlinearity. The resultant harmonic is then amplified and extracted from the drain through a circulator by tuner 2 in the frequency selective tuner.

Fig. 9.21 shows, for example, the output power measured for two oscipliers at various doubling frequencies.

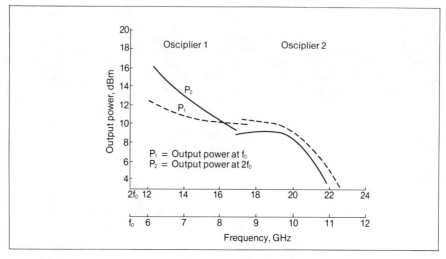

FIG. 9.21.   Power Performance of Two Dual-Gate MESFET Oscipliers

6. PULSED OSCILLATORS

During recent years solid state microwave devices have achieved considerable technical importance in radar applications in the measurement of range and velocity.

Doppler frequency shift provides an accurate technique for the measurement of target velocity for many applications. In CW Doppler radar systems the Doppler shift is proportional to the axial component of the velocity and is therefore related to the difference between the transmitted and received frequencies. Pulsed RF systems are capable of providing both the target velocity and range information. The transmitted and received signals when mixed generate the IF signal with the appropriate Doppler modulation impressed upon it. One transient effect which can cause problems in pulsed r.f. systems is the frequency change during the RF pulse due to changes within the microwave device. This phenomenon is commonly termed 'chirp' and has to be minimized if it is a significant fraction of the receiver bandwidth. Extremely elaborate and expensive techniques have been employed to minimize 'chirp' in Gunn diode oscillators, for example.

The GaAs dual gate FET can be employed in an oscillator circuit such as the common gate configuration of Fig. 9.22 and a property of the device exploited producing a low chirp pulsed r.f. output. This property is best illustrated by considering the model of the dual gate FET as the cascode connection of a common source device used as the element for obtaining steady state oscillations followed by the common gate FET operating as a high speed switch to obtain pulsed r.f.

output. This is easily done by applying a negative pulse to the second gate.

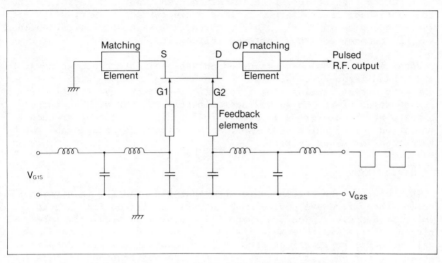

FIG. 9.22.   Schematic of Pulsed Dual-Gate FET Oscillator

TABLE 9.2.   Input Reflection Coefficient S_{11} and Forward Gain S_{21} for a Dual-Gate GaAs FET as a Function of V_{G2}

V_{G2}, volts	S_{11}		S_{21}	
	MAG.	ANG.	MAG.	ANG.
0	.56	−42.0	1.15	−58.5
−0.4	.58	−42.6	0.93	104.8
−0.9	.61	−46.4	0.57	116.7
−3.2	.62	−95.3	0.03	171.0
V_{G1} = −1.0V V_{DS} = 5 volts				

Table 9.2 shows the magnitude and phase of the gate 1-to-source and gate 1-to-drain scattering parameters, S_{11} and S_{21} of a chip 1μm by 300μm GaAs FET for different gate 2 bias levels. It can be seen that as the second gate voltage is increased from 0 to −0.9V the phase of the input reflection coefficient remains virtually unchanged while

the forward gain of the device decreases by as much as 6 dB. Minimal phase change in S_{11} maintains the operating frequency. The phase change beyond $-0.9V$ is quite rapid as the device reaches pinch-off but the gain of the device has already dropped to such an extent that oscillations can no longer be sustained. An important factor in the dual-gate oscillators favour is that drain current is virtually zero when the device is 'off'. This results in less heat dissipation in the FET channel and thus improves reliability.

Joshi et al (1980) have reported pulsed oscillators operating at X and J band frequencies with pulse widths ranging from 20 nS to 1μsec and duty cycles ranging from 1 to 25%. The frequency variation for the duration of the pulse was approximately 0.3 MHz with an output spectra which was virtually symmetrical indicating linear frequency variation during pulse duration. Such a figure is considerably better than that for diode sources.

7. CONCLUSIONS

This chapter has attempted to show that the GaAs FET is a versatile solid state microwave device much as various silicon FETs are at lower frequencies. The dual-gate FET exhibits particularly interesting properties which enable its use in the circuits discussed as well as other applications such as oscillator/mixers with Doppler sense. The single gate FET is also finding applications in tunable filters (Presser, 1979) when combined with varactor diodes. The author is certain that the GaAs FET will find a dominant role in many systems applications during the next few years.

8. BIBLIOGRAPHY

Ayasli, Y., Pucel, R.A., Fabian, W. and Vorhaus, J.L. A monolithic X-band single pole, double throw bidirectional GaAs FET switch. Research Abstracts of the 1980 GaAs IC Symposium, Las Vegas, USA, Paper 21.

Brandwood, D.H., Cross coupled cancellation system for improving cross-polarization discrimination. Antennas and Propagation - IEE Conference Publication, No. 169, Part 1, November 1978, pp.41-45.

Chu, A.S. and Chen, P.T. An osciplier up to K-band using dual gate GaAs MESFET. IEEE MTT-S International Microwave Symposium Digest, May 1980, pp.383-386.

Fabian, W., Vorhaus, J.L., Curtis, J.E. and Ng, P. Dual gate FET switches. GaAs IC Symposium, Research Abstracts, Lake Tahoe, Sept. 1979, Paper 28.

Furutsaka, T., Tsuji, T. and Hasegawa, F. IEEE Trans. on Electron Devices, Vol. ED-25, 1978, p.563.

Garver, R.V. Broadband PIN diode phase shifters. IEEE Trans. 1972, MTT-20, pp.314-323.

Gaspari, R.A. and Yee, H.H. Microwave GaAs FET switching. IEEE
MTT-S International Microwave Symposium Digest, Ottawa, Canada,
1978, pp.58-60.

Hicks, D. and Raymond, G. Adaptive arrays and sidelobe cancellers
for communications and radar applications. Proceedings of the 1978
Military Microwaves Conference, London, Oct. 1978, pp.366-378.

Higashisaka, A., Furutsuka, T. et al. Power GaAs MESFETs with a
graded recess structure. Proceedings of the 11th Conference on
Solid State Devices, Tokyo, 1979, Japanese Journal of Applied
Physics, Vol. 19, pp.339-343.

Joshi, U.S. and Pengelly, R.S. Ultra low chirp GaAs dual-gate FET
microwave oscillators. IEEE MTT-S International Microwave Symposium
Digest, May 1980, pp.379-382.

McLevige, W.V. and Sokolov, V. A monolithic microwave switch using
parallel-resonated GaAs FETs. Research Abstracts of the 1980 GaAs
IC Symposium, Las Vegas, USA Paper 20.

McLevige, W.G. - private communication.

Pengelly, R.S. and Suckling, C.W. The application of gallium arsenide FETs in microwave signal control circuits. IEE Colloquium,
London, December 1980, Digest No. 1980/81.

Pengelly, R.S. and Suckling, C.W., to be published.

Pengelly, R.S., Suckling, C.W. and Turner, J.A. Performance of Dual-
gate GaAs MESFETs as phase shifters. 1981 IEEE International Solid
State Circuits Conference, New York, Feb. 1981, Session XI on
Microwave Circuits.

Pengelly, R.S. GaAs monolithic microwave circuits for phased array
applications. IEE Proc. Vol. 127, Pt.F, No. 4, August 1980,
p.301-311.

Pengelly, R.S. and Suckling, C.W. - private communication.

Pengelly, R.S. Broad and narrow-band frequency discriminators using
dual-gate GaAs field effect transistors. Proceedings of the 9th
European Microwave Conference, Brighton, England. Sept. 1979,
pp.326-330.

Presser, A. High speed, varactor-tunable microwave filter element.
IEEE International Microwave Conference MTT-S Digest, May 1979,
pp.416-418.

Schiek, B. and Kohler, J. A method of broadband matching of microstrip differential phase shifters. IEEE MTT Trans. p.666-671,
Aug. 1977.

338

Schiffman, B.M.   A new class of broadband microwave 90-degree phase shifters.   IRE MTT Trans. p.232-237, April 1958.

Tsai, W.C., Paik, S.F. and Hewitt, B.S.   Switching and frequency conversion using dual gate FETs.   Conference Proceedings of the 9th European Microwave Conference, Brighton, Sept. 1979, pp.311-315.

Tsai, W.C., Tsai, T.L. and Paik, S.F.   PSK modulator using dual-gate FETs.   1979 IEEE International Solid State Circuits Conference Digest, pp.164-165.

Tsironis, C. and Harrop, P.   Dual gate GaAs MESFET phase shifter with gain at 12 GHz.   Electronics Letters, 3rd July 1980, Vol. 16, No. 14, pp.553-554.

Tsironis, C., Stahlmann, R. and Pouse, F.   A self oscillating dual gate MESFET X-band mixer with 12 dB conversion gain.   Proceedings of the 9th European Microwave Conference, Brighton, 1979, pp.321-325.

White, J.F.   Diode phase shifters for array antennas.   IEEE Trans. MTT-22, 1974, pp.658-674.

Yamamoto, R., Higashisaka, A. and Hasegawa, F.   IEEE Trans. on Electron Devices, Vol. ED-25, 1978, p.567.

CHAPTER 10
Gallium Arsenide
Integrated Circuits

1. INTRODUCTION

The superior physical properties of gallium arsenide over those of
silicon have led researchers worldwide to use the material as a semiconducting and semi-insulating substrate for both high frequency
analogue and high-speed digital integrated circuits. Much activity
is now concentrating on the ability to integrate large numbers of FETs
into chip areas of less than 4 mm^2 to perform sophisticated functions
such as code generation and random access memories at speeds of up to
4 GBits/sec. Analogue ICs which contain complete receiver front-ends,
phase and amplitude coded transmitters, analogue to digital converters
and synthesisers, for example, are presently being developed following
the encouraging results produced during the last five years with
single function circuits such as low-noise amplifiers and oscillators.

In parallel with the utilization of both silicon IC circuit techniques at the lower frequencies (i.e. up to approximately 4 GHz) and
the ingenious use of FETs in circuits up to 35 GHz or so, sophisticated technologies are emerging to cope with the demands of the
circuit designers.

This chapter is divided into four main sections. The first section
deals with the philosophy behind and design approaches of analogue
circuits. This is followed by an equivalent account for digital GaAs
ICs. A review is presented of the various approaches and technologies
that have been developed including the use of ion implantation and
'planar' IC processing. A considerable number of circuit examples
are included to indicate to the reader the range of activities being
pursued in this exciting field of microwave solid state device development. An attempt has been made to include up-to-date circuit
examples which aim to demonstrate the way in which gallium arsenide
ICs - whether they be monolithic microwave or digital - are rapidly
increasing in complexity.

2. MONOLITHIC MICROWAVE CIRCUIT DESIGN

As may be appreciated from the rest of this book, GaAs is an excellent material for field effect transistors and Schottky diodes. Because it can also be a low loss dielectric in its semi-insulating form it has become the basic material for microwave integrated circuits where both active and passive elements are combined on the same chip. The passive elements take the form of either distributed or lumped elements. The two most popular transmission lines used on GaAs are microstrip and coplanar waveguide. Of these two, microstrip has become the most exploited since the advantage put forward for coplanar waveguide, that of accessible ground planes on the top surface of the chip is only useful for simple circuits. With the advent of 'via' technology microstrip is much more flexible. Lumped elements produce more circuit design flexibility, too, provided that the equivalent circuits of these components (i.e. the component parasitics and loss) can be accurately modelled (Footnote). In many cases these models are most easily predicted by the use of transmission line theory, either single or coupled. Lumped components are particularly useful at frequencies up to J-band. However, in certain cases (for example, where the chip needs to be thin for thermal dissipation reasons) the loss factor of these lumped components can be low (i.e. 20 or so). Much of the early work on GaAs monolithic circuits used lumped elements to gain their advantages in broad-band circuits.

A. LUMPED COMPONENTS

(i) Inductors

The lumped inductor may take one of three forms; a single loop, a straight ribbon or a multiturn 'spiral'. On GaAs single loop inductors rarely exceed a few nanohenries whilst multiturn spiral inductors have been fabricated on GaAs up to 50 nH (Suffolk, 1980).

Single loop inductor calculations use the formula of Grover (1946) which is:-

$$L = 12.57a\left[2.303\log_{10}(\frac{8\pi a}{W})-2 + \mu\delta\right] \text{ nH} \qquad 10.1$$

where L is the inductance in nanohenries

    a is the mean radius

    W is the strip width

    μ is the relative permeability of the strip and

    δ is the skin depth.

At microwave frequencies, δ is small and equation 10.1 reduces to:

$$L = 12.57a\left[2.303\log_{10}(\frac{8\pi a}{W})-2\right] \quad \text{nH} \qquad 10.2$$

FOOTNOTE. A lumped element at microwave frequencies is usually defined as a component whose maximum dimension is, at most, a tenth of a wavelength.

It is interesting to note that a straight conductor has a greater inductance than the single loop inductor having the same total length, since

$$L = 2\ell \left[2.303 \log_{10} \left(\frac{2\pi\ell}{W} \right) - 1 + \frac{W}{\ell\pi} \right] \quad nH \qquad 10.3$$

where ℓ is the ribbon length in cm.

For a typical ribbon inductor on GaAs at X-band, for example, $W/\ell = 0.1$ and therefore $W/\ell\pi$ can be neglected.

Equations for spiral inductors range from the original expressions by Terman (1943) through the expressions of Grover (1946) to the more recent expressions of Dill (1964) and Greenhouse (1974).

Grover's formula for square multiturn spirals is given as:

$$L = .008N^2S \left[\log_e \frac{S}{b} + 0.726 + 0.1776 \frac{b}{S} + \frac{1}{8} \frac{b^2}{S^2} + \ldots \right] \qquad 10.4$$

where N is the number of turns,

    S is mean side of the spiral

and $b = pN$

where p is the pitch of the spiral.

Many of the formulae used in the past are not applicable for very small coil sizes (which may be typically less than 0.5 mm in outside dimension) on GaAs and the most accurate formulae are due to Greenhouse (derived from Grover and taking into account the negative mutual inductance between turns). Experimental measurements and theoretical calculations agree within 10% (Greenhouse, 1974). The loss component of inductors should be as small as possible. It may be shown that the loss factor of an inductor is proportional to the inductance and strip width and the (frequency of operation)[1.5] (Aitchison et al, 1971). On GaAs, because the metal thicknesses are usually close to a skin depth the Q is also dependent on the metal thickness. Thus, Figures 10.1(a) and (b) show the Q for single loop inductors at X-band as a function of metallization thickness and geometry (Pengelly et al 1977). In many cases the GaAs chip thickness will only be 100 to 200μm and thus the effect of the proximity of a ground plane under the chip is to reduce the effective inductance and increase the loss factor of the inductor.

The characteristic impedance Z_0 of the transmission line formed by an inductor and the ground plane can be written simply as:

$$Z_0 = L.v \qquad 10.5$$

where v is the velocity of propagation and L is the inductance.

The effective inductance L_e in the presence of a nearby ground

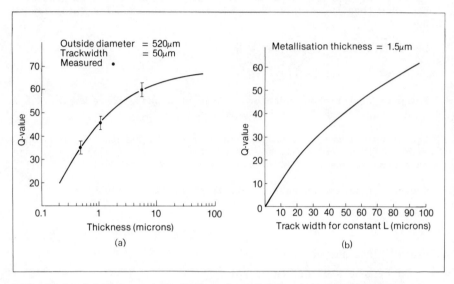

FIG. 10.1. (a) Measured Variation in Q-Value with Metallisation
Thickness of 1.0 nH Single Loop Inductor at 10 GHz
(b) Measured Q-Value vs Track Width for 0.5 nH Single
Loop Inductor at 10 GHz

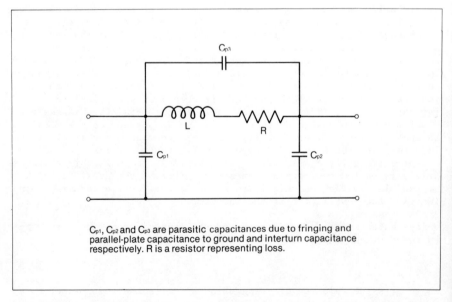

C_{p1}, C_{p2} and C_{p3} are parasitic capacitances due to fringing and
parallel-plate capacitance to ground and interturn capacitance
respectively. R is a resistor representing loss.

FIG. 10.2.   Equivalent Circuit of Multiturn Spiral Inductor

plane, can be computed in terms of the effective impedance of a short-circuited line

$$Z_{IN} = \omega L_e \simeq Z_0 \gamma \ell \qquad 10.6$$

where γ is the propagation constant and ℓ is the physical length. Thus from 10.6

$$L_e = \frac{Z_0 \ell}{v} \qquad 10.7$$

When $v = \dfrac{3 \times 10^{10}}{\sqrt{\varepsilon_R}}$ cm/sec, we have

$$L_e = \frac{Z_0 \sqrt{\varepsilon_R}}{30} , \qquad \text{being in nH/cm.} \qquad 10.8$$

Inductances are of the order of 10 nH per cm length for spiral inductors and therefore from Equation 10.8

$$Z_0 \simeq \frac{300}{\sqrt{\varepsilon_R}}$$

Thus the inductor will be very close to the value calculated in equation 10.4 as long as the equivalent impedance is greater than $300/\sqrt{\varepsilon_R}$. For GaAs, $\varepsilon_R = 13$ and therefore $Z_0 = 83$ ohms. It is possible to prove that for substrate thicknesses of 100 to 200μms and strip widths of 15 to 20μms this condition is met. Thus, the method which is used to calculate the parasitic shunt capacitance present in a multiturn square spiral, for example, is based on coupled transmission line theory (Smith, 1978). Interturn capacitances can be calculated by using coupled line theory but generally it has been found more accurate to produce an empirical formulae from measured results (Rickard, 1978). Calculations result in equivalent circuits for inductors which take the general form of Fig. 10.2.

(ii) Capacitors

There are two basic types of capacitor that can be used on GaAs — these are the interdigital capacitor (Fig. 10.3(a)) and the parallel plate or overlay capacitor (Fig. 10.3(b)). For capacitances greater than a few pF's it is convenient to use overlay capacitors employing a thin film of dielectric such as silicon nitride (Si_3N_4). Film thicknesses are usually in the range of 1000 to 3000Å.

The interdigital capacitor is formed as the result of the fringing fields between two sets of metal fingers as shown in Fig. 10.3(a). Lim and Moore (1968) have derived a closed form expression for the capacitance of such periodic structures. The theory neglects the effect of corners in the structure and assumes the fingers to be of zero thickness. Alley (1970) has shown that an equivalent two port

matrix may be calculated for interdigital capacitors but his work
covered only capacitors in shunt with a transmission line. A more
general and accurate method has been introduced by Hobdell (1979)
which is based on the computation of the even and odd mode impedances
of coupled line triplets. (Smith 1971). The interdigital structure
is considered as a pair of coupled parallel lines shunted by the
finger admittances. The method allows the derivation of the various
components in the equivalent circuits shown in Fig. 10.3(c).

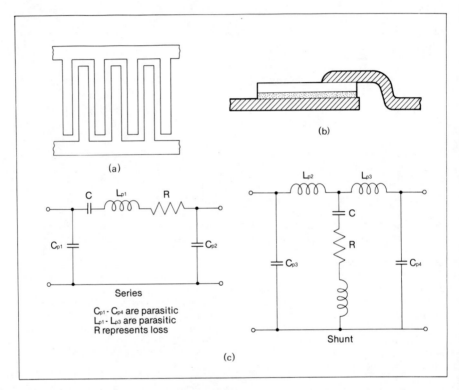

FIG. 10.3.(a) Interdigital Capacitor
         (b) Overlay Capacitor
         (c) Equivalent Circuits

    Fig. 10.4 shows the Q factor of a 0.25 pF capacitor versus finger
width to gap ratio at a test frequency of 10 GHz. As may be seen Q
value is low until the aspect ratio of the capacitor becomes such that
the capacitor is wide compared to its length. Fig. 10.5 shows the
dependence of interdigital capacitor loss on the metallisation thickness. It can be seen that, as the metal thickness decreases below one
skin depth (approximately 1μm for gold at 14 GHz) the Q factor of the
capacitor decreases quite rapidly.

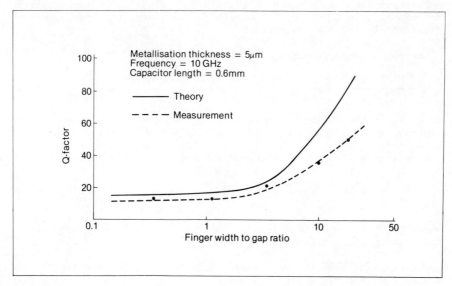

FIG. 10.4.   Q-Factor as a Function of Finger Width to Gap Ratio for Interdigital Capacitors

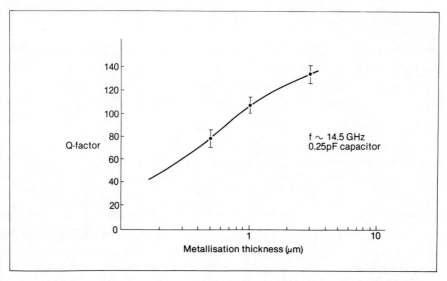

FIG. 10.5.   Measured J-Band Interdigital Capacitor Q vs Metallisation Thickness

For applications in which large values of capacitance are required, for example, in the case of decoupling capacitors, the overlay capacitor of Fig. 10.3(b) is employed.

The capacitance is calculated from

$$C = \frac{8.842\varepsilon_R \ ab}{t} \quad pF \qquad\qquad 10.9$$

where ε_R is the relative permittivity of the dielectric, a and b are the dimensions of the electrodes and t is the thickness of the dielectric. The predominant loss in overlay capacitors tends to be the resistive loss of the electrodes. The Q value is given by

$$Q = \frac{1}{\omega C R_S} \qquad\qquad 10.10$$

where R_S is the resistive loss of the electrodes, C is the capacitance and ω the angular frequency.

Tserng et al (1981), for example, have reported Q's greater than 50 at X-band for small 1 pF overlay capacitors on GaAs.

(iii) <u>Resistors and Active Loads</u>

Resistors for gallium arsenide ICs can be made by employing either the intrinsic resistivity of the n-type layer (which for a carrier concentration of 10^{17} donors cm^{-3} is approximately 300 ohms per square) or the resistivity of thin deposited films. The parameters that determine the choice of resistor type include:-

1. Current density handling;

2. Power handling:

3. Temperature coefficient of resistance;

4. Reproducibility; and

5. Reliability.

Current handling is determined by the amount of current that can be passed through the resistor structure per unit thickness and width. Typical values are in the mid 10^4 A cm^{-2} for composite thin film materials such as tantalum nitride (Ta_2N). Thin metal films may have higher current handling properties but usually have considerably lower resistivities. The temperature coefficient of resistance is important in that it effects, for example, the bias point of transistors. Since GaAs FETs have a positive temperature coefficient of resistance, i.e. their drain current decreases with increasing temperature it is desirable to use resistors with a negative temperature coefficient. Such resistor materials will be discussed later in this chapter.

In many applications bulk GaAs resistors can be used (even though these have high positive TCR). As we have already seen in Chapter 2 such resistors will have a saturation characteristic which will depend not only on the carrier concentration but also on the geometry of the resistor. This saturation characteristic can be used to advantage in logic circuits for example where a non-linear pull-up resistor is required to load inverter circuits (Mun, 1980). However, for linear circuits such saturation characteristics are undesirable. The use of active loads in the drains of common source connected FETs (Fig.10.6 (a)) rather than resistor loads results in better large signal performance for a given small signal gain (shown in Fig. 10.6(b)). This is because the r.f. signal resistance of the active load is substantially higher than the d.c. bias resistance. With the active load it is possible to choose a quiescent d.c. operating point near, say, half the value of the I_{dss} of the common-source FET. This circuit can, therefore, source or sink nearly equal currents into capacitive loads – an improvement over the poor sourcing capability of the resistive load.

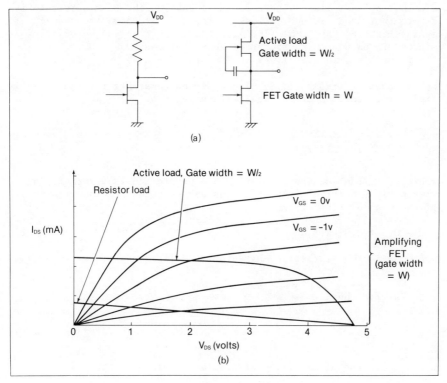

(a)

(b)

FIG. 10.6(a)   Resistor and Active Loads
      (b)   Comparison of Resistor Load to FET Active Load

B. DISTRIBUTED COMPONENTS

Two different transmission line techniques have been considered for application to GaAs monolithic microwave circuits. These are the microstrip approach and the coplanar approach. In the microstrip approach the ground plane is on the backside of the insulating GaAs substrate material. This presents grounding difficulties (for components on the top of the GaAs) which have been overcome by the use of via technology as explained later in this chapter. In the coplanar transmission line approach the ground plane is on the same side of the GaAs as the active devices. In certain cases this can be more convenient than a microstrip approach but the ground planes occupy valuable GaAs area and also result in topographical difficulties.

The insertion loss of transmission lines on GaAs have been measured by various workers (Courtney, 1977; Ch'en et al, 1979) and some of the results are shown in Fig. 10.7(a,b,c). Fig. 10.7(a) compares the loss of GaAs microstrip and coplanar transmission lines related to an alumina coplanar line. The additional loss on the GaAs substrate is due to increased dielectric loss. Fig. 10.7(b) shows the results of measurements by Chu et al (1981) on the dispersion characteristics of microstrip on GaAs with frequency where the GaAs was 200μm thick. A 254μm thick alumina result is shown for comparison purposes.

Getsinger's formula (1973) with modified constants for the effective dielectric constant agrees well with experiment

$$\varepsilon_{eff} = \varepsilon_s - \frac{(\varepsilon_s - \varepsilon_{eo})}{1 + G(f/f_p)^2} \qquad 10.11$$

where ε_s is the substrate relative dielectric constant, ε_{eo} is the microstrip effective dielectric constant at zero frequency. G is given by an emprical formula

$$G = 0.5 + 0.01 \; Z_0 \qquad 10.12$$

where Z_0 is the microstrip characteristic impedance at zero frequency, and

f_p is given by

$$f_p = \frac{Z_0}{2\mu_o h} \qquad 10.13$$

where μ_o is the permeability of free space and h is the GaAs thickness.

Attenuation as a function of frequency is shown in Fig. 10.7(c) for a line of 40 ohm characteristic line impedance on a 200μm thick semi-insulating GaAs substrate. The solid line corresponds to the

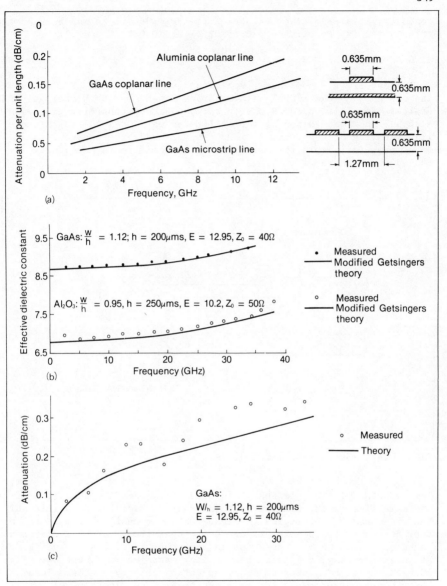

FIG. 10.7.(a) Measured Insertion Loss of Coplanar and Microstrip
Transmission lines on GaAs Substrates
(b) Effective Dielectric Constant of Microstrip on GaAs
and Al₂O₃ Substrates as a Function of Frequency
(c) Attenuation as a Function of Frequency for a 40Ω
Microstrip Line on a Semi-Insulating GaAs Substrate

conductor loss predicted by Pucel's theory (1968). As may be seen the attenuation is approximately 0.15 dB/cm at 10 GHz.

C. GaAs PLANAR DIODES

The rest of this book has given extensive details on MESFET theory and design. As the MESFET is a major component for both analogue and digital GaAs IC's up to 20 GHz or so, the Schottky barrier diode is also an important component for millimeter wave GaAs IC's. An idealized drawing of the type of diode a number of workers have been developing for use in the millimeter and submillimeter frequency ranges is shown in Fig. 10.8.

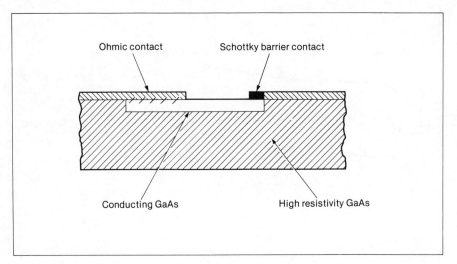

FIG. 10.8.   Surface-Oriented Diode

The important feature is that both diode terminals are accessible from the same side of the GaAs wafer. In order to keep the parasitics of the diode small, the region of conducting GaAs must be limited to only that necessary to support proper device operation. Several different techniques have been adopted for making Schottky diodes including selective epitaxy, molecular beam epitaxy and ion implantation. Fig. 10.9(a) shows the construction of selective epitaxy diodes. Texas Instruments were amongst the earliest to produce diodes using this technique (Mehal et al, 1968; Shaw, 1966) whilst Standard Telecommunication Laboratories (Antell, 1971; Allen et al, 1973) used selective epitaxy to produce 5 to 7μm diameter mixer diodes giving conversion losses of 7.5 dB at 70 GHz. Vapour phase epitaxy has also been used in Japan (Sato et al, 1975; Araki, 1978) to produce diodes having a cut-off frequency of greater than 700 GHz.

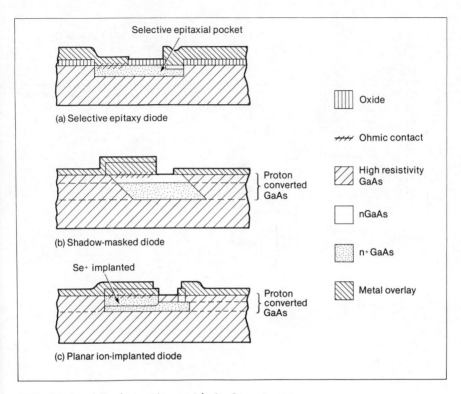

FIG. 10.9.   Various Planar Diode Structures

Immorlica and Wood (1978) have developed a fabrication technique which exploits carefully controlled shadow masking of a proton and an evaporated beam to produce a Schottky barrier junction at the edge of the conductive region in a self-aligned manner (Fig. 10.9(b)).

Ion implantation can also be utilized to produce surface oriented diodes as shown in Fig. 10.9(c). The devices are made with material in which two epitaxial layers of GaAs, an n^+ layer followed by an n layer, are grown upon a semi-insulating substrate using the $AsCl_3$-Ga-H_2 system. The n^+ layer was approximately $3\mu m$ thick with a carrier concentration of 3×10^{18} cm^{-3} and the n layer was $0.2\mu m$ thick with a 1 to 2×10^{17} cm^{-3} carrier concentration. Se ion implantation was used to decrease the resistance of the alloyed Au-Ge ohmic contact. Proton bombardment was used to isolate the devices from other parts of the circuit (Murphy et al, 1978).

352

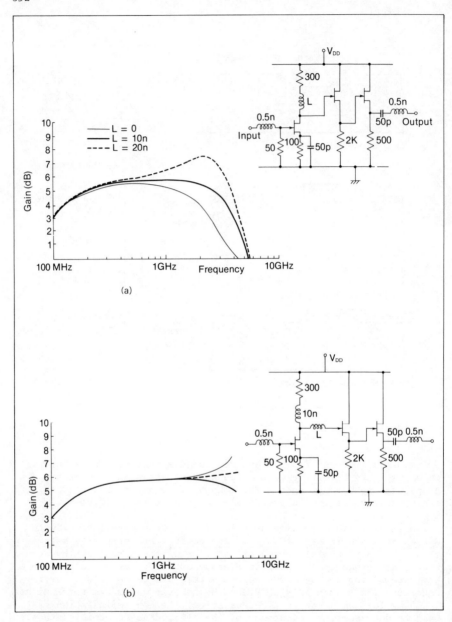

FIG. 10.10.(a) Common Source, Source Follower Amplifier and Effect
of Inductive Peaking
         (b) Common Source/Source Follower and Effects of
Interstage Inductive Peaking

D. LOW FREQUENCY CIRCUIT TECHNIQUES

(i) Amplifiers

At frequencies below approximately 4 GHz the reactances associated with the parasitics of MESFETs are small enough to enable them to be used without the need for matching components. Fig. 10.10(a) shows the circuit of a 50 ohm input impedance amplifier which uses three FETs, one in common-source the other two in source-follower mode to produce a 50 ohm output impedance with stable operation (Wilcox, private communication). As may be seen from Fig. 10.10(a) the 1 dB bandwidth of the amplifier is approximately 1.5 GHz. By introducing some inductive peaking into the drain load of the input stage the bandwidth is extended to 3 GHz. Fig. 10.10(b) shows the effect of introducing inductive peaking at the interstage between the common source and source follower FETs. A bandwidth of well over 5 GHz with 5.5 dB gain results. Obviously, the presence of the shunt 50 ohm resistor at the input of the amplifier will increase the noise figure. However, for many IF applications the high input impedance of the amplifier (without the 50 ohm resistor) is desirable enabling the circuit to work directly from the output impedance of either diode or FET mixers. As may be appreciated from Fig. 10.10 the amplifiers discussed are d.c. coupled. It is therefore, necessary that the peak output voltage from one stage drives the subsequent stage correctly - accurate bias voltages on the FETs are therefore essential. The use of ion implantation results in more uniform FET characteristics thus minimizing this potential problem.

Fig. 10.11(a) shows another approach to wideband amplifiers at low frequencies (Hornbuckle et al, 1981).

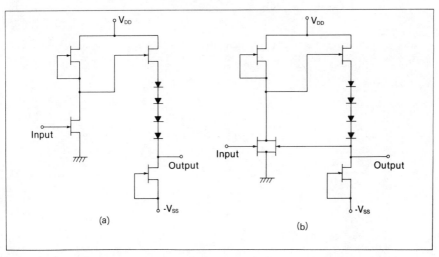

FIG. 10.11.(a)   Open Loop Amplifier Utilizing Level Shift Diodes
        (b)   Internal Feedback Amplifier

A common source FET is used as the basic amplifying element. Each common source stage has a buffer stage following it consisting of a source-follower and level shifting diodes. For example, the level shifting diodes are arranged such that a +3 volt drain voltage at the common source FET will result in a −1V level at the output of the source follower which is suitable for direct coupling to the next amplifying stage. The circuit of Fig. 10.11(a) has approximately 14 dB gain and a 2.7 GHz bandwidth when lightly loaded (Hornbuckle et al, 1981). Feedback can be added to such an amplifying stage to produce the circuit of Fig. 10.11(b). Active feedback using a FET can be used with the advantage that this type of feedback is superior to resistive feedback as will be seen.

The gain of the two types of amplifier shown in Fig. 10.11 can be calculated.

The voltage gain of the circuit of Fig. 10.11(a) is given by

$$G_1 = g_m \cdot R_{LOAD} = \frac{g_m}{(g_d + \frac{g_d}{2})} = \frac{2g_m}{3g_d} \qquad 10.14$$

where g_m is the amplifying FET's transconductance, g_d is the amplifying FET's drain conductance and $g_d/2$ is the active load drain conductance. The gain is independent of the gate width of the FET as long as the active load FET has a gate width which is half the width of the amplifying FET.

$$\therefore \quad G_1 = 20\log \left(\frac{2g_m}{3g_d}\right) \quad dB \qquad 10.15$$

The g_m per mm gate width of GaAs MESFETs with 1μm gate lengths is approximately 90 mS and g_d per mm gate width is approximately 10 mS.

Thus from Eq. 10.15

$$G_1 = 15.6 \ dB$$

The gain of the buffer/level shift stage, G_2, is given by:-

$$G_2 \simeq \frac{g_m R_S}{1 + g_m R_S} \qquad 10.16$$

where $R_S = \frac{1}{2g_d} \qquad 10.17$

Thus $G_2 = -1.74 \ dB$.

Series diode resistance should also be taken into account. This is usually of the order of 0.2 dB per diode. Thus the calculated gain for the configuration of Fig. 10.11(a) is 13.06 dB.

The gain of the circuit of Fig. 10.11(b) can be determined by calculating the effect of the feedback FET as an extra load on the amplifying drain. The added conductance is $g_{df} + G_2 \cdot g_{mf}$ where g_{df} and g_{mf} are the drain conductance and transconductance of the feedback FET respectively.

If the feedback FET has a gate width which is 25% of the amplifying FET then the latter's gain, G_3, is given by:

$$G_3 = \frac{g_m}{g_d + \dfrac{g_d}{2} + 0.25\ g_d + G_2 \cdot 0.25 \cdot g_m} \qquad\qquad 10.18$$

$$= \frac{g_m}{1.75\ g_d + 0.2\ g_m}$$

$$\therefore\ G_3 = 8.52\ dB.$$

Subtracting the 2.54 dB loss due to the buffer and diodes results in a calculated gain of 5.98 dB per stage.

The gain, G_4, for a four stage amplifier such as that shown in Fig. 10.12 will therefore be

$$G_4 = 13.06 + 5.98 + 5.98 + 3.1 \;=\; \underline{28.12\ dB}$$

where the last term represents the gain of the common-source FET for operation into a 50 ohm load, the FET having a gate width of 300μm.

(ii) Mixers

There are a variety of ways in which the amplifying and switching properties of FETs can be combined to achieve mixing. These have been more fully covered in Chapter 6, including the use of dual-gate structures. The disadvantage of the dual-gate FET mixer is that with broadband IF operation a large local oscillator (LO) feedthrough can occur which will overload the following IF amplifier. Since the IF amplifier is integrated with the mixer on the same chip, LO rejection is needed. The three FET mixer shown in Fig. 10.13(a) can be used where the FET driven by the LO signal switches the path to ground from the amplifying FET's source. A constant current source is also incorporated, which is connected to a negative supply, to bias the amplifying FET's source to zero voltage resulting in no current flow when the switching FET is 'on'. The LO feedthrough is therefore reduced but the conversion efficiency of this mixer is also lower than the dual gate FET mixer. However, greater conversion efficiency and further suppression of RF and LO feedthrough can be achieved by the

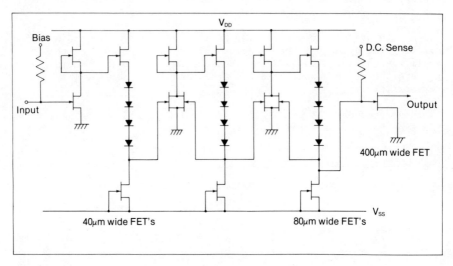

FIG. 10.12.   Multistage GaAs IC Feedback Amplifier

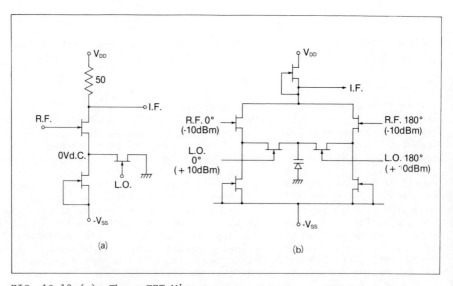

FIG. 10.13.(a)   Three-FET Mixer
           (b)   Active Double Balanced Mixer

use of the configuration shown in Fig. 10.13(b). In this balanced
approach the outputs of two mixers are summed in an active load. The
central node between the LO FETs is capacitively decoupled to ground.
Since the two LO and two RF signals are 180° out of phase the LO and
RF signals are suppressed at the IF sum port.

E. HIGH FREQUENCY CIRCUIT TECHNIQUES

So far we have considered frequencies where the FET's parasitic
elements do not dominate the device and thus the FETs can be considered as being transconductance devices with known drain conductances.

Provided the geometries of the FETs are sufficiently small monolithic circuit techniques can be used to produce feedback amplifiers to
much higher frequencies than hybrid circuits employing discrete FETs.
Thus, FETs can be made as parts of monolithic circuits which have
much lower values of gate-to-source and drain-to-source capacitance
than discrete versions because of the lack of bonding pad areas and
the much more compact nature of their geometry.

The design principles of feedback amplifiers can be extended to much
higher frequencies and bandwidths by using low parasitic, high transconductance FETs. For example, the high g_m FET with the equivalent
circuit of Fig. 10.14 has been used to produce a monolithic 4 to 18
GHz amplifier. In Fig. 10.14 the figures shown in parenthesis are
those associated with a normal discrete FET having the same g_m. As may
be seen there is a substantial reduction in the gate-source, C_{GS} and
drain-to-source, C_{DS}, capacitances over a discrete FET.

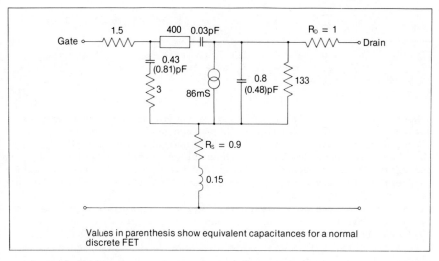

Values in parenthesis show equivalent capacitances for a normal
discrete FET

FIG. 10.14.   High g_m, Low Parasitic GaAs FET Equivalent Circuit

358

Fig. 10.15(b) shows the performance of the circuit of Fig. 10.15(a)
which uses such a FET. Such low parasitic FETs mean that many of the
circuit techniques currently used up to 4 GHz are becoming available
up to 20 GHz.

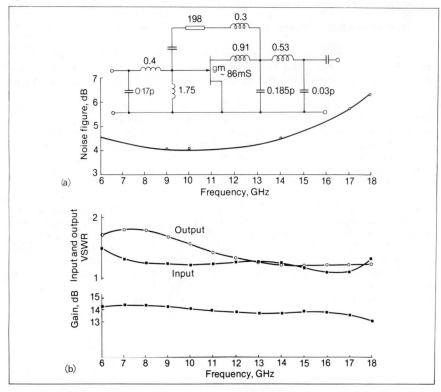

(a)

(b)

FIG. 10.15(a) Broadband 6-18 GHz Monolithic GaAs FET Gain Block
        (b) Performance of 2 Stage 'High g_m, Low Parasitic' FET
   Feedback Amplifier

    Most analogue monolithic microwave circuits of today, however, use
conventional passive techniques to produce such circuits as amplifiers, mixers and oscillators. Such techniques are very similar to
those used in the generation of conventional hybrid circuits. However
one of the main advantages to be gained from monolithic techniques is
the reproducibility of fabricated circuits and their total lack of any
need for adjustment. Such an advantage results partly from a judicious choice of the possible configurations available to implement a
particular circuit function. The analysis and optimization of a circuit must be accompanied by a sensitivity analysis which will take
into account the tolerances associated with both active and passive
elements as well as predict the effect of component values and parasitics only being known to a certain accuracy.

The following example illustrates the technique (Pengelly, 1981).

(i) Sensitivity Analysis

The design of monolithic low noise receiver front-ends has shown that, with careful consideration in the use of GaAs FETs and passive components, circuits can be designed which are much more tolerant of component variations from batch to batch and the absolute accuracy of those components in circuit realisations than in other circuits which produce similar performance.

Two design examples are presented which offer two solutions to a monolithic broadband low noise amplifier circuit at S-band. The extent to which these circuits meet certain requirements is measured in terms of sensitivities of reflection coefficient, gain flatness and noise figure to on-chip passive component and GaAs FET S-parameter variations. Such sensitivities determine the variation that will be seen from batch to batch. Also the sensitivity of the circuits to prime component values and parasitic component values is assessed. The monolithic GaAs FET preamplifier forms part of a complete S-band image reflection receiver front end covering the frequency range 2.7 to 3.5 GHz to be realised on a GaAs chip approximately 6 mm square.

Monolithic Circuit Design Principles

The first solution consists of a passively matched preamplifier using a sub micron gate length FET.

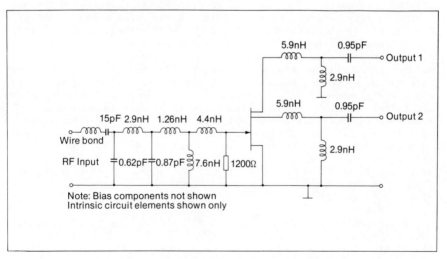

FIG. 10.16.   Circuit Diagram of Low Noise Passively Matched Splitter Amplifier

The circuit diagram is shown in Fig. 10.16, the circuit producing a gain of 9 dB with a 3 dB noise figure, where the FET is stabilised using resistive loading, rather than inductive loading in the source of the FET, the former technique leading to a circuit which is less sensitive to matching component variations. The circuit consumes 50 mW power and operates from a 5 volt supply. The IC measures 3.5 to 4.5 mm. The GaAs IC uses a considerable number of passive matching components to produce flat gain over the operating frequency range. Hence the Q values of the lumped elements used need to be as high as is feasible to minimise the effect of their loss on gain and noise figure. At S band most inductors are produced using multiturn spirals which use 15µm wide conductors (assuming a 150µm thick GaAs chip with a ground plane on its backside). Since the IC was designed initially to use VPE material a number of 'select-on-test' bias resistors are required (to enable the device to be operated at approximately 15% I_{DSS} for low noise operation) where the variation in I_{DSS} and pinch-off voltage, V_P are of known distribution over a GaAs wafer.

The behaviour of such a circuit is shown in Fig. 10.17(a),(b) and (c). The input VSWR is not particularly low because the device is matched for noise figure in a common source configuration. Figs. 10.18(a), (b) show the results of a Monte Carlo sensitivity analysis on the circuit of Fig. 10.16 where the components are varied by up to $\pm$ 10% in all their values in a normally distributed fashion. From these figures it may be expected that the variation in gain of the chip design with random variations in component value will be of the order of $\pm$ 1.3 dB (standard deviation about the mean value of 8.7 dB at 2.7 GHz). In reality it is likely that major component variations will be systematic - for example, resistors will be either all high or low in value due to the resistivity of the films being incorrect. For example (From Figs. 10.17(a)), if all the interdigital capacitors are calculated to be 20% too low in value there will be a reduction of 1 dB in gain over that expected. More importantly if element values are calculated incorrectly (either due to the use of incorrect formulae or component equivalent circuit models failing in accuracy) the circuit response will be affected. For example, if all inductors are calculated to be 50% lower in value than later measurements confirm, Fig. 10.17(c) for example, predicts that the noise figure will increase to over 9 dB at 3.5 GHz as against the nominal value of 3 dB. Obviously such discrepancies are unlikely but accuracies to within 10 to 15% are normal for spiral inductors, for example, when using the formulae of Grover (1946) or Greenhouse (1974). Thus, such changes in performance are excellent indicators of the circuit sensitivity.

The second circuit design shown in Fig. 10.19 uses a common gate, common source, source-follower cascade to produce a low noise preamplifier. The common gate input stage provides an almost simultaneous power gain and noise figure match for a FET having a 20 mS transconductance. Although the noise figure and associated gain of a common gate stage are higher and lower respectively than a common source stage a low I/P VSWR can be achieved over wide bandwidths without the need for balanced stages.

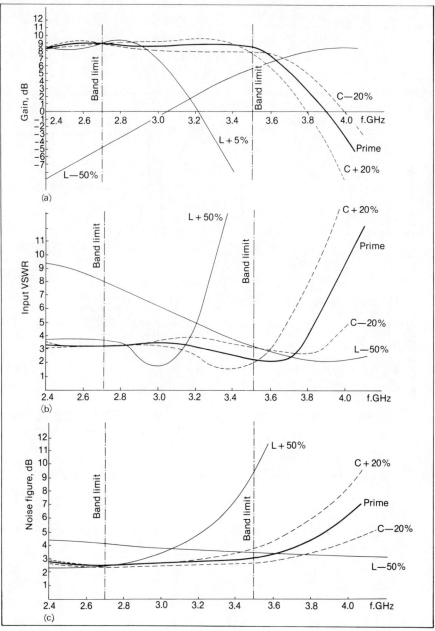

FIG. 10.17. Performance of Passively Matched Low Noise Amplifier
  (a) Variation of Gain
  (b) Variation of Input VSWR
  (c) Variation of Noise Figure

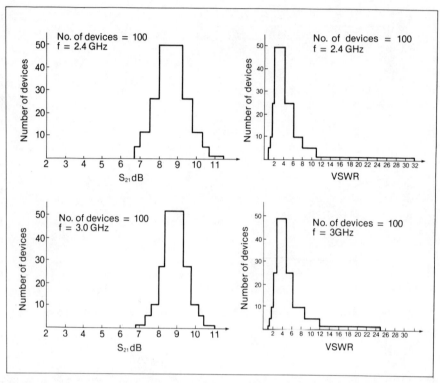

FIG. 10.18.   Monte Carlo Results for Passively Matched Low Noise
    Splitter Amplifier

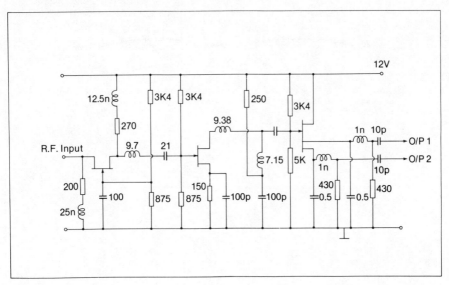

FIG. 10.19.   Circuit Diagram of Low Noise Actively Matched Splitter
Amplifier

The FETs used, employ submicron gate lengths. The GaAs chip
measures 2.2 mm by 3 mm. This circuit produces 19 dB gain over the
2 to 4 GHz frequency range with a noise figure of <4 dB over the 2.7
to 3.5 GHz band.

Applying the same Monte Carlo sensitivity analysis as in the first
design, indicates that this preamplifier is three times less sensitive
to component value changes, even though there are more FETs, resistors etc. Fig. 10.20(a),(b),(c) show the gain, input VSWR and noise
figure change with $\pm$ 50% changes in inductors and $\pm$ 20% changes in
capacitors. As may be seen the design is virtually insensitive to
capacitance changes of this order and indeed the $|S_{21}|$ and $|S_{11}|$ are
much more 'well behaved' where in fact a 50% decrease in inductance
only decreases the gain by 2 dB in the 2.7 to 3.5 GHz band. Noise
figure is also relatively well behaved for inductance values 50% lower
than the optimised values.

Thus the actively matched solution appears to offer a design which,
with expected $\pm$ 10% variations in component parameters, will produce a
monolithic circuit with high reproducibility. The use of implanted
material also means that I_{DSS} and V_P variations will be considerably
smaller than for a smaller VPE wafer, thus lowering the cost of the
complete IC.

For the case of likely variations in component values, based on
measurements made using test structures, prime element values are

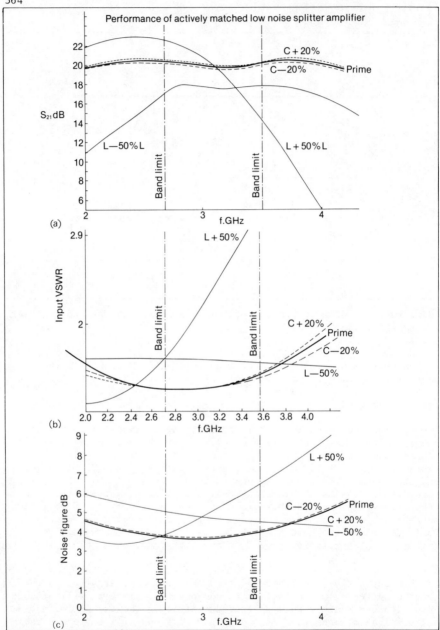

FIG. 10.20(a) Variation of S21 with Matching Component Change
        (b) Variation of Input VSWR with Matching Component Change
        (c) Variation of Noise Figure with Matching Component
             Change

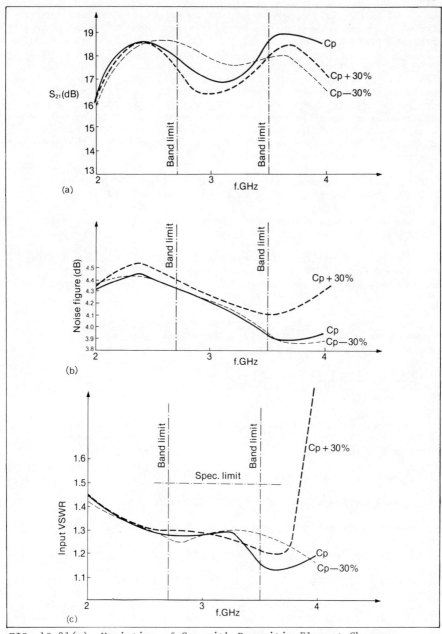

FIG. 10.21(a) Variation of S₂₁ with Parasitic Element Changes
   (b) Variation of Noise Figure with Parasitic Element
        Changes
   (c) Variation of Input VSWR with Parasitic Element Changes

known to within 15%. Such variations in capacitor, resistor and inductor values show that the circuit is well behaved with acceptable performance variations.

So far the effect of variations in the parasitic elements associated with the lumped components has been neglected. Extensive measurements on such components has shown that the parasitic elements can be predicted to within 30%. Thus, the circuit of Fig. 10.19 has been assessed in its performance from the viewpoint of changing the parasitics from the values used for circuit design (which are actually those found by fitting the equivalent circuit models of the component to their measured S-parameters) by $\pm$ 30%.

Fig. 10.21(a) shows a mean gain value of 17.75 dB over the 2.7 to 3.5 GHz frequency range which is a worst case value for probable component value errors and loss. If all the parasitic capacitances are varied by $\pm$ 30% a worst case variation in gain of $\pm$ 1 dB is observed.

Fig. 10.21(b) and (c) show, respectively, the variation in noise figure and input VSWR for the case where the active and passive component values are set for the expected worse case performance and the parasitic elements are varied by $\pm$ 30%.

The actively matched solution, uses approximately 480 mW of d.c. power, with a 12 volt supply rail (because of the utilization of resistive loading of the FETs rather than r.f. chokes).

3. DIGITAL CIRCUITS

A. INTRODUCTION

The principal requirements of a digital integrated circuit technology to make possible the development of ultra high speeds and large scale integration are:

1. A low chip area per logic gate.

2. A low gate power dissipation.

3. An extremely low speed-power product.

4. A very low gate propagation delay and,

5. A very high yield.

Since the dynamic switching energy or speed-power product, $P_D \tau_d$, is the minimum energy that a gate can dissipate during a logic transition and, assuming two transitions per gate, the power dissipation for a chip with N gates with an average clock frequency of f_c is given by:

$$P = 2 N_{f_c} (P_D \tau_d) \tag{10.19}$$

Table 10.1 demonstrates equation 10.19 in a form where the maximum allowable number of gates per chip or the maximum speed power product is calculated for a 2 Watt chip dissipation. It may be seen that for a clock frequency of 1 GHz and a complexity of 1000 gates per chip a speed-power product of 1 pJ per gate is needed.

TABLE 10.1. Maximum Allowable Gate Dynamic Switching Energies ($P_D \tau_d$) for Various Average Gate Clocking Frequencies (Assuming 2 Watt Chip Dissipation)

Gates/Chip		Clock Frequency					
		0.1 MHz	1 MHz	10 MHz	100 MHz	1 GHz	10 GHz
ULSI	10^5	10	10	1	10^{-1}	10^{-2}	10^{-3}
VLSI	10^4	10^3	10^2	10	1	10^{-1}	10^{-2}
LSI	10^3	10^4	10^3	10^2	10	1	10^{-1}
MSI	10^2	10^5	10^4	10^3	10^2	10	1
SSI	10	10^6	10^5	10^4	10^3	10^2	10
DEVICE	1	10^7	10^6	10^5	10^4	10^3	10^2

All energies are in picojoules

Eden et al (1979) have analyzed the dependence of the MESFET characteristics on the performance of the device as a high speed switch.

The drain current of a FET of channel width W is given by

$$I_{DS} = W Q_c v_d \qquad \text{10.20}$$

where Q_c is the charge per unit area in the channel and v_d is the drift velocity of the carriers in the channel. Q_c and v_d are bias dependent.

The charge Q_c in the FET channel and the average control field E_c are both proportional to the gate voltage above pinch off $(V_{GS}-V_P)$ (see Chapter 2).

$$I_{DS} = K(V_{GS}-V_P)^2 \qquad \text{10.21}$$

where $\quad K = \dfrac{\varepsilon \mu_n W}{2a \, L_g}$

where a is the channel depth, μ_n is the electron mobility in the channel and L_g is the gate length.

Correspondingly, the transconductance g_m is given by:

$$g_m = \frac{dI_{DS}}{dV_{GS}} = \frac{\varepsilon \mu_n W}{a \, L_g} \, (V_{GS} - V_P) \qquad 10.22$$

For short gate length FETs Shockley's expression (Eq. 10.22) no longer holds since the electron velocities are saturated and g_m is given by:

$$g_m = \frac{\varepsilon v_{sat} W}{a} \qquad 10.23$$

Now the switching speed (or propagation delay) of a transistor with N similar devices connected to it (its 'fanout') is limited by the device's current gain-bandwidth product f_T to a value given by:

$$\tau_d = \frac{N}{\pi f_T} \qquad 10.24$$

now $f_T \simeq \dfrac{g_m}{2\pi C_{GS}}$ (see Chapter 2)

so that

$$\tau_d = \frac{2N \, C_{GS}}{g_m} \qquad 10.25$$

This equation, however, does not reflect the non linear behaviour of the FET when it is switching from its 'on' to 'off' states. Consider the simple inverter circuit of Fig. 10.22 which consists of a common source inverting FET with a depletion mode active load.   In order to produce the lowest $P_D \tau_d$ product the FET must switch between the 'hard-off' state at $V_{GS} = V_P$ and the 'hard-on' state at $V_{GS} = V_P + V_M$ where V_M is the (peak-to-peak) logic swing. In most FET logic circuits, the $I_{DS} - V_{GS}$ characteristics of the FET and active loads define the switching limits (i.e. ground and the drain voltage, V_{DD}). For simplicity we assume that the gate input range, like the output range is positive which for depletion mode FETs means that some constant level shift voltage is needed in series with the actual FET gate to give the apparent positive threshold.

Let us consider the output transition time for a step input voltage applied to the gate. If we choose the value of the load current, I_0, to be half the maximum FET current (i.e. $I_0 = \frac{1}{2}I_{DM}$) then the negative and positive transition times for the step input will be equal and given by:

$$\Delta T = \frac{2C_L V_M}{I_{DM}} \qquad 10.26$$

where C_L is the capacitance loading on the gate and V_M is the logic swing.

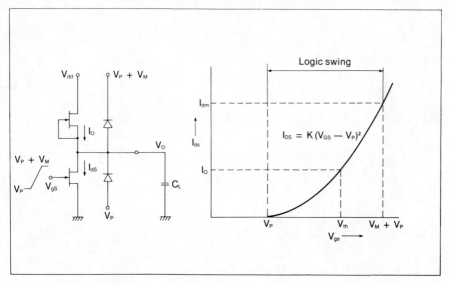

FIG. 10.22.   Simple FET Inverter Circuit and Drain Current as a
Function of the V_{GS} Characteristic Assumed for the Switching FET

The logic propagation delay for chains of inverters is related to
this output slew time ΔT. For maximum FET currents of $K(V_{GS} - V_P)^2$,
the logic propagation delay is given by:

$$\tau_d = \frac{2}{3} T \qquad\qquad 10.27$$

Substituting Eq.10.26 into 10.27 gives

$$\tau_d = \frac{4C_L V_M}{3I_{DM}} \qquad\qquad 10.28$$

The peak power dissipation for this circuit is given by

$$P_D = V_{DD} I_0 = V_{DD} \frac{I_{DM}}{2} \qquad\qquad 10.29$$

For normal 50% duty cycle switching,

$$P_D = \frac{V_{DD} I_{DM}}{4} \qquad\qquad 10.30$$

therefore from Eq. 10.28 and 10.30

$$P_D \tau_d = \frac{1}{3} C_L V_{DD} V_M$$

Now V_{DD} will in practice be approximately twice V_M so that

$$P_D \tau_d = \frac{2}{3} C_L V_M^2 \qquad\qquad 10.31$$

Now, from Eqn. 10.21

$$I_{DM} = K V_M^2 \qquad\qquad 10.32$$

Substituting in Eq. 10.2 we obtain

$$\tau_d = \frac{4C_L}{3K V_M} \qquad\qquad 10.33$$

therefore

$$V_M = \frac{4C_L}{3K\tau_d}$$

Substituting into Eq.10.31 gives:

$$P_D \tau_d = \frac{32C_L^3}{27K^2 \tau_d^2} \qquad\qquad 10.34$$

This latter equation shows the penalty that has to be paid for high switching speed for a particular technology.

K can be increased from Eq. 10.21 by going to a higher mobility material (e.g. GaAs instead of Si) or by reducing the gate length L_g. C_L is to a large extent dominated by the effect of substrate capacitance of interconnecting lines etc. Here again GaAs has an advantage over Si. For VLSI circuits it may be argued that there will be so many interconnections that fringing and cross-over capacitances predominate, these being independent of the material used. However, from Eq. 10.31 it may be appreciated that a method of overcoming this is to reduce the logic swing V_M whilst increasing the gate width of the driving FETs to maintain the required τ_d.

B. GaAs DIGITAL CIRCUIT TECHNIQUES

By far the most work on GaAs digital IC's has been done by using MESFETs as the active devices.

The depletion mode MESFET (D-MESFET) is the most easily fabricated device but, because the device is normally-on, the logic gates employed require two power supplies and they also need to contain some form of voltage level shifting, the level of integration that can be achieved will only be up to the LSI (1000 gate) level.

Table 10.2 shows the maximum obtainable 'flip-flop' clock frequencies that have been demonstrated for various circuit techniques and technologies at 1980. D-MESFET logic has the highest attainable frequency of operation of all the semiconductor approaches.

TABLE 10.2. Maximum Obtainable Flip-Flop Clock Frequencies as a Function of Technology

Technology		Structural Dimensions	Gate Power (mW)	Clock Frequency GHz	Remarks
Si			2	1.25	Differential logic
Current mode		2µm	1	1.11	ECL/EC^{2}L
logic		Emitter	1	1.67	ECL (simulation)
M.O.S.		0.4µm channel	1	0.9	
		0.8µm channel	0.3	0.65	
		1µm channel	2	0.83	punch through
			10	1.64	buried channel
GaAs	DMES	0.5µm gate	5.6	3.25	
			40	7.86	
		0.75µm gate	30	4	
		1µm gate	10	2.67	
	SDFL	1µm gate	0.5	1.89	
			1.5	1.9	counter ct
	EMES	0.6µm gate	1	2.6	counter ct
			1.9	3.33	
		0.8µm gate	1	1.3	
			0.02	0.61	
	JFET	1µm gate	0.5	0.93	
			0.2	1.03	
			0.25	1.6	counter ct

(i) Depletion Mode Logic

Early work on D-MESFET digital circuits was carried out using a buffered FET logic approach (Liechti, 1977; Van Tuyl et al, 1977). The circuits for an inverter and several basic logic functions (NAND and NOR) are shown in Fig. 10.23.

An n-channel D-MESFET requires a negative gate voltage to turn it off - thus level shifting must be introduced at some point so that the output logic levels match the input levels. In buffered FET logic (BFL), the negative voltage swings are produced by level shifting the positive drain voltages at the gate output. This is accomplished by level shifting diodes in the source-follower output stage of the gate.

As shown in Fig. 10.23(a) and (b) the NOR and NAND functions are performed by FET devices in parallel and series respectively. By combining series and parallel FETs higher level logic functions are produced (Fig. 10.23(c)) and (d)). Depletion mode GaAs BFL NAND/NOR (or NAND/WIRED-AND) gates with up to two NAND input terms and up to two of these NAND functions connected together having propagation delays as low as τ_d = 110 pS have been reported (Van Tuyl et al,1977). Four such gates have been utilized to implement a fast complementary clocked divider which gave toggle frequencies up to 4.5 GHz for τ_d = 111 pS effective gate delays at a P_D of approximately 40 mW per gate.

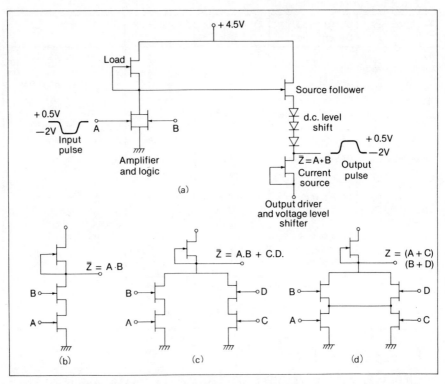

FIG. 10.23. (a) Basic BFL NOR circuit, (b) NAND, (c) Combined NAND + NOR, (d) Combined NAND/WIRED-AND

Examples of digital logic circuits employing the buffered D-MESFET approach are given in Section 5 of this chapter.

Schottky diode FET logic (SDFL) is a cousin of BFL and has been the logic architecture which has produced the first LSI level circuit (Lee et al, 1980). In the SDFL approach the basic 'OR' logic function is performed by diodes and level shifting is provided by the logic diode and an extra level shifting diode in some implementations at the input stage. (Fig. 10.24(a)). This latter feature enables the SDFL circuit to dissipate considerably less power than the BFL approach. Because the diodes used for the logic function are very small compared to the FETs in the BFL approach the SDFL technique leads to dense circuits.

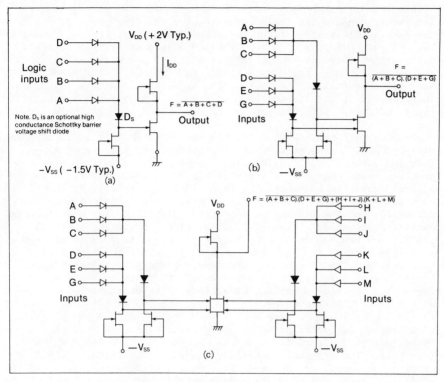

FIG. 10.24. (a) Schottky Diode FET Logic (SDFL) NOR Gate
(b) SDFL OR/NAND Gate
(c) SDFL OR/NAND/WIRED-AND Gate

The SDFL circuit is implemented using a selective ion implantation process such that the doping profiles for the diodes and transistors can be individually optimized (Eden et al, 1977). A benefit of this fabrication process is that FETs with very low pinch-off voltages can be produced leading to power dissipations which are nearly as low as enhancement-mode devices. The high density of gates available with SDFL make it possible to provide high switching speeds (τ_d between 75 and 150 psec) with low power levels (P_D between 100μW to 1 mW per gate) for speed-power products $P_D \tau_d$ as low as 16 fJ (femtojoules).

SDFL has been used with multilevel logic gate configurations. Circuits with both 2 level OR/NAND (Fig. 10.24(b)) and 3 level OR/NAND/WIRED-AND (Fig. 10.24(c)) gates have been fabricated using the planar 1μm gate length GaAs IC process pioneered by Rockwell (Eden et al, 1980).

(ii) Enhancement Mode Logic

Normally-off enhancement-mode MESFETs (E-MESFETs) offer circuit
simplicity because the logic gates require only one power supply.
The permissible voltage swing is rather low however because Schottky
barrier gates on GaAs cannot be forward biased above approximately
0.8V, without drawing excessive current. A logic swing of 0.5V,
therefore, although being a desirable goal, leads to the need for
very tight control in the fabrication of very thin active layers so
that the FET is totally depleted at zero gate bias but has a high
transconductance per unit gate width when the device is 'on'.

The more difficult fabrication procedure of E-MESFET logic compared
with D-MESFET logic can be partially overcome by the use of a p-n
junction gate FET (JFET) since the larger built-in voltage of the p-n
junction means that the JFET can be biased to approximately 1 volt
without excessive current being taken (Zuleeg et al, 1978; Immorlica
et al, 1977). First attempts with diffused p-n junctions were not
successful but the introduction of ion implantation has made the task
considerably easier. Fig. 10.25 illustrates the cross sectional
profiles of the McDonnell-Douglas EJFET process (Zuleeg et al, 1980).
A silicon nitride (Si_3N_4) layer is firstly deposited on to a Cr
doped wafer. The n^+ contact layers are produced by selectively
implanting with selenium. The n-channel and resistor loads are
formed by a selective implant of silicon. The wafers are then capped
with a 1500Å thick layer of nitride and annealed at 775°C to activate
the implanted ions. A gate mask is now used to define the p^+ gate
region which is formed by selectively implanting with Mg ions. The
Mg implant is designed to produce a precise junction depth which in
turn yields the desired threshold voltage. After further nitride
capping and annealing the Ge-Au-Ni contact metal is deposited and the
contact pattern defined. The second metal layer consists of Pt-Au to
form the low resistance gate-metal and first level interconnects.
After a 4000Å thick dielectric layer is deposited and vias defined
for interconnects the final metal layer is deposited. This process
results in circuits such as that shown in Fig. 10.26 which is a
divide-by-two circuit using the master-slave flip-flop technique.
FET gate lengths are 1μm with FET gate widths of 5 to 10μm. The
circuit operates at 500 MHz with a total power dissipation of 1.6 mW.
As may be appreciated from Fig. 10.26 the EJFET process results in a
planar IC.

The E-MESFET approach is potentially an excellent candidate for LSI
or VLSI circuits at G bit s^{-1} rates. Table 10.2 again shows the
results obtained to date and the performance projection for the technology.

The circuit implementations of either E-MESFET or EJFET logic are
identical. A basic NOR gate is shown in Fig. 10.27(a) where it is
noted that the FETs are directly coupled. The normally off FETs
start to conduct as soon as their gate voltage becomes positive, a
logic '0' corresponds to a voltage near zero below the switching

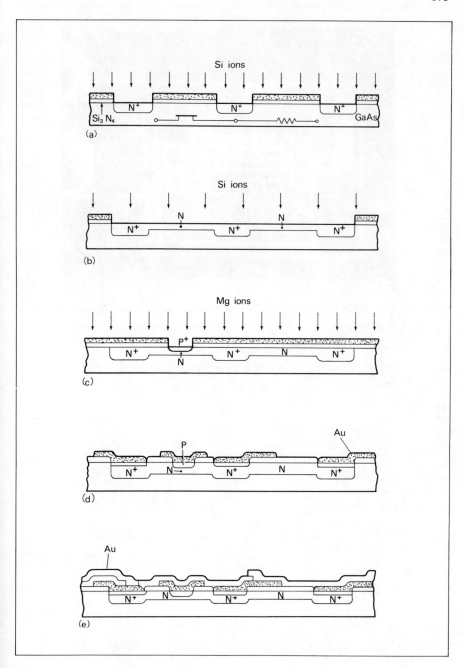

FIG. 10.25.   McDonnell-Douglas EJFET Process

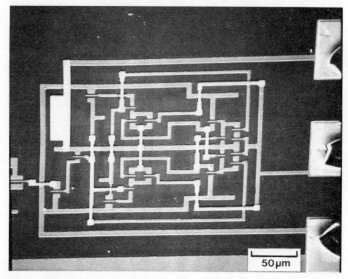

FIG. 10.26.   EJFET Divide-by-Two Circuit (courtesy McDonnell-
Douglas)

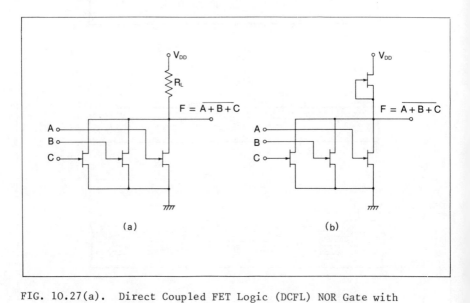

FIG. 10.27(a).   Direct Coupled FET Logic (DCFL) NOR Gate with
Resistive Load R_L
         (b). DCFL Circuit using Depletion Mode FET Active Load

threshold whilst a logic '1' corresponds to a positive voltage somewhat above this threshold.

A significant improvement to the directly coupled logic gate is to
substitute a depletion mode active load for the load resistor R_L as
in Fig. 10.27(b). This active load sharpens the transfer characteristic and improves the speed and speed-power product of the circuits.
However, the fabrication of the depletion mode active load requires a
carrier concentration profile which is different from that of the
enhancement-mode devices.

One of the most interesting approaches to the direct coupled FET
logic approach is the use of enhancement and depletion mode MOSFET
structures. The main advantage of such a scheme is that it is a
direct GaAs analogue of the very successful silicon MOS architecture -
thus this technology can directly benefit from the extensive computer
aided design software packages which already exist. These include
circuit and logic simulations, test vector generation and verifica-
tion, design rule checking and continuity checking.

Lack of a stable native oxide on GaAs has hampered the development
of such logic but work is well underway to realise InP devices where
the number of interface states is lower than in GaAs (see Chapter 11).
GaAs MOSFETs have been integrated into a ring oscillator by Yokoyama
(1980) to produce a τ_d = 110 pS with a P_D = 18 mW per gate. τ_d's of
70 pS with P_D's of 1 mW per gate are projected for 0.5μm structures.

Fujitsu (Yokoyama, 1981) have also reported a self-aligned source/
drain planar MESFET structure using enhancement mode FETs. Ti/W
gates are used to define the areas where n^+ is to be implanted.
Considering Fig. 10.28, the Ti/W gates are formed on n-type GaAs
which is formed using a selective ion implantation into Cr-doped
semi-insulating GaAs. A high dosage Si^+ implantation is then made
using the Ti/W gate as the implantation mask. This n^+ implant is
annealed at 850°C for 15 minutes without alloying the Ti/W gates
which stay unalloyed up to at least 860°C. The Ti/W gates are deposited using d.c. sputtering to a thickness of 0.5μm and etched using
a $CF_4 + O_2$ gas plasma. Fabrication is completed by ohmic metallization with Au-Ge-Au (Fig. 10.28(d)). In the resulting structure the
Schottky gate reverse breakdown voltage depends mainly on the donor
density profile of the n^+ regions, since the n^+ regions are in direct
contact with the gate electrode at their boundaries. The ion implantation energy and Si^+ dose are chosen so as to maintain a reverse
breakdown of at least six volts and a peak carrier density of 10^{18}
cm^{-3}.

Ring oscillator results for such a technology using 1.5μm gate
length FETs show that the minimum propagation delay is 50 pS at a
speed-power product of 287 femtojoules. The lowest speed-power
product is 14.5 fJ at a propagation delay of 98 pS. When compared to
SDFL the propagation delay is two times smaller at any power dissipation and the speed-power product is smaller by one order of magnitude for a propagation delay of 100 psecs.

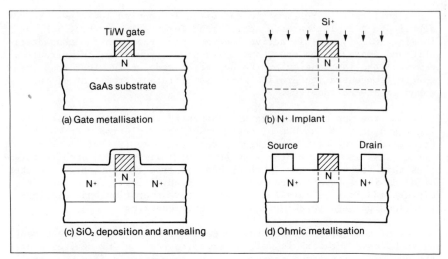

FIG. 10.28.   Fabrication of Self-Aligned Enhancement Mode GaAs
    MESFET

(iii) Quasi-Normally off Logic

The major problem with enhancement-mode MESFET logic is the manufacture of such circuits with an acceptable yield at the MSI integration level. Fig. 10.29(a) shows, for example, the basic DCFL d.c.
transfer curves computed for different values of FET pinch-off
voltages, V_p. It may be seen that the noise margins of the basic
normally-off circuit depend on the actual V_p value. Hence for conventional E-MESFET logic a V_p tolerance of only approximately 200 mV is
allowable. The present status of implantation technology into GaAs,
particularly from run-to-run is close to this tolerance so that a
circuit architecture which produces a larger allowable V_p variation
is attractive.

An alternative circuit approach has been adopted by Thomson CSF
(Nuzillat et al, 1980) which tolerates a wider range of pinch-off
voltages. The approach uses a so-called 'quasi-normally-off' logic,
i.e. transistors operating as enhancement mode devices with a pinch-
off voltage close to zero, but whose actual value can be either
positive or negative. The basic quasi-normally-off logic gate
(Fig. 10.29(c)) has been derived from a buffered version of normal
E-MESFET logic. The addition of the Schottky diode operating in the
current-switched mode in the second stage leads to several advantages
related to the d.c. transfer characteristics as may be seen by
comparing Fig. 10.29(a) or (b) with (c):-

(i) An increase in the acceptable pinch-off voltage range to -0.2 volts to +0.2 volts;

(ii) A higher logic swing (0.85 volt compared to 0.5 volt); and

(iii) A low logic level almost equal to ground (10 mV typically).

Ring oscillator results using such a logic architecture have resulted in minimum propagation delays of 95 to 135 pS (depending on the circuit configuration) and speed-power products, $\tau_d P_D$, of 200-250 fJ, using MESFETs with 1 x 35μm gates operating at a drain voltage of +2.5 volts.

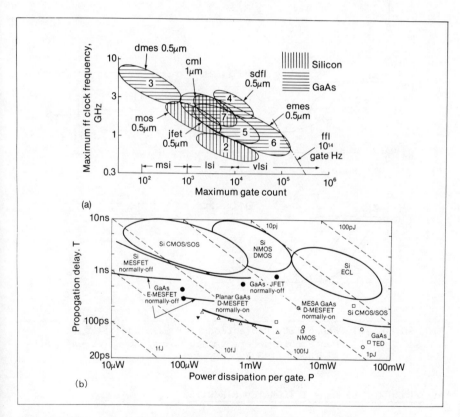

(a)

(b)

FIG. 10.29.   Circuit Diagrams of Various 2-Input NOR Gates Implemented with Enhancement Mode FETs and Dependence of the DC Transfer Curve on the FET Pinch-off Voltage
(a) Basic DCFL Logic Circuit, (b) Buffered Logic Gate with Normally-off MESFETs, (c) Basic Quasi-Normally-off Logic Gate

Fig. 10.30(a) shows the projections for the future of GaAs logic compared to those expected for Si. It may be seen that for 0.5 to 1μm gate length FETs all the logic techniques discussed above will reach the 1000 gate level excepting BFL which will, however, reach the highest speeds of close to 10 GHz. Fig. 10.30(b) shows the contemporary technology performance comparisons for the various GaAs and Si architectures.

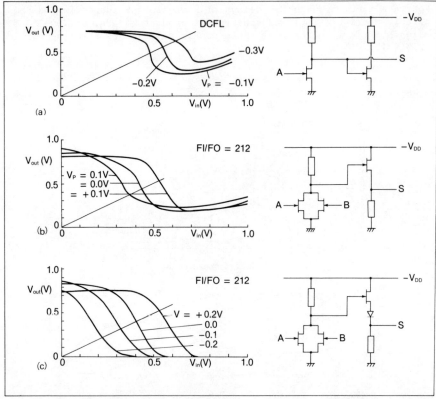

FIG. 10.30. (a) Capability Projections for GaAs and Si Logic
(b) Contemporary Performance Comparisons for GaAs and Si Logic (1980)

4. TECHNOLOGY OF GaAs INTEGRATED CIRCUITS

In order to give the reader an understanding of the relative complexity of GaAs integrated circuit manufacture when compared to discrete FET devices, the approaches being taken to define high yield processes are reviewed. Because many laboratories developing GaAs ICs have already considerable experience in the fabrication of discrete FETs there tends to be a marked resemblance between the two fabrication processes particularly where epitaxial layers are used.

To realise low noise and high gain FETs, the devices are optimised
with respect to material parameters and device geometries as described
in Chapter 2. Apart from the use of low resistance ohmic contacts
usually utilizing AuGe/Ni, the FETs have multiple parallel gates for
analogue applications to reduce gate resistance and as thick a gate
metal as is possible. Photolithographic techniques are usually
employed to define gate lengths down to 0.5μm. Below that value
electron beam lithography is used using such equipment as the Cam-
bridge EBMF2 (Lawes, 1979). Electron beam lithography is used for
gate lengths greater than 0.5μm when a fast turn-around time is
required (since no high definition gate mask is required) by directly
exposing a resist using the electron beam. The electron beam machine
automatically aligns the gates to alignment marks usually defined
close to source and drain contacts which have been previously defined.
Gate lengths of 0.2μm can be successfully defined.

In monolithic circuits device bonding pads generally do not occur
as direct connection to the circuit is made thus reducing the parasitic capacitances (such as C_{DS}). n^+ contact layers are generally
only used where the lowest noise figures are needed and the FET
devices employ a channel recess to minimise this factor. A recessed
gate structure can also improve the power handling capabilities of
the FETs as has already been discussed in Chapter 3.

When using ion implanted material the implants are arranged to give
the required FET or diode performance characteristics. Implantation
may range from simple Gaussian profiles involving one implant to more
complicated implantations involving several energies and doses as
well as different species. For example, Hewlett Packard, in their
BFL logic IC's use a dual implant of Se and Si to produce the n, n^+
regions required for low forward resistance diodes and low channel
resistance FETs.

A. SOME IC TECHNOLOGIES - A REVIEW

Fig. 10.31 shows the silicon like planar ion-implanted process
technology developed by Rockwell in order to realize the advantages
of ion implantation in GaAs (Zucca et al, 1980). The figure shows
the device process steps necessary to fabricate a planar GaAs integrated circuit incorporating a FET, a level shifting and a switching
Schottky diode. Fabrication of the device wafer is started by the
deposition of a Si_3N_4 (silicon nitride) layer onto a flat, qualified,
semi-insulating GaAs substrate (qualification of substrates is
described in Chapter 4). This layer of Si_3N_4 remains on the slice
during all subsequent fabrication steps. A first photoresist stage
defines the device and circuit areas which require low dose implants
such as the channels of FETs. A shallow Se implant (Fig. 10.31(a))
is then followed by a deeper n^+ implant (Fig. 10.31(b)) for device
contact areas or Schottky barrier switching diodes after exposure of
a photoresist. The implantation steps are followed by encapsulation
of the slice with Si_3N_4 and annealing at $850^\circ C$ in an H_2 atmosphere.
(Fig. 10.31(c)). This high temperature anneal converts the shallow

382

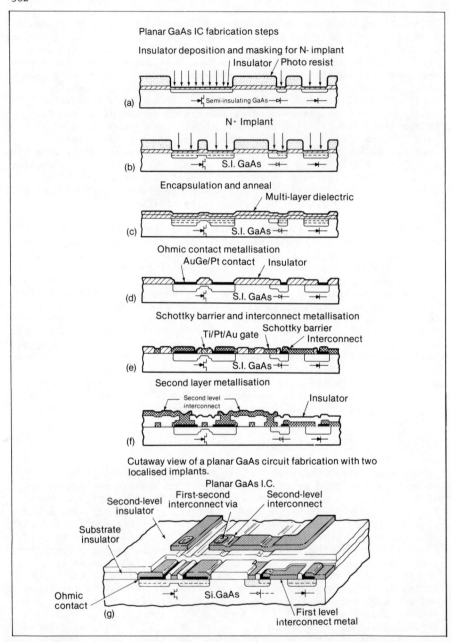

Planar GaAs IC fabrication steps

Insulator deposition and masking for N- implant

(a)

Insulator / Photo resist

Semi-insulating GaAs

N+ Implant

(b) S.I. GaAs

Encapsulation and anneal

Multi-layer dielectric

(c) S.I. GaAs

Ohmic contact metallisation

AuGe/Pt contact   Insulator

(d) S.I. GaAs

Schottky barrier and interconnect metallisation

Ti/Pt/Au gate   Schottky barrier
Interconnect

(e) S.I. GaAs

Second layer metallisation

Second level interconnect   Insulator

(f)

Cutaway view of a planar GaAs circuit fabrication with two localised implants.

Planar GaAs I.C.

Second-level insulator   First-second interconnect via   Second-level interconnect

Substrate insulator

Ohmic contact   Si.GaAs   First level interconnect metal

(g)

FIG. 10.31.   Planar GaAs Circuit Fabrication Process Using Ion Implantation Directly into Qualified Semi-Insulating Substrate Material Adopted by Rockwell

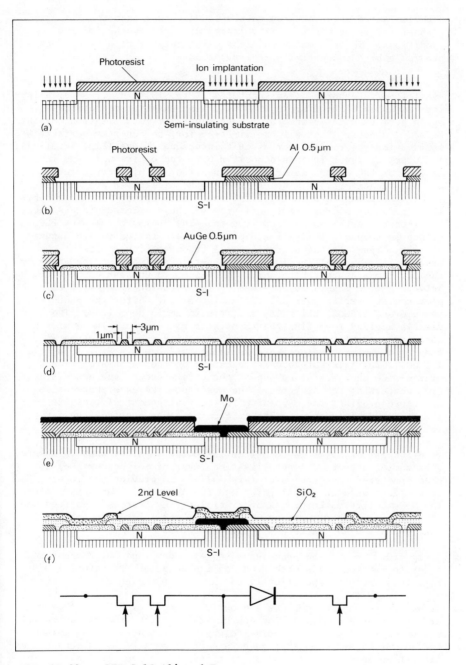

FIG. 10.32.   LEP Self-Aligned Process

n-Se-implanted areas (approximately 1500Å thick) into the FET active
channel layers with pinch-off voltages dictated by the implant
conditions. The n^+ implanted areas using sulphur at a higher dosage
provide a high conductivity region.

Device ohmic contacts are defined next with a standard photoresist
and lift-off technique (see Chapter 4) (Fig. 10.31(d)). After alloying the ohmic contact areas at $450^{\circ}C$, a photoresist operation is
again performed to define the Schottky barrier metallization for the
FETs and diodes. Rockwell use Ti-Pt-Au for the Schottky barriers as
well as for the first layer circuit interconnections (Fig. 10.31(e)).
A dielectric layer is then deposited onto the entire wafer as insulation for the second layer interconnection and dielectric for
circuit capacitors. Via holes through the dielectric are used for
interconnections between the first and second layers of metals. Fig.
10.31(g) shows a cross-sectional view of the completed process where
the interconnects can be clearly seen. The approach taken by LEP,
France is to use a self-aligned process where the gates are automatically aligned to the n^+ regions. A cross-sectional view of the
process is shown in Fig. 10.32. The n-type active layer is grown by
the low temperature VPE technique discussed in Chapter 4. Isolation
between active layers is obtained by implantation with boron using a
photoresist barrier over all the active areas. After the photoresist
is removed, a 5000Å thick layer of aluminium is deposited. The gate
mask is applied to define the transistor gate areas. The 3 micron
gates are undercut beneath the 2 micron-thick, photoresist barrier
leaving a 1 micron effective gate length as shown in Fig. 10.32(b).
A 5000Å thick layer of AuGe is deposited on the substrate. The 3
micron-wide photoresist on top of the gate prevents the AuGe layer
from contacting the Al gates. The mushroom-like gate/photoresist
structure provides the mechanism for self-alignment of the gate to
the source/drain regions. The AuGe metal layer provides the first
level of interconnections.

An insulating layer is deposited following an alloy step to ensure
good ohmic contacts. A metallisation layer of molybdenum (Mo) is
then sputtered onto the wafer (Fig. 32(e)) to provide the interconnections between the Al gates and the AuGe ohmic contacts. After
the Mo metal is patterned, another dielectric layer is deposited and
contact vias etched. The second level of metal interconnections is
then deposited and patterned (Fig. 10.32(f)).

Fig. 10.33 shows the processing sequence developed by Plessey in
order to fabricate ICs either on ion-implanted or VPE material.
Fig. 10.33 relates specifically to the use of implanted material.

The active channel regions and ohmic contacts of FETs, for example,
are formed by a dual dose and energy implantation schedule which will
produce devices with low source to drain resistance. This implantation may be achieved through a thin Si_3N_4 mask to avoid ion channel-
ling which leads to poor characteristics in the profile 'tail'. The
wafer after implantation is then capped with reactively sputtered

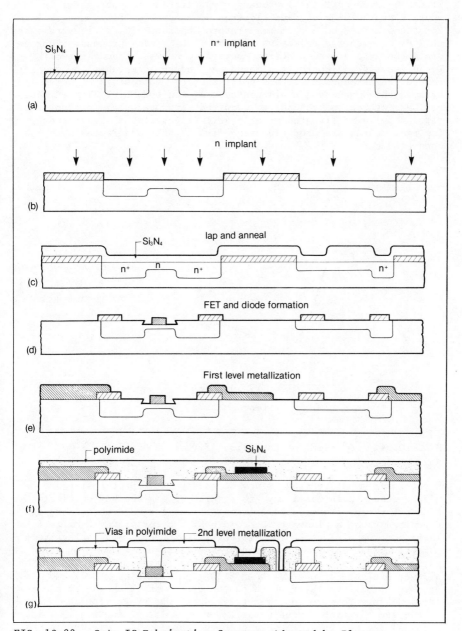

FIG. 10.33. GaAs IC Fabrication Sequence Adopted by Plessey

Si_3N_4 and annealed at $850^\circ C$ in an H_2 atmosphere for 15 minutes. Fig. 10.34 shows the result of doing this using Se as the species. It can be seen that the low energy implantation is successful in increasing the near surface electron concentration to around 10^{18} cm^{-3} for the 5×10^{13} cm^{-2} room temperature implant. The active areas are defined by either a selective implant or by a mesa process. Source-drain contacts are then defined using a photoresist and lift-off technique and the contacts alloyed to produce acceptably low specific contact resistances. At this point the I_{SAT} current is measured between the source and drain contacts of monitor FETs. The FET channels are etched to produce the required saturated drain current the gates then being situated at or near the peak of the higher energy implant. Gates are produced using Ti-Al. FETs and diodes have now been produced.

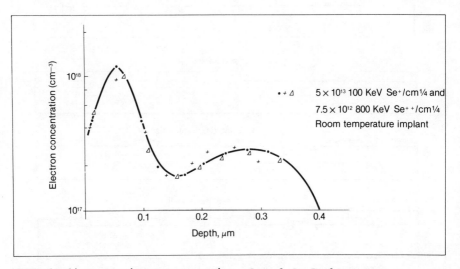

FIG. 10.34.   Carrier Concentration of Dual Se Implant

First layer metallization is now defined using a photoresist and lift-off process – this metal may, for example, form the bottom electrodes of overlay capacitors. The silicon nitride for such capacitors is deposited using a plasma enhanced deposition technique (Commizoli, 1976) and the dielectric (of dielectric constant = 5.5) selectively removed using a plasma etching process (Tolliver, 1980). At this point a polyimide material is spun onto the wafer. This material has a lower dielectric constant than Si_3N_4 (polyimide ε_R = 3.5) and produces a uniform layer which can be varied in thickness from a few thousand Angstroms to approximately 10 microns by varying the spin speed, the dilution of the material or by multiple coatings. This material is then cured to form an excellent low loss dielectric which is used in several ways. Firstly it is used as the dielectric between

metallization layers; secondly, it is used as a dielectric for overlay capacitors; thirdly, it is used as a 'stress' relief barrier
between the GaAs and certain other layers, such as resistive films and
finally it is used as an ion-milling barrier. In the latter case,
since GaAs is removed rapidly by the ion-milling technique used to
define certain metallizations (particularly where the lift-off technique is inappropriate), the polyimide film (which is not milled)
acts as a protective layer.

Following polyimide deposition, vias are opened up in the polyimide
and this is followed by the interconnect metallization which also
forms the top electrodes of overlay capacitors. Thin film resistors
are put down on the top of the polyimide film and interconnected to
the metallizations. The polyimide film provides protection for the
active and passive components on the IC.

In many cases it is necessary to provide low capacitance interconnections between components such as FET sources or the centres of
spiral inductors to other circuit elements. These are produced by
using an 'air bridge' technology which is the same as that explained
in Chapter 3. These air-bridges are usually 2 to 4µm in thickness.
Where current handling is of importance and where circuit losses need
to be reduced, the metal thickness has to be increased. This can be
achieved in two ways - either by plating-up the evaporated metal or
by depositing a thick layer of metal with subsequent definition by
ion milling.

Vias have been used successfully in discrete FETs to make connections between the source contacts of the FETs and the ground plane on
the back of the GaAs wafers (D'Asaro et al, 1977). In GaAs integrated
circuits it is often desirable to ground various circuit components
effectively without the need for top surface grounding areas which
are subsequently connected to the IC package ground. Such a method
(still used for simple single-function ICs) is a severe constraint to
the GaAs IC circuit designer where more than a few active devices are
present or where the frequency of operation is above 10 GHz or so.

Monolithic power FET amplifiers which rely upon microstrip transmission lines for matching circuitry, for example, present a design
trade-off between thermal considerations and low-loss circuitry
(Driver et al, 1981). The thickness of the GaAs is often chosen to
be 100µm to give acceptable losses in the circuit, while maintaining
satisfactory heat-transfer for the power FETs. With discrete FETs
the GaAs wafers are usually reduced in thickness to 25µm and then
plated up to a greater thickness following the etching of the vias.
With ICs this substrate thickness is not possible over the entire
wafer for the reasons given above.

Two procedures have, therefore, been developed for producing vias
in monolithic circuits (Driver et al, 1981). In the first method,
large areas (or 'tubs') (typically 300µm by 1500µm) are first etched
in the back of the GaAs wafer using an acid-hydrogen peroxide based

etch. These areas are aligned to the FET areas, for example, on the front of the wafer using infrared techniques. A second photoresist layer is used as an etching mask to produce 100μm diameter holes to meet the source pads of the FETs. These holes, together with the back of the wafer are subsequently metallized to form the via connections and ground plane respectively. Fig. 10.35 shows the results of such a method.

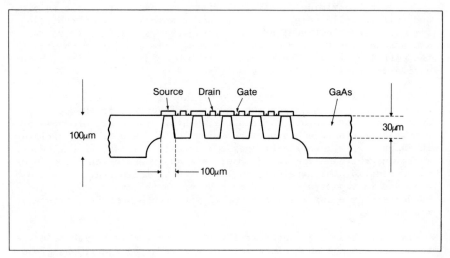

FIG. 10.35.   Etched Tub and Vias

In the second method, termed the sparse 'via' approach only a few large area (i.e. 250 by 250μm) vias are used. These may occur, for example, at the ends and the middles of transistors for source grounding with the source effectively connected together by air-bridges on the top surface. Shunt connected components can also be connected to ground using such a technique. Elsewhere on the chip the GaAs is of the required thickness for distributed or lumped elements.

Raytheon (Vorhaus et al, 1981) have developed a method of producing plated gold integral beam leads which extend over the edge of the chip typically by some 150μms. Such beam leads simplify the mounting and bonding of the GaAs chips into test fixtures or packages. These beam leads are formed using the same process as is used to produce 'air-bridges'. Development of the integral beam lead approach is aided by 'dicing' the wafers, following IC manufacture, from the backside using chemical etching. This back side etching makes possible the use of structures on the front side, such as leads, which extend beyond the individual chip boundaries through what would normally be the scribe grid. The chemical dicing also introduces much less stress than mechanical techniques and also conveniently allows the fabrication of non-rectangular chips.

B. RESISTOR TECHNOLOGY

It is possible to make resistors from gallium arsenide by etching a mesa in the highly n doped wafer surface and then making ohmic contacts to this feature. Such resistors have a sheet resistivity of about 300 $\Omega/\square$ for a typical GaAs FET layer. This resistivity is suitable for many resistor requirements, but mesa resistors have two disadvantages which have led to their replacement by other resistive films.

At certain current levels, the resistance of such resistors deviates from ohmic behaviour due to electron velocity saturation. This effect involves the limited velocity of conduction electrons within the material at a given applied electric field, and cannot be avoided. The effect can, in fact, be used to advantage in digital circuits but is undesirable in linear circuits. Highly doped gallium arsenide resistor films also exhibit a positive temperature coefficient of resistance (TCR), which combines with the positive TCR of certain active components to give unacceptable parameter changes with temperature. For these reasons it is desirable to utilise other resistor systems which will give truly ohmic behaviour, a negative TCR, and also will allow various resistivities to be achieved without wide variations in thickness. It is also essential that the deposition and processing conditions are compatible with the components already on the wafer, and thus materials requiring high temperature deposition or annealing are precluded. Two suitable resistor systems that have been evaluated are tantalum nitride and chromium-silicon monoxide cermets.

Tantalum nitride is deposited by sputtering tantalum in an argon-nitrogen mixture. As the partial pressure of nitrogen in the plasma is increased the deposited film changes from pure tantalum to Ta_2N and subsequently to TaN. This change is accompanied by an increase in resistivity by a factor of about 10, and the TCR decreases from +1000 ppm $^{o}C^{-1}$ to about -100 ppm $^{o}C^{-1}$.

It is a useful property of such reactively sputtered films that the resistivity and TCR vary with the nitrogen partial pressure over a limited range. If the nitrogen concentration is between 1% and 10% of the sputtering gas, both parameters vary only slightly even though the structure and composition of the film changes (Gerstenberg et al, 1964). This variation is shown in Figure 10.36.

A serious drawback to the use of tantalum nitride resistors has been the tendency of sputtered films to craze, indicating that the films were in compression. Attempts were made to overcome this problem by sputtering onto heated substrates and also by heat sinking the wafers to a water cooled pallet. Various annealing procedures were also investigated but no satisfactory solution to the problem has yet been found.

Tantalum nitride has a resistivity of 2.5×10^{-4} $\Omega.cm$ which corresponds to a sheet resistivity of 25 $\Omega/\square$ at a thickness of 0.1μm.

This is felt to be the minimum thickness compatible with stability
and reproducibility, even at this thickness the sheet resistivity is
rather low for many circuit requirements. In addition it is necessary to consider the current density limitations of the resistive
films in view of the need to pass currents of the order of tens of
milliamps. Resistor films are in general limited to current densities
in the range 1-10 x 10^4 A.cm^{-2} and thus circuit requirements severely
restrict the degree to which resistors can be reduced in thickness
and width. Increasing the resistor length is an inefficient solution
since it not only uses more wafer area, but also increases the complexity of the circuit and the risk of parasitic effects at high
frequency operation.

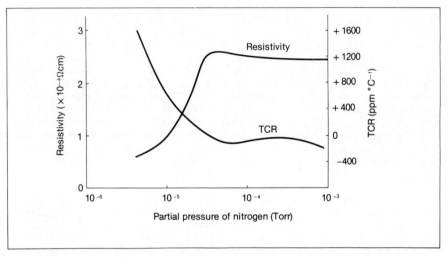

FIG. 10.36.   Resistivity and TCR as a Function of Nitrogen
   Partial Pressure for TaN Resistors

Cermet resistors have an advantage over many other resistor systems
because their resistivity can be adjusted over a wide range by varying the material composition.

Chromium-silicon monoxide cermets have been used extensively as
resistive films (Maissel, 1970) and have been deposited by co-
evaporation or RF sputtering.

The problems of evaporation have been well documented (Ostrander,
1962; Braun, 1966 and Pitt, 1967), for GaAs IC's it was considered
that the improved reproducibility gained by RF sputtering a composite
target would be of great value.

Initial assessment of the cermet was carried out using a silicon
monoxide target partially coated with chromium, leaving exposed areas

of silicon monoxide. This technique allowed the composition of the
sputtered films to be varied by altering the relative areas of
chromium and silicon monoxide exposed to the plasma. The sputtered
films were assessed initially by Auger analysis (AES) and measurement
of sheet resistivity by a four-point probe instrument.

In practice cermet resistors are sputtered from targets consisting
of an intimate mix of chromium and silicon monoxide powders. A film
resistivity of 1 x 10^{-2} Ω.cm allows resistors of 300 $\Omega/\square$ to be fabricated at a thickness of 0.33μm, this corresponds to an atomic concentration of about 55% chromium in the film. The sputtering rate of
chromium is less than that of silicon monoxide, and by applying a
substrate bias during sputtering it is possible to modify the chromium content of the film and thus make small adjustments to the
resistivity from deposition to deposition. Auger analysis has indicated that bias sputtering induces chemical changes in the film and
this effect needs to be further investigated.

Films of 50-60% atomic chromium have been found to give TCR's of
between -400 and -250 ppm $^{o}C^{-1}$, which is acceptable for circuit
stability. Limiting current densities of 3 x 10^{4} A.cm^{-2} are achieved.
The films are stabilised by annealing in air at 300°C immediately
after sputtering.

A summary of resistor materials is given in Table 10.3.

C. CAPACITOR TECHNOLOGY

Capacitors are required at various stages in RF circuits and in the
case of decoupling capacitors the values required can be several
hundred picofarads. Such values can only be achieved if large areas
are employed or if very thin dielectric films are used. Capacitors
of values less than one picofarad can be made successfully by using
an interdigital structure.

The two dielectrics used in current Plessey designs, for example,
are polyimide and silicon nitride which have dielectric constants of
3.5 and 5.5 respectively. Silicon nitride has the advantage that it
can be applied in thinner layers than polyimide and thus it is useful
for the larger values of capacitors.

(i) Polyimide Dielectric Layers

A major advantage of the use of polyimide layers is the ease and
cheapness of fabrication of such films. The starting material may be
a polyimide precursor resin which on curing undergoes a condensation
reaction to form an imidised, cross-linked structure. Alternatively
an already imidised material can be used, where curing is used to
remove a carrier solvent and to promote limited cross-linking.

Several resins of both types have been evaluated for application to
GaAs IC's. The most suitable is a condensation polymer which forms

TABLE 10.3. Resistor Material Summary

Material	TCR ppm/°C	Ω/□	Manufacturing Technique	Remarks
Ti	+2500	10	Filament or EB evaporation	Excellent adhesion to GaAs. Good stability after annealing
NiCr	<200	90 Typically	Sputtering from target in inert gas	Ω/□ dependent on annealing schedule. Good stability after annealing
Ta2N	-100 Typically	90 Typically	Reactive sputtering	Good adhesion onto polyimide
Cr	+3000	1.5	Evaporation	Low Ω/□ ∴ not very suitable for GaAs ICs. Good adhesion to GaAs
CrSiO (CERMET)	-300 to +100 depending on composition and annealing	50 to 500	Sputtering from a composite target	Ω/□ variable over good range. Annealing schedule for wanted TCR
Au/SiO2 composite	-500 to +500 depending on Ω/□	0.1 to 1000	r.f. sputtered from composite target	Difficult to reproduce Ω/□ accurately
Bulk GaAs	+3200	300 for 10^{17}/cc material	Epitaxial or ion implanted	Current saturation a disadvantage. Limited Ω/□ without selective implants
CrGe	Close to 0	100	Evaporation	Difficult to reproduce Ω/□ stability questionable

an amide-imide structure on curing. Curing temperatures are limited
to below 350°C (to avoid prolonged exposure of the active devices to
high temperatures), and adequate polymerisation occurs at these
temperatures.

Polyimide films are applied to the wafer by spinning in the same
way as photoresist layers and confer good step coverage of underlying
structures. For this reason polyimide layers are also used as inter-
layer insulation in multilevel devices, the conformal nature of the
spun film gives a quasi-planar surface suitable for further processing steps. Polyimide films allow sputter deposition and ion milling
of conductor and resistor films without detriment to other device
structures. In addition these interlayers provide good stress
barriers between thin film materials which are not normally compatible.

Fabrication of polyimide layers by spinning a fluid is a satisfactory process for producing films of a few microns thickness. For
films of thickness much less than a micron, pinholing of the spun
film becomes a problem, and in order to reduce the physical size of
high value capacitors, another dielectric is required with the capability of forming high quality layers approximately 0.1µm thick.

(ii) Silicon Nitride Layers

Silicon nitride is a suitable dielectric material for microwave use,
having a dielectric constant higher than that of polyimide, and
capable of being deposited in the requisite thin films. Of the
various methods of deposition available, the one chosen is plasma
enhanced chemical vapour deposition (PECVD). This allows high
quality films to be produced at substrate temperatures below 350°C.

The deposition is carried out in a parallel plate reactor in which
silane, ammonia and nitrogen flow across the wafers in a vacuum chamber maintained at a pressure of about one Torr. The wafers rest on a
heated pallet which forms the earthed lower electrode of the parallel
plate system. An RF discharge at 13.56 MHz sustains the reaction and
deposition rates of about 10 $nm.min^{-1}$ are achieved at power levels of
50 mW cm^{-2}.

Films are assessed by measurement of refractive index by ellipso-
metry and measurement of etch rate in buffered hydrofluoric acid
(BHF). It is possible to vary the refractive index from 1.85 to 2.15
by adjustment of the flow ratio of silane to ammonia, and this ratio
is usually set to give films having a refractive index of 2.00. The
BHF etch rate of the films decreases with increasing refractive index
as the films become progressively enriched in silicon, as is shown in
Figure 10.37(a).

An essential advantage of the PECVD method of deposition is that
wafer temperatures are low enough not to damage the active areas of
the devices. In order to see how far the deposition temperature
could be reduced, films were deposited under otherwise identical

394

conditions on substrates at temperatures between 150°C and 300°C. Figure 10.37(b) shows that the BHF etch rates of the low temperature films were in excess of ten times greater than those of films deposited at 300°C, even though the films had similar refractive indices. It is considered that high etch rates indicate poor quality films. Capacitor dielectric layers are therefore routinely deposited at the higher temperature of 300°C.

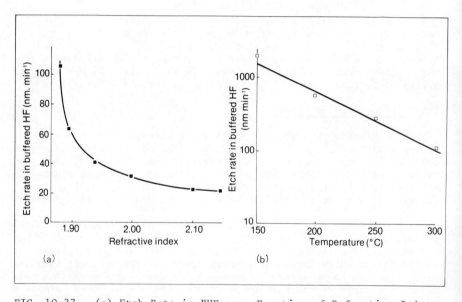

FIG. 10.37. (a) Etch Rate in BHF as a Function of Refractive Index for Silicon Nitride

(b) Etch Rate in BHF as a Function of Silicon Nitride Deposition Temperature

Reproducibility and uniformity are very important parameters for routine production of high tolerance dielectric films, and the process has been established with this in mind. The area available for deposition is sufficient for four 3 inch diameter wafers. Uniformity under normal deposition conditions is better than 6% over the whole deposition area, and better than 4% over individual wafers. Reproducibility from run to run is excellent, measured refractive indices varying by ± 2%, which is within the measurement error of the ellipsometer. Deposition rates are consistent to within ± 7%, and equipment modifications are expected to reduce this significantly in the near future.

Of all the factors affecting the deposition process, it has been found that power density is one of the most relevant to uniformity. For this reason depositions for capacitor dielectrics are carried out at power densities below 20 mW cm^{-2}. Less critical depositions (such as passivation layers) can be carried out at higher powers in order to reduce deposition times (Figure 10.38).

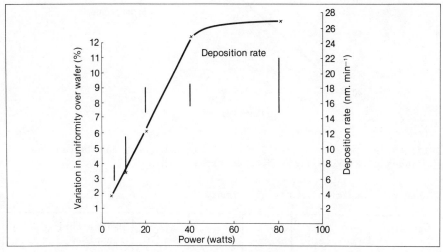

FIG. 10.38. Variation of Deposition Rate and Uniformity as a
    Function of Applied RF Power for PECVD Silicon Nitride

Auger analysis of films has been used to measure the silicon to nitrogen ratio, although this measurement is complicated by the different removal rates of silicon and nitrogen by the Auger spectrometer. The PECVD films are compared to a silicon nitride standard of composition Si_3N_4. The results are shown in Figure 10.39.

It is recognised that the films will contain significant quantities of hydrogen derived from the gases used in the deposition process. The use of an electron microprobe analyser in conjunction with Auger spectroscopy has enabled an estimate to be made as follows of the atomic composition of a typical film: silicon 45%, nitrogen 35%, hydrogen 20%.

Since this instrument requires much thicker films than are normally grown, microprobe analysis is not considered suitable for routine analysis of film composition.

A summary of capacitor materials is given in Table 10.4.

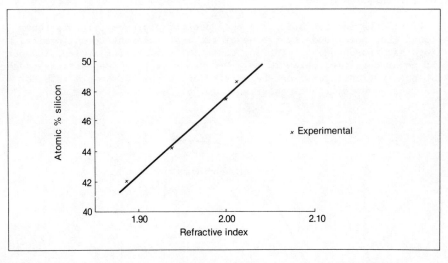

FIG. 10.39.   Relationship Between Silicon Content and Refractive
Index of PECVD Silicon Nitride Films

TABLE 4.   Capacitor Material Summary

Material	Relative Dielectric Constant	C/A pF/mm^2	Ω	TCC ppm/$^\circ$C	Manufacturing Technique
SiO$_2$	4–5	400	Good	50	Sputtering with a SiO$_2$ target
Si$_3$N$_4$	5.5	485	Good	25	Reactive plasma sputtering
Al$_2$O$_3$	6 to 10	795	Good	100–500	Anodization of evaporated film
Polyimide	3 to 4.5 depending on material	31	Good to V. good	–500	Spun and cured film
Varactor diode	–	1000	Good	–	Evaporated Schottky

D. PLASMA ETCHING

Both polyimide and silicon nitride are plasma etched (see for example, Tolliver, 1980) to define dielectric structures. Polyimide is etched in oxygen at about 1 Torr in a system with fixed parallel plate electrodes. At a power density of 3.5 $W.cm^{-2}$ polyimide films are etched at a rate of about 50 nm min^{-1} which is similar to the removal rate of photoresist. Thus thicknesses of polyimide up to several microns may be satisfactorily masked with photoresist layers.

Silicon nitride films are plasma etched in a parallel plate reactor using carbon tetrafluoride-oxygen mixture at 500 mTorr and a power density of 0.5 $W cm^{-2}$ at an etch rate of 50 nm min^{-1}. Positive photo-resist layers are used as masks without problems caused by rapid removal of photoresist by the oxygen component of the etchant gas mixture. This is because 0.1μm silicon nitride films can be adequately masks by photoresist films thicker than 0.5μm. The effect of electrode separation has been found to be very important in optimising etch rate ratios between the mask and the material to be etched.

E. ION MILLING

Ion milling (see for example, Bollinger and Fink, 1980) is carried out using an uncollimated argon ion beam, which operates at a current density of about 0.5 mA cm^{-2}. This power enables conductor and inductor metallisations to be ion milled at a rate exceeding 25 nm min^{-1}. The chief advantages of this technique are the elimination of undercutting of the metal being patterned and the ability to achieve precise dimensional control features of a few microns wide. Control of conductor profiles is achieved by careful adjustment of the baking cycle applied to the photoresist layers used as ion-milling masks (Brambley and Vanner, 1979).

The removal rate of gallium arsenide by an argon ion beam at the current density quoted, is much higher than the removal rates of the conductors. In order to avoid unwanted etching of the substrate, circuits are designed so that the gallium arsenide substrate is always protected by a polyimide layer during ion-milling. Polyimide has been found to be an excellent material for this purpose with a low ion-milling rate. After ion-milling, photoresist residues are removed by a brief oxygen plasma etch followed by a solvent clean. This has been found to cause no adverse effects to the polyimide surface.

Wet etching of tantalum nitride or chromium-silicon monoxide cermets poses additional problems. Tantalum etchants often attack gallium arsenide, and most cermets do not etch satisfactorily in chromium or silicon monoxide etchants (Glang and Gregor, 1979). The use of ion milling for patterning resistors made of these materials avoids all of the problems involved in developing satisfactory wet etchants, and confers more precise dimensional control.

F. INDUCTORS

Single turn loop inductors for inductance values up to 1 nH are easily formed using gold films. The requirements for low frequency circuits are for larger value inductors necessitating the use of a spiral structure. The entire inductor can be covered by a polyimide film and the centre contact to the spiral made through a via. Alternatively, an airbridge technique can be used as shown in Fig. 10.40 where, in order to avoid the use of a long airbridge which may be mechanically unsound, the connection from centre to outside is made underneath the spiral. At frequencies below a few gigahertz the skin depth of the gold metallization becomes such that simple lift-off techniques cannot be used to define the 3 to 5µm thick inductors. Thus, the inductors are either plated-up or are defined using ion beam milling.

G. INTERCONNECTIONS

A reliable and reproducible interconnection technology is one of the key factors in successful monolithic circuit fabrication. FETs use Schottky contacts formed from metals such as Ti-Al, or Al which although providing good performance usually require an interconnect scheme to gold for reliability. Because there is no physical bond the oxide film that forms on the top of aluminium, for example, prevents an ohmic contact between the aluminium and the interconnect metal. Titanium, if sintered, will leach out the oxide to form conducting titanium oxide. A further barrier metal such as nickel is used to separate the titanium from the gold which would otherwise form intermetallic compounds. Fig. 10.41 shows examples of interconnections where in Fig. 10.41(a) connection is made between Ti-Al gates and a Au second level metallization and in Fig. 10.41(b) between Au first level and second level metallizations.

5. INTEGRATED CIRCUIT EXAMPLES

This part of the chapter is broadly divided into descriptions of analogue and digital GaAs circuits with the aim of giving the reader an appreciation of the performances being obtained at the present time (1981) and the likely circuit complexities and performances which will be produced in the next few years.

A. SMALL SIGNAL AMPLIFIERS

Much of the work reported on GaAs monolithic small signal amplifiers to date has concentrated on the realisation of broadband low noise units intended for electronic warfare applications. Indeed Pengelly et al (1976) reported the first monolithic GaAs IC to employ GaAs FETs which was a broadband X-band amplifier covering 8 to 12 GHz. More recently single stage amplifiers embracing the input and output matching networks as well as self-bias etc. have been fabricated by Stubbs (1980) resulting in gains of 6 dB from 6 GHz to 18.5 GHz. These ICs employed a conventional mesa-epitaxial process together

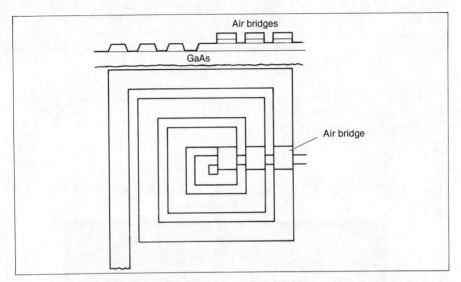

FIG. 10.40.   Example of Airbridge Technique Applied to Spiral
Inductor

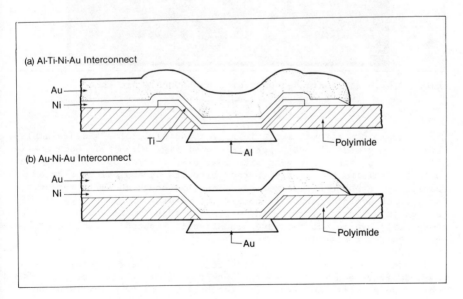

FIG. 10.41.   Interconnect Schemes

with mesa 'bulk' resistors and polyimide overlay capacitors (Fig. 10.42). This integrated circuit forms the basis for more complex multistage amplifiers employing 0.6μm gate length FETs.

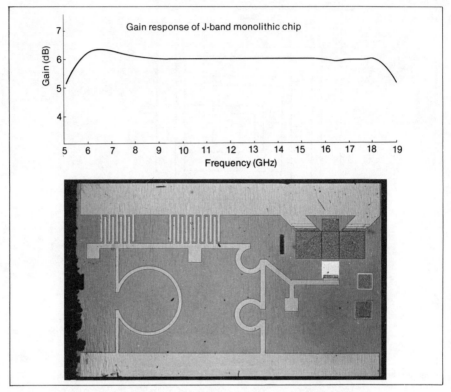

FIG. 10.42.　(a)　Gain Response of J-band Monolithic Chip
　　　　　　　(b)　Broadband J-band Amplifier

For example the circuit shown in Fig. 10.43(a) is of a 4 stage amplifier which yields the gain performance of Fig. 10.43(b). Such amplifier chips provide flat gain with low noise figures over 8 to 18 GHz with the ability to cascade further circuits by employing balanced amplifier techniques (Chapter 5). Thus, although an amplifier of, say, 70 dB overall gain will consist of several monolithic gain stages in a hybrid circuit, the extensive adjustments and build of a completely hybrid approach are dramatically reduced leading to smaller size and lower cost.

In order to remove the need for broadband couplers several workers have designed and fabricated feedback amplifiers (Niclas et al, 1980) which use reactive and resistive feedback to maintain low input and output reflection coefficients over a very wide bandwidth.

In general, such techniques allow the direct cascading of amplifier stages without the need for considerable numbers of passive matching components which consume large GaAs areas. However, such circuits require more d.c. power than the balanced approach owing to the need for large FET transconductances.

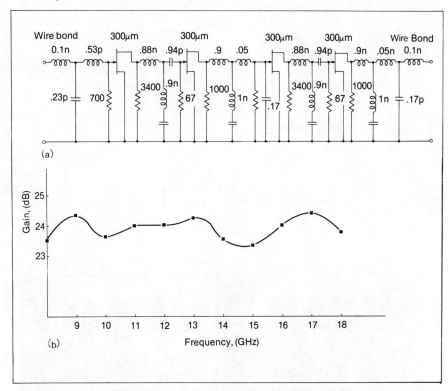

FIG. 10.43(a). 4 Stage Monolithic J-band Octave Bandwidth Amplifier
        (b). Performance of 4 Stage Monolithic J-band Octave
    Bandwidth Amplifier

An alternative approach in low-noise amplifier design which is applicable to frequencies less than 6 GHz at present is the use of common gate input stages and source follower output stages to provide the matched conditions. Fig. 10.19 shows, for example, the circuit of a three stage amplifier consisting of a common gate 300μm wide FET, followed by a simply matched common source stage cascaded with a source follower. All stages are biased at the low-noise condition to produce the circuit performance of Fig. 10.44(a). Fig. 10.44(b) shows the layout of this amplifier which also provides an in-phase power division.

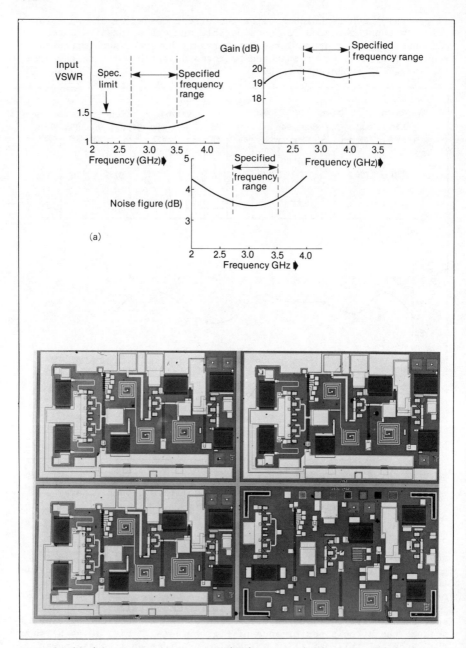

FIG. 10.44.(a) Performance of CG/CS/CD Monolithic Preamplifier
(b) Low Noise Common Gate, Common Source, Source
Follower S-band Active Splitter

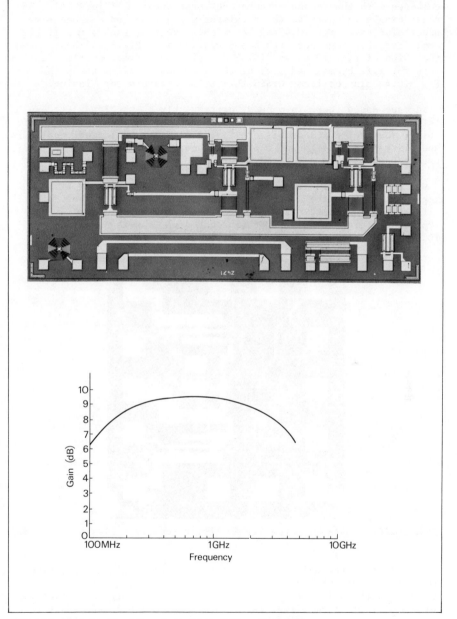

FIG. 10.45.(a)   Broadband (4 GHz) Video Amplifier
         (b)   Common Source-Source Follower Amplifier-Frequency
   Response

The techniques of common gate and source follower stages can be combined with the feedback technique to produce amplifiers which are considerably insensitive to transistor variations and component value uncertainties. Fig. 10.45(a) shows the layout of such a monolithic amplifier chip together with the performance in Fig.10.45(b). Both the chips of Fig. 10.44 and 10.45 use silicon nitride overlay capacitors for r.f. bypass and d.c. blocking purposes whilst the chip of Fig. 10.44 also utilizes CrSiO thin film resistors for biasing.

In many cases there is a need, particularly at the lower frequencies (of up to several gigahertz), to produce ICs involving balanced circuits. Such circuits as differential amplifiers, balanced modulators and mixers are well known in Si IC technology and an example of their use on GaAs will be described later.

One of the most important constraints of differential amplifier implementation is the d.c. voltage match between the two halves of the differential circuit, often referred to as the offset voltage. A typical differential amplifier fabricated on GaAs is shown in Fig. 10.46.

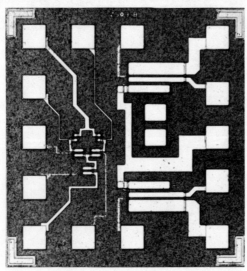

FIG. 10.46.   GaAs Differential Amplifier (courtesy Plessey Co. Ltd.)

On the left of the chip is the differential amplifier using active loads and a current source as well as a pair of larger gate width devices (with their active loads) used as output buffers on the right. This differential amplifier is produced on ion implanted GaAs to reduce the differences in the d.c. characteristics of the FETs to an acceptable level. Potential applications for such differential amplifiers include logarithmic amplifiers and modulators. In many

applications, for example radar, the input signals can have a very
wide dynamic range in excess of the dynamic range of a typical linear
amplifier. This problem may be overcome by the use of the logarithmic
amplifier (Hughes,   ). In order to achieve a maximum logarithmic
error of less than 1 dB the linear gain of each stage of the logarithmic amplifier must be 10 dB or less.

Differential amplifiers on GaAs have been designed employing feedback to achieve 10 dB gain over a 4 GHz bandwidth (Wilcox, 1980).

Considerable interest is at present being focused on gigahertz
biphase shift keyed (BPSK) modulation and demodulation. A BPSK modulator can be produced using the circuit of Fig. 10.47.

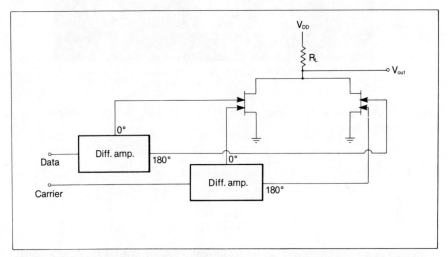

FIG. 10.47.   GaAs Dual Gate Modulator

Dual gate FETs are used to perform modulation of the r.f. carrier and
the 0° and 180° phase shifts for the data and carrier are performed
by differential amplifiers.

The circuit techniques described so far have tended to be broadband
in nature. However, GaAs monolithic circuits have also been produced
which are designed for narrow bandwidth applications where noise
figure is of prime importance for example. Fig. 10.48 shows an
example of a narrow band amplifier fabricated by Hughes Aircraft
Company which has a 12% 1 dB bandwidth centred at 10 GHz. The circuit uses both lumped and distributed elements in a chip size of
2.67 by 1.78 mm. Vias are used through the 100μm thick chip to
ground the input matching interdigital capacitor, the gate and drain
bias network bypass capacitors and the FET sources themselves. The
chip which uses ion implanted GaAs has a gain of 7 dB and uses a

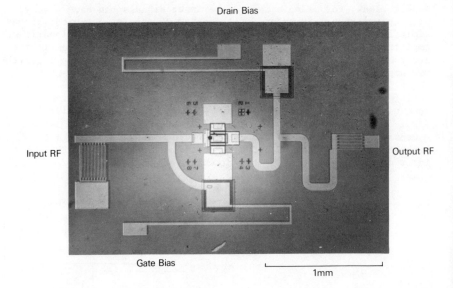

Drain Bias

Input RF

Output RF

Gate Bias

1mm

FIG. 10.48. X-band Medium Power FET Amplifier (courtesy Hughes Aircraft Co.)

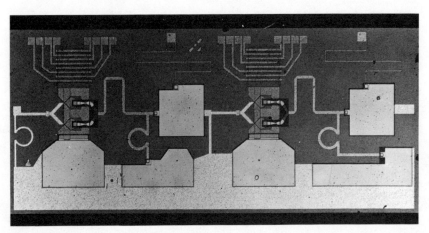

FIG. 10.49. Direct TV Preamplifier

a 600μm by 1μm FET.

Fig. 10.49 shows a two stage low noise preamplifier fabricated by Plessey which is designed to cover the TV satellite band of 11.7 to 12.5 GHz with a noise figure of less than 4 dB and 15 dB gain. The chip uses 0.6μm gate length FETs, polyimide capacitors and 'bulk' resistors.

B. POWER AMPLIFIERS

Monolithic FET power amplifiers have received considerable attention by a number of workers (Driver et al, 1981; Tserng et al, 1981; Vorhaus et al, 1981). The approaches that have been taken vary considerably ranging from push-pull amplifiers at X-band to feedback amplifiers at S-band. Three approaches to GaAs monolithic power amplifier design are discussed.

(i) Two Stage Push-Pull Amplifier

Texas Instruments have designed and fabricated a two stage push-pull amplifier at C-band together with a differential or 'paraphase' amplifier to drive this push-pull configuration (Sokolov et al, 1980). Based on experience with discrete power FETs having 100μm thick substrates monolithic circuits are fabricated on substrates of similar thickness. The chips are mounted on to gold plated copper test fixtures using a 96% Sn/4% Ag solder (having a melting point of 217°C) to provide a good heat sink.

A schematic diagram of the two stage, four transistor, push-pull design is shown in Fig. 10.50.

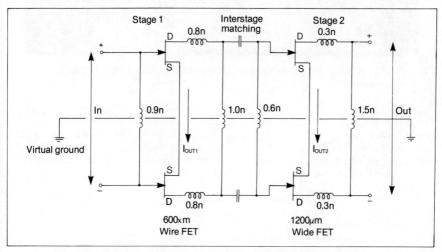

FIG. 10.50.   Schematic Diagram of Two-Stage X-band Push-Pull Amplifier

Fig. 10.51 shows a chip photograph of the GaAs IC. The series inductors and capacitors are integrated 'on-chip' whilst the shunt inductors are realized with 25μm diameter bond wires allowing the centre
frequency of the amplifier to be adjusted.

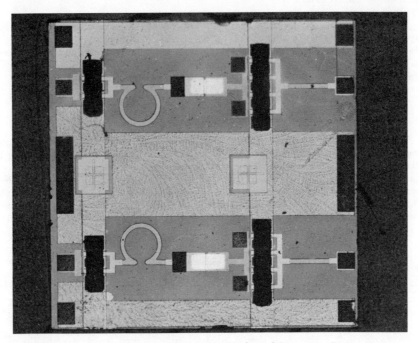

FIG. 10.51.   Two Stage Push-Pull Amplifier (courtesy Texas
   Instruments)

The output stage consists of a pair of 1.2 mm gatewidth FETs with the
input stage using a pair of 600μm FETs. Each of the multi cell FETs
uses a plated 'airbridge' for source interconnections. Grounding is
achieved in this design using top ground planes which are connected
to the test fixture using wire mesh.

Two advantages of the push-pull configuration are used in the
design. Firstly, the push-pull arrangement is essentially a series
connection corresponding to a total gate width at the output of
2.4 mm. This means that the impedance matching problem is halved
over the simpler single-ended approach. Secondly, current at the
fundamental frequency flows between the source of each transistor and
the source of its push-pull counterpart. Coplanar grounding

metallization connects all sources, so that these r.f. currents are primarily confined to the chip's surface. If each transistor pair is closely matched, then only a small fraction of the total r.f. source current needs to flow to true chassis ground. Therefore, connection of the chip's grounding metallization is not as crucial as in the case of a single-ended amplifier.

Since the amplifier operates in the push-pull mode, a virtual ground exists symmetrically between the two halves of the circuit (see Fig. 10.50). The design of the amplifier merely consists of a single-ended approach with the shunt inductors doubled in value for the push-pull implementation.

As may be seen from Fig. 10.51 a series overlay capacitor is used for d.c. blocking between first and second stages.

The push-pull amplifier needs to be driven with antiphase signals and then the signals combined in another circuit providing a 180° phase shift. A three transistor monolithic X-band differential or 'paraphase' amplifier was developed by Texas Instruments to enable the input antiphase signals to be generated without the need for a conventional passive (and, therefore, large) 'rat-race'. The 'paraphase' amplifier is the solid state equivalent of circuits produced at lower frequencies using valves (Seely, 1958). Fig. 10.52 shows a schematic of the paraphase circuit.

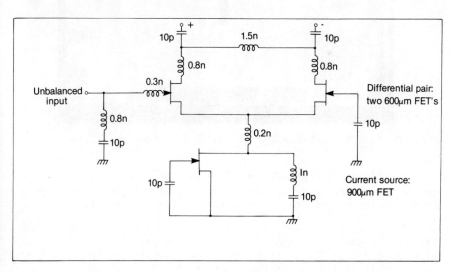

FIG. 10.52.   Schematic Diagram of X-band Paraphase Amplifier

Two 600μm gatewidth devices are used for the differential pair. The unbalanced input is applied to one of the gates whilst the other is r.f. short circuited. The balanced output is taken via matching

410

circuits. The transistor that functions as the current source is realized by a FET having a total gatewidth of 900μm. The gate of this transistor is r.f. short-circuited to ground and a shunt inductor is used at the drain to resonate the drain-to-source capacitance.

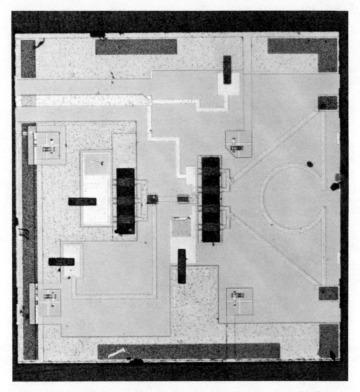

FIG. 10.53.   Microphotograph of X-band Paraphase Amplifier

Fig. 10.53 shows the realization of this amplifier. The dark bars are plated airbridge source interconnects. The lighter areas are the 10 pF overlay capacitors together with airbridge connections to avoid capacitor short circuits. The push-pull amplifiers have achieved over 1 watt output power with 10 dB gain at 9 GHz. Fig. 10.54 shows the small signal and large signal gain characteristics as a function of frequency from 8.5 to 11.5 GHz. 1 dB gain compression at 9.5 GHz was 0.9 watt with a small-signal gain of 14 dB at 9.2 GHz. Power added efficiencies of between 17% and 19% have been measured.

The separate paraphase amplifier has a gain of approximately 2 to 3 dB over a 1 GHz bandwidth centred at 8.3 GHz with a phase difference at the outputs of between 140° and 180°. Further circuit and fabrication improvements can improve the performance and bandwidth of this circuit considerably.

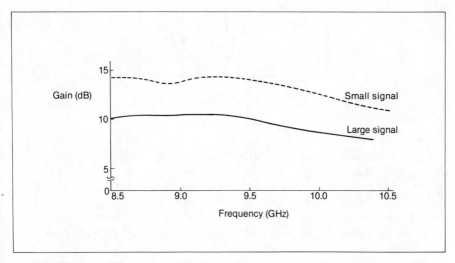

FIG. 10.54.   Gain Versus Frequency for a X-band Two-Stage Push-Pull
    Amplifier

(ii) Single-Ended Amplifiers

   Several laboratories have reported monolithic single ended power
amplifiers (Driver et al, 1981; Vorhaus et al, 1981). Most IC power
FET amplifier work is concentrated at X-band where bipolar transistors cannot be used. Both distributed and lumped element passive
matching is utilized with the FETs having gate widths up to 2.4 mm.

   For example, the equivalent circuits for 300 and 1200μm gate widths
FETs shown in Fig. 10.55 can be used together with the large signal
output resistance R_D to design multistage amplifiers. Fig. 10.56
shows, for example, the optimized results of computer calculations
for a 3 stage amplifier utilizing high impedance transmission lines.
A photograph of two completed amplifier chips is shown in Fig. 10.57.
These amplifiers have 3 or 4 stages with the first two being 300μm
gate width devices and the third a 1200μm gate width device. The
last stage of the four stage amplifier uses a 2400μm gate width FET.
The 300μm gate width devices have 75μm unit width gate fingers whilst
the 1200 and 2400μm FETs have unit gate widths of 150μms, thus reducing the number of FET cells. The amplifier chips are 1 mm by 4 mm
in dimension. The various loops in the photograph are all modelled
as transmission lines. The FET sources are all interconnected by
plated Au 'airbridges' which are grounded by the bars along the chip
edges. Si_3N_4 overlay capacitors are used for both d.c. blocking and
r.f. bypass. Gains of over 27 dB have been obtained with -3 dBm

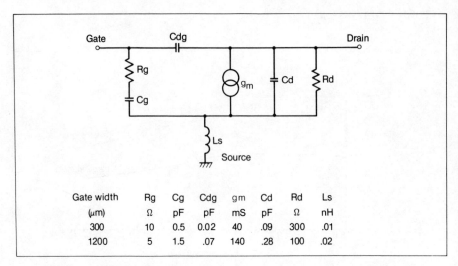

FIG. 10.55.   Simplified Power FET Model for Monolithic Amplifier
Design

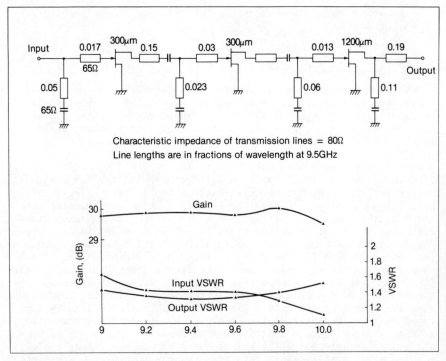

FIG. 10.56.   Circuit Topology and Calculated Gain-Frequency Response
of a 3-Stage Monolithic Amplifier using Distributed Matching

input and 290 mW output at 9.2 GHz for the 3 stage amplifier.
Increasing the input drive resulted in a 400 mW output with 23 dB
gain corresponding to a power-added efficiency of 15%.

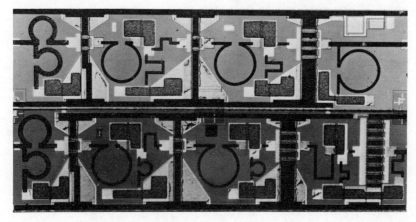

FIG. 10.57.   Three and Four Stage Monolithic Power FET Amplifiers
(courtesy Texas Instruments Ltd.)

An output power of 800 mW was obtained from the four stage design
with 32 dB gain at 8.7 GHz. With increased drive, 1 watt output was
obtained with 27 dB gain at 8.9 GHz with a 1 dB bandwidth of 8.6 to
9.2 GHz. By modifying the chip shunt inductor values only it was
possible to produce 2W output power with 28 dB gain and a power added
efficiency of 37% at 3.5 GHz with a 1 dB bandwidth of 1 GHz.

Driver et al (1981) have reported similar results using a distributed lumped element matching approach on GaAs where, unlike the
approach of Tserng et al (1981) above, interdigital capacitors have
also been used. Two stage amplifiers have been reported, contained
on chips measuring 2 by 4.75 mm which use via hole technology for the
FETs and shunt connected interdigital capacitors. This design has
produced 28 + 0.7 dBm output power over 5.7 to 11 GHz with 6 + 0.7 dB
gain. The power added efficiency of the amplifier was 8 to 12% over

414

the band. Vorhaus et al (1981) recently described the fabrication of
a two stage monolithic X-band power amplifier exhibiting a power
output of 565 mW with 8 dB gain over 8.2 to 10.5 GHz. This circuit
is somewhat simpler than the approaches of Tserng or Driver. Power
added efficiencies were 16%.

(iii) Feedback Amplifiers

   Pengelly et al (to be published) have investigated the use of feedback techniques to produce large percentage bandwidth monolithic
power FET amplifiers in the 2 to 4 GHz region. At these frequencies
conventional broadband matching circuits would require rather large
value inductors which consume large areas of GaAs as well as having
low Q. In order to reduce the number and magnitude of matching components the high transconductance of the power FET can be exploited by
employing feedback. This technique also reduces the sensitivity of
the resultant amplifier to transistor variations and improves the
intermodulation performance of the circuit.

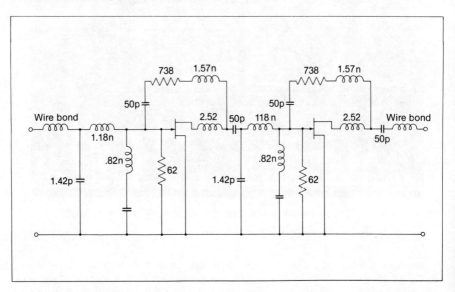

FIG. 10.58.   Circuit Diagram of Two-Stage Monolithic S-band Power
   FET Amplifier

Fig.10.58(a) shows the integrated circuit layout for a two stage
monolithic amplifier having 22 dB gain and a 1 watt O/P power at the
1 dB gain compression point over the 2.5 to 3.5 GHz frequency range.
Fig. 10.58(b) shows the equivalent circuit of the chip where each FET
has a total gate width of 2.4 mm, thus enabling high transconductance
per stage. The input and output VSWRs are less than 2:1. The chip
uses silicon nitride capacitors for r.f. bypass, d.c. blocking and
matching together with CrSiO cermet resistors and a low dielectric
constant dielectric to separate first and second level metallizations.
Chip size is 2.3 x 4.5 mm which is not significantly larger than the
X-band power amplifiers previously described.

C. OSCILLATORS

Both fixed and varactor diode tuned oscillators have been implemented monolithically. The first monolithic oscillator was reported by
Joshi (1979) which used a GAT5 FET geometry in common gate configuration. The oscillator circuit is shown in Fig. 10.59(a).

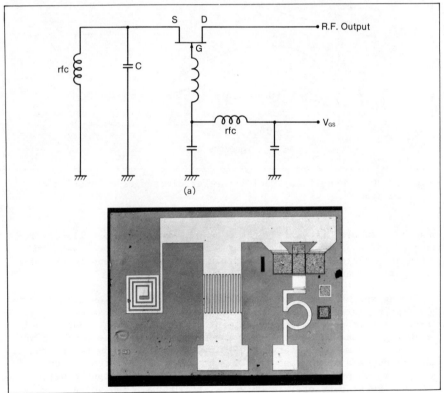

FIG. 10.59(a). Common Gate Circuit Configuration used for First
Monolithic Oscillator
(b). Photomicrograph of Monolithic Oscillator

The computed output reflection coefficient for the circuit of Fig. 10.59(a) is greater than one for frequencies above 8 GHz reaching a maximum of 9.3 at 12 GHz. Fig. 10.59(b) shows the monolithic implementation of the oscillator circuit. The gate feedback inductance is realized using a single loop inductor and the capacitive source termination using an interdigital capacitor. A spiral inductor is connected to the source of the FET to act both as an r.f. choke and as a d.c. return for the source. The chip measures 1.8 by 1.2 mm. The output power, frequency and d.c. to r.f. conversion efficiency as a function of gate bias are shown in Fig. 10.60.

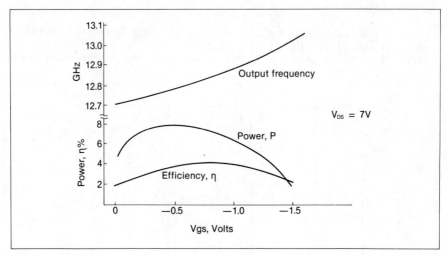

FIG. 10.60.   Monolithic Oscillator Performance as a Function of Gate Bias

As the gate bias is varied from 0 to −1.5 volt, the oscillator frequency varies from 12.7 to 13 GHz. Maximum output power was 8 mW with an efficiency of 4%. The drain port had no tuning to maximise the power transfer to the load.

Tserng and Macksey (1981) have produced monolithic oscillators with 'off-chip' varactor tuning diodes. A schematic diagram of the oscillator circuit is shown in Fig. 10.61. The FET has been operated in common gate and in common drain modes. In common gate mode a power output of 20 mW was observed at 10.8 GHz with a narrow tuning range. Efficiency was approximately 8%. In common drain a tuning range of 4 GHz was observed (16 to 20 GHz) with an average output power of 10 mW. Worst case efficiency was 2% at 20 GHz. The voltage swing of the hyperabrupt varactor was only 9 volts.

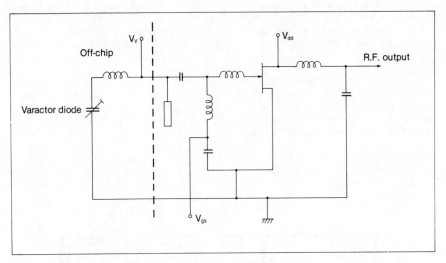

FIG. 10.61.   Monolithic Common Drain FET VCO Circuit

D. SWITCHES

The principles of operation of GaAs FET switches have been covered
in Chapter 9.

McLevige et al (1981) have produced monolithic SPDT configurations
using parallel resonated series-connected FETs at X-band.   The chip
dimensions for the switches was 1.78 by 0.76 mm where the FETs were
0.7μm gate length devices ranging in widths from 900 to 1200μm.
Pinch-off voltages of approximately 6 volt were used to give a compromise between low 'on'-resistance and low gate-leakage.   The SPDT
switches have given 28 dB isolation at 10.2 GHz with an insertion
loss of 0.7 dB.   A minimum isolation of 20 dB was maintained over a
0.6 GHz bandwidth.   The d.c. power dissipation due to gate leakage is
typically 200 to 400μW.   This is a distinct advantage over the use of
PIN diodes as used in conventional switches.   Similar devices have
switched 1W of power with an insertion loss of 1.0 to 1.4 dB with
20 dB minimum isolation over a 0.4 GHz bandwidth centred at 10.4 GHz.

Broader band monolithic SPDT switches have been reported by Ayashi
et al (1980) where the drain circuit of the switch is incorporated
into the microstrip lines used.   The drain to source capacitance of
the FET, which is shunted to ground, and the drain overlay inductance
are treated as part of a transmission line.   Thus when no gate bias
is applied the FET shunts the microstrip line with its low resistance
and with negative gate bias applied the channel is pinched-off the
FET then acting as a low impedance transmission line with low insertion loss.   The switches were fabricated on 100μm thick GaAs substrates

418

with multifingered 1600μm gate width FETs designed specifically for
low drain to source resistance. OFF-state isolations of 40 dB were
measured at 8.5 GHz with 0.5 dB ON-state insertion loss. At 11.8 GHz
the insertion loss is 0.6 dB with 33 dB isolation. Complete X band
coverage has been achieved with 1 dB insertion loss and 30 dB isolation.

Suckling et al (1981) have designed SPDT switches at S-band which
provide over 30 dB switching range with 0.5 dB insertion loss. These
monolithic circuits are shown in Fig. 10.62 and consist of lumped
inductors and capacitors which when combined with the reactances of
the FETs form a matched transmission structure over the 2.5 to 3.5
GHz frequency range. Such circuits are capable of switching 1.5
watts of power. Similar circuits have been designed which use optimized FET structures capable of switching over 3.5 watts of r.f.
power. The chips which are 150μm thick are approximately 2.8 by 3.3
mm and contain both polyimide isolated second level metal as well as
'airbridged' gate connections for the control voltage.

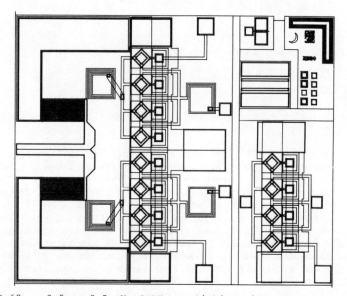

FIG. 10.62.   2.5 to 3.5 GHz SPDT Monolithic Switch

E. MIXERS

The area where GaAs monolithic circuits have had the greatest impact
in the millimeter wave region is mixers where GaAs FETs are used to
amplify the IF signals which occur at a few gigahertz.

Chu et al (1980) have designed a mixer which downconverts from
31 GHz to 2 GHz. The monolithic chip is shown in Fig. 10.63.

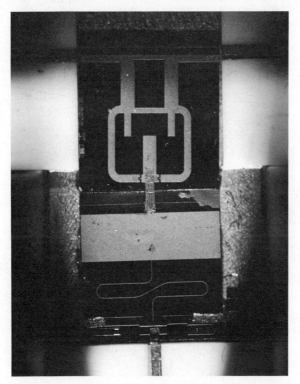

FIC. 10.63.   Millimeter Wave Monolithic Mixer (Lincoln Labs, MIT)

The configuration of the balanced mixer is similar to that of Vendelin (1968). The 90° branched arm hybrid coupler provides a 20% bandwidth over which the amplitudes of the two output arms match to within 0.25 dB. The 90° phase relationship is also maintained to within + 6° over this bandwidth. The stubs on the output arms of the hybrid are used to peak the isolation between the hybrid coupler ports. The quarter wave stub between the Schottky diodes provides an r.f. short.

The structure of the input matching circuit for the MESFET amplifier is also shown in Fig. 10.63. The matching circuit consists of a shunt open-circuited stub followed by a length of high-impedance line. The sizes of the left-hand and right-hand chips are 2.7 by 2.7 mm. The use of distributed matching results in a simple one-level metalliza-tion.

The overall gain and noise figure of this millimetre wave receiver are shown in Fig. 10.64. With an LO of 29 GHz, the conversion gain from r.f. to IF varies from 5 to 3 dB in the signal frequency range of 31 to 31.5 GHz, corresponding to an IF frequency of 2 to 2.5 GHz. The noise figure was 11.2 dB at 31 GHz.

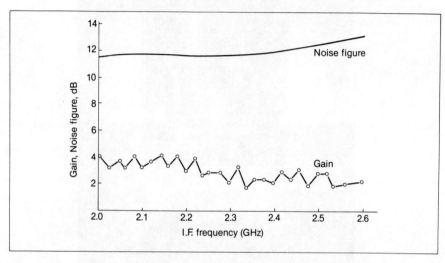

FIG. 10.64. Noise Figure and Gain of 31 GHz Monolithic Receiver as a Function of IF

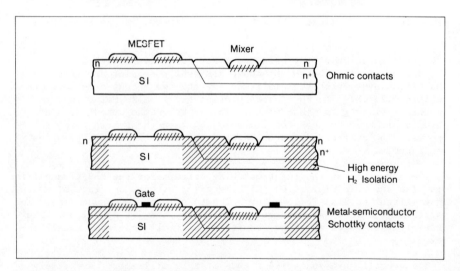

FIG. 10.65. Fabrication Process for Schottky Diodes and FETs Using Selective Epitaxy and Proton Isolation

The fabrication of this millimetric receiver combines the processing requirements for MESFETs and high frequency Schottky diodes. The fabrication steps are shown in Fig. 10.65. Ohmic contacts were fabricated using AuGe. The n-type material is removed by etching from the ohmic contact areas of the planar diodes to expose the n$^+$ layer. The contact materials are evaporated through photoresist windows onto the n-type GaAs for the MESFETs and onto n$^+$ type GaAs for the diodes. The ohmic contacts are then alloyed. High energy proton bombardment was used to isolate the devices as shown in Fig. 10.65. The devices are completed by evaporating the Schottky metal through photoresist openings to form the gates of the MESFETs and the anodes of the diodes.

Lower frequency mixers have been produced monolithically using GaAs FETs in balanced structures by Van Tuyl (1981) and this circuit has been covered in some detail in Section 2D. Novel mixer circuits using common gate connected FETs to combine the LO and r.f. signals have been used along with a common-source mixer FET over the 2 to 4 GHz frequency band. The common-source mixer FET is operated in a self-biased mode with an active load which itself has feedback applied between source and gate (Fig. 10.66).

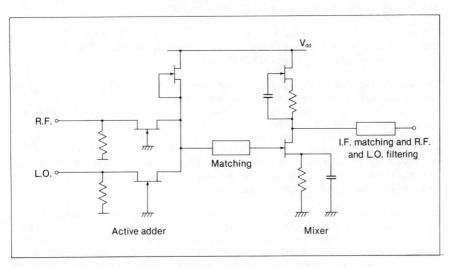

FIG. 10.66.   Novel Common Gate/Common Source Mixer Circuit

The common source FET is operated in a switching mode. Rees and Greenhalgh (to be published) have analysed such a circuit to show that the conversion gain of the mixer is constant over a relatively large change in LO power. Some matching is needed between the common gate stage and the mixer FET. The circuit has an input impedance of 50 ohms and output impedance close to 1 Kohm at an IF of 350 MHz.

The conversion gain of such an arrangement is 8 dB with a noise figure of 6 dB.

F. FURTHER LEVELS OF INTEGRATION

The level of integration of circuits is increasing at a rapid rate. Van Tuyl (1981) has demonstrated the realisation of a signal generator using heterodyne techniques which consists of a tunable oscillator, on/off modulator, mixer, IF and buffering amplifiers, all fabricated on a 600μm by 650μm chip with a 16 bonding pad layout. The active circuit area occupies 350μm by 400μm and contains 29 FETs, 8 diodes, 6 varactor diodes and 1 resistor. The only off-chip components are the tuning inductors for the oscillator, a low pass filter and amplifier. The circuit generates signals of up to 6 dBm power over a 1.4 GHz bandwidth.

LEP (Harrop et al, 1980) have produced a monolithic receiver for direct TV satellite reception for the 11.7 to 12.5 GHz band which uses a microstrip approach. The chip includes a common drain FET oscillator which is stabilized 'off-chip' with a dielectric resonator (see Chapter 7), a 2 stage low-noise preamplifier and a dual-gate FET mixer. An image rejection filter is also incorporated on a chip which measures 10 mm square and 300μm thick.

Overall conversion gain of the chip is 15 to 16 dB over 12 to 12.25 GHz with a 4.5 dB noise figure. The technique adopted for fabrication was a single level metallization.

Plessey (Suffolk et al, 1980) have designed an S-band image rejection receiver which consists of a low noise preamplifier which divides the incoming r.f. signal into 2 in-phase channels which are applied to FET mixers. The LO is amplified and divided into two quadrature channels to feed the balanced mixer arrangement. Further quadrature phase shifting is achieved using active circuits at IF followed by IF signal addition and amplification. This receiver operates over the 2.5 to 3.5 GHz band and is designed to occupy a chip size approximately 8 mm by 8 mm.

Integrated transmitters are also being developed by a number of companies including Texas Instruments, Raytheon and Plessey. These circuits incorporate small signal and power amplifiers up to 1 watt level as well as switches and phase shifters. Table 10.5 gives an indication of the types of analogue circuits being developed together with an indication of the level of integration in 1980 and that perceived in 1985.

G. DIGITAL CIRCUITS

The levels of integration associated with analogue circuits is only at the SSI level, i.e. the number of FETs rarely exceeds 30 or so. Many of the early circuits fabricated to demonstrate the performance levels of various logic architectures contained a similar number of

TABLE 5. Analogue GaAs ICs

Circuits	1980	1985	Companies at present include
Receivers (L through J band)	8 GHz* 5 chips	20 GHz 2 chips	Plessey, Rockwell, LEP
Transmitters (1 watt or greater)	8 GHz* 4 chips	20 GHz 2 chips	Plessey, Raytheon, Texas Instruments
Phase shifters	-	20 GHz 1 chip	Plessey, Texas Instruments, Raytheon
Switches (1 watt or greater)	X-band* 1 chup	20 GHz 1 chip	Plessey, Texas Instruments, Raytheon
Very broadband amplifiers	Hybrid* 6 to 18 GHz 1 chip	0.5 to 18 GHz, 1 chip	Plessey, Watkins Johnson, Rockwell
Active filters	-	4 GHz 1 chip	Plessey
Oscillators	X,J band 1 chip*	VCO Jband 1 chip	Plessey, Rockwell, TI
A/D flash converters	3 bits 500 MSps	7 bits 500 MSps	TRW, Rockwell
Sample and hold	1 chip* 250 MSps	1 chip 1 GSps	Plessey, TRW
PLL Synthesisers	5 GHz 15 chips	10 GHz 15 chips	Plessey, Rockwell Hughes
Logarithmic amplifiers	-	1 GHz 1 chip	Plessey
Mixers	35 GHz* 1 chip	Lower noise figures	MIT, Honeywell

*Existing chip functions

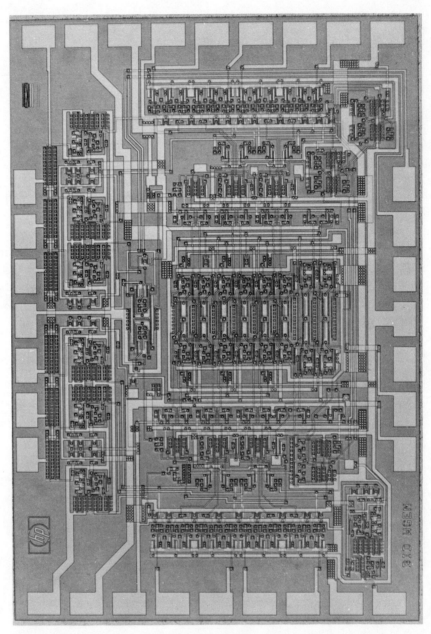

FIG. 10.67.   Photomicrograph of High Speed GaAs Word Generator
(courtesy of Hewlett Packard)

FETs. However, the growth in the complexity of digital circuits fabricated on GaAs has been quite remarkable (particularly since 1978) and the LSI level was reached in 1980 by Rockwell with a 8 by 8 multiplier using SDFL.

It is predicted that the 10,000 gate level will be reached by 1984 using similar logic architectures. The area of GaAs digital logic is a rapidly expanding one and this section deals with a few examples to illustrate the achievements to date.

Fig. 10.67 shows the microphotograph of a word generator fabricated by Hewlett Packard (Liechti, 1981). The circuit contains 400 transistors and 230 diodes contained on a chip size of 1.6 by 1.1 mm. Fig. 10.68 shows the basic IC construction in which a transistor, a diode and various interconnect lines are shown. The IC uses a dual implant of Si and Se into an LPE undoped buffer layer to produce the active and ohmic contact regions respectively. Silicon dioxide is used as the dielectric layer separating first and second level metallizations. The word generator is capable of generating bit streams at data rates of up to 4 GBit/second. Buffered depletion mode logic is used with typical flip-flop frequency divider output transition times of 80 psecs.

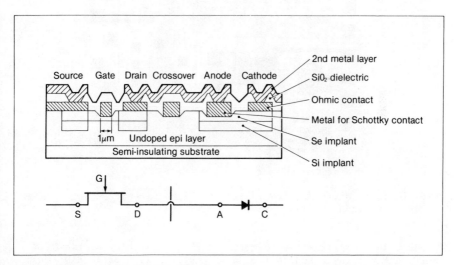

FIG. 10.68. Basic IC Fabrication Process used in Word Generator (courtesy Hewlett Packard)

Lee et al (1980) have reported the operation of an 8 by 8 multiplier using SDFL. This circuit has 1008 gates with over 6000 active components. The chip measures 2.25 by 2.7 mm and has a peak power dissipation of 2.08 watts. The multiply time is 5.3 nanoseconds using FETs of 1μm gate length. A similar circuit occupying a 1.125 by

1.35 mm chip size has provided 5 by 5 multiplication.

The fastest Si circuit has an 8 by 8 multiply time of 45 nsec.

Fig. 10.69 shows the microphotograph of ÷2 and ÷8 frequency dividers fabricated using EJFET technology (Zuleeg, 1980). The dark areas of the photograph show output buffers enabling signals to be taken off chip into 50 ohm terminations for measuring purposes. The divide-by-two circuit when operated at 500 MHz consumed 1.8 mW of d.c. power. The area occupied by one EJFET gate is $500\mu m^2$ and assuming that 75% of the chip area is used for interconnections results in a 1000 gate circuit (LSI) requiring a chip of 1.44 by 1.44 mm. Total power consumption of such a circuit is 0.2 watts. For VLSI (10,000 gates) such designs would require an area of 4.5 by 4.5 mm and would dissipate 2 watts. One of the areas for exploitation of such a technology is in static random access memories.

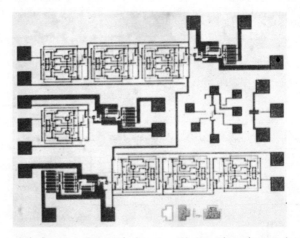

FIG. 10.69. Divide by Two, Divide by Eight Circuits using EJFETs (courtesy McDonnell-Douglas)

Table 10.6 outlines the performance characteristics of two GaAs EJFET RAM designs based on 1µm gate length FETs. The 1K RAM is designed for high speed whilst the 6K RAM is designed for lowest power dissipation. Such RAMs are at least 5 times faster than state-of-the-art Si circuits.

Fig. 10.70 shows the chip layout of high speed dividers utilizing BFL which operate to 4 GBit/sec (Wilcox, 1980). The FET gate lengths are 1µm and it is expected that 0.5µm gate length FETs will give operation up to 7 GBit/sec. The divider which uses the fast complimentary clock principle, is similar to that reported by Cathelin

TABLE 10.6.   Performance Characteristics of Two GaAs EJFET RAM
Designs (Gate Length = 1μm)

	1K static RAM	4K static RAM
Access time	910 pS	6 nS
Cycle Time	960 pS	12 nS
Memory stack power	125 mW	10 mW
Total power	600 mW	100 mW
Cell power	120 W	2 W
Cell size	70 x 54 m^2	32 x 61 m^2
Chip size	2.7 x 2.6 mm^2	3.5 x 3.5 mm^2
Organisation	256 x 4	4096 x 1

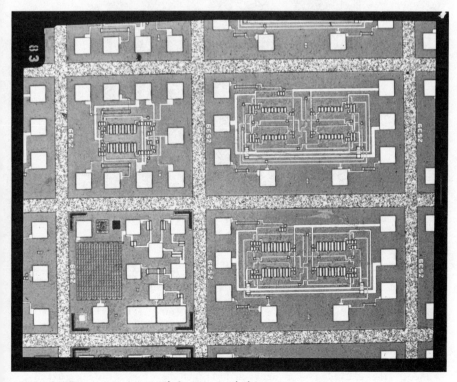

FIG. 10.70.   Array of High Speed Dividers

et al (19), this latter circuit having a maximum toggle frequency
of 5.5 GHz using 0.6μm gate length FETs. The divider is contained
together with the complimentary clock generator on a chip measuring
1 by 2.8 mm and dissipates 400 mW of which 240 mW is due to the
buffer stages used to drive the signals into 50 ohms for test purposes.

Enhancement mode, normally-off logic is also being actively researched by various European laboratories. STL (Mun et al, 1980) have
reported the use of both resistive and velocity saturation loads, the
latter enabling increased noise margins and output voltage swings to
be produced. Fig. 10.71 shows a microphotograph of an 11 stage ring
oscillator fabricated using mesa isolated 1.5μm gate length FETs.
The active layer is VPE grown GaAs of 0.12μm thickness which is ano-
dically etched from an initial 3μm thickness down to the required
thickness in the areas of the gates only. Typical standard deviation
in pinch-off voltage is 90 mV for wafers approximately 20 mm square.

FIG. 10.71.   Microphotograph of 11 Stage Ring Oscillator using
   Enhancement Mode FETs (courtesy STL)

In the SDFL approach (Eden et al, 1980) (Fig. 10.24) the logic
functions are supplied by 1 by 2μm Schottky diodes which have junction
capacitances of 2 femto Farads and series resistances as low as 300
ohm. These diodes require a deep (0.5μm) low sheet-resistance implant
for their fabrication whereas the low power high transconductance
MESFETs require a shallow (0.1μm) higher sheet-resistance implant.
Extra fan-in is provided by additional logic diodes with little
degradation in speed because of the low reverse-bias diode capacitance. These diodes also provide half the level shifting required
between the drain and gates in depletion mode logic. The pull down

active load provides the bias current for the logic diodes and most of the current to turn off the gate of the inverter. Fan-out of the basic NOR gate is limited to 3 or 4 by the ratio of the pull-down and pull-up currents. Fan-out is extended by the use of source followers as buffers. Ring oscillators consisting of chains of odd numbers of logic gates are used to evaluate propagation delay (τ_d) and power-delay product ($P_D\tau_d$). 5μm NOR gates have yielded power-delay products as low as 27 fJ/gate with 156 pS propagation delay, while 20μm NOR gates have provided 75 pS propagation delay and 170 fJ/gate power-delay product.

Three stage D flip-flop dividers (÷8) containing 25 NOR gates have been fabricated in SDFL and operated up to 1.9 GHz. Power dissipation ranged from 45 to 145 mW for these dividers.

An MSI combinational circuit consisting of a 3 x 3 bit parallel multiplier with a total gate count of 75 is useful as an example of what can be achieved using SDFL. Very high speed multipliers are an important part of many signal processing and computer systems because the data rate is often determined by the time to perform a complex multiplication. In a N x N parallel or array multiplier, N(N-2) full adders and N half adders are used to sum the partial product bits in (N-1) sum delays plus (N-1) carry delays. Thus in a large parallel multiplier the speed will be determined by the full adder cell propagation delay.

By using a NOR implemented full adder as shown in Fig. 10.72 a $3\tau_d$ delay is incurred for the sum and a $2\tau_d$ delay for the carry with 12 gates being required. The full adder (F) and half adder (H) cells are interconnected as in Fig. 10.73. For a 3 bit multiplier, 3 half and 3 full adders are needed. Consider an input code (111 x 10S = $S\overline{SSS}$ SS) where a square wave input S is applied to the A_0 input. (Fig. 10.74). Upon this A_0 pulse input the product outputs P_2, P_3 and P_4 will go to $\overline{S}$ and the other outputs to S. The total power dissipation for such a 3 x 3 multiplier has been measured at 31.5 mW for V_{DD} = 1.77 volt and V_{SS} = -0.94 volt corresponding to a power dissipation of 420μW per gate. The highest speed devices have yielded 3 x 3 multiplication times of 1.5 nS.

Enhancement mode logic has been demonstrated by Mizutani et al (1980) to have a considerably better performance at liquid nitrogen temperatures. A gate propagation delay of 51 psec with a power dissipation of 1.9 mW has been achieved with a 20μm gatewidth ring oscillator. The associated power delay product was 97 fJ. This is compared to a propagation delay some 50 to 60 percent greater at room temperature for the same power-delay product.

6. CONCLUSIONS

It has been seen in this chapter that the GaAs field effect transistor is playing a most important role in the development of integrated circuits. Analogue microwave circuits fabricated on

430

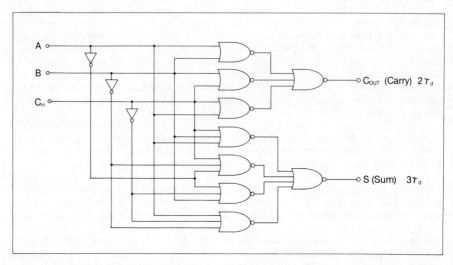

FIG. 10.72. Full Adder Cell for Array Multiplier as Implemented with 12 SDFL NOR Gates. Carry delay is $2\tau_d$ and Sum Delay is $3\tau_d$ where τ_d is the Basic NOR Gate Delay

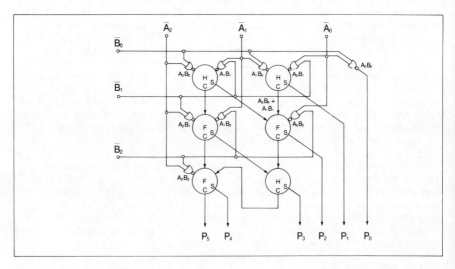

FIG. 10.73. Logic Diagram of 3 x 3-bit Parallel Multiplier. Full Adder Cells (F) are as shown in Fig. 10.72. Half-adder Cells (H) are Implemented with 5 SDFL NOR Gates. Input Inverters, Output Buffers, and on-chip Feedback Connections are not shown.

gallium arsenide have a performance which is unattainable using silicon. Digital logic circuits on GaAs have a five times speed advantage over silicon at the present time but the massive investment in the very high speed IC (VHSIC) programme in the USA will undoubtedly narrow the gap. However, there are many promising logic architectures based on GaAs MESFETs, JFETs and MOSFETs which will progress rapidly as the technology matures. The bibliography below is intended as a guide to further reading and papers not referred to in the text are also included. The complexity of both analogue and digital GaAs ICs is increasing rapidly. Markets for GaAs ICs are becoming well defined and a $1500 million market in the USA alone is anticipated by 1990, which assuming modest requirements in the early 1980's infers a growth in the market in excess of 100% per annum over the next decade.

7. BIBLIOGRAPHY

Aitchison, C.S., Davies, R., Higgins, I.D., Longley, S.R., Newton, B.H., Wells, J.F. and Williams, J.C. Lumped microwave circuits. Design Electronics, October 1971, pp.30-39.

Allen, R.P.G. and Antell, G.R. Monolithic mixers for 60-80 GHz. Proc. 1973 European Microwave Conference, Paper A153.

Alley, G.D. Interdigital capacitors and their application to lumped element microwave integrated circuits. IEEE Trans. Microwave Theory Tech., Vol. MTT-18, pp.102--1033, Dec. 1970.

Antell, G.R. Monolithic mixers for millimeter wavelengths. Proc. 1971 European Microwave Conference, p.C2/3:1.

Araki, T. A high cut-off frequency planar Schottky diode with a stripe geometry junction. IEEE Trans. on Electron Devices, Vol. ED-25, 1978, p.1091.

Ayashi, Y., Pucel, R.A., Fabian, W. and Vorhaus, J.L. A monolithic X-band single pole, double throw bi-directional GaAs FET switch. IEEE GaAs IC Symposium Research Abstracts, Las Vegas, USA, Nov. 1980, paper 21.

Bollinger, D. and Fink, R. A new production technique: Ion milling. Solid State Technology, November 1980, pp.79-84.

Brambley, D.R. and Vanner, K.C. IPAT 1979 Proceedings, LEP Consultants, Ediburgh 1979, pp.47-54.

Braun, L. and Lood, D. Proc. IEEE, Vol. 54, 1966, pp.1521-1527.

Cathelin, M., Durand, G., Garant, M. and Rocchi. 5 GHz binary frequency division on GaAs. Electronics Letters, Vol. 16, No. 14, pp.535-536.

Ch'en, D.R. and Eisen, F.H. Ion implanted high frequency high speed GaAs integrated circuits technology. J. Vac. Sci. Technol, Vol. 16 No. 6, Nov/Dec. 1979, pp.2054-2062.

Chu, A., Courtney, W.E. and Sudbury, R.W. A 31 GHz monolithic GaAs mixer/preamplifier circuit for receiver applications. IEEE Trans. on Electron Devices, Vol. ED-28, No. 2, Feb. 1981, pp.149-154.

Commizoli, R.B. RCA Review, Vol. 37, December 1976.

Courtney, W.E. Complex permittivity of GaAs and CdTe at microwave frequencies. IEEE Trans. Microwave Theory Tech., Vol. MTT-25, pp.697-701, Aug. 1977.

D'Asaro, L.A., Di Lorenzo, J.V. and Fukui, H. Improved performance of GaAs microwave field effect transistors with via connections through the substrate. Int. Electron Devices Meeting, Digest Tech. Papers, pp.370-371, 1977.

Dill, H.G. Designing inductors for thin film applications. Electronic Design, February 1964, pp.52-59.

Driver, M.C., Wang, S.K., Przybysz, J.X. et al. Monolithic Microwave amplifiers formed by ion implantation into LEC gallium arsenide substrates. IEE Trans. on Electron Devices, Vol. ED-28, No. 2, Feb. 1981, pp.191-196.

Eden, R.C., Welch, B.M. and Zucca, R. Low power GaAs digital ICs using Schottky diode FET logic, 1978, Int. Solid State Circuits Conf. Digest of Tech. Papers, Feb. 1977, pp.68-69.

Eden, R.C., Welch, B.K., Zucca, R. and Long, S.I. The prospects for ultra high-speed VLSI GaAs digital logic. IEEE Journal of Solid State Circuits, Vol. SC-14, No. 2, April 1979, pp.221-239.

Eden, R.C., Lee, F.S., Long, S.I., Welch, B.M. and Zucca, R. Multi-level logic gate implementation in GaAs ICs using Schottky diode - FET logic. 1980 Int. Solid State Circuits Conf. Digest of Technical Papers, Feb. 1980, pp.122-123.

Gerstenberg, D. and Calbick, C.J., J. Appl. Phys. Vol. 35, 1964, p.402.

Getsinger, W.J. Microstrip dispersion model. IEEE Trans. Microwave Theory Tech. Vol. MTT-21, No. 1, pp.34-39, Jan. 1973.

Glang, R. and Gregor, L. Handbook of Thin Film Technology, McGraw-Hill, New York, 1970, Section 7-36.

Greenhalgh, S.G., Rees, G. and Pengelly, R.S. A novel MESFET mixer using Monolithic GaAs circuit techniques. To be published.

Greenhouse, H.M. Design of planar rectangular microelectronic inductors. IEEE Trans. on Parts, Hybrids and Packaging, Vol. PHP-10, No. 2, June 1974, pp.101-109.

Grover, F.W. Inductance Calculations. Van Nostrand, Princeton, N.J. 1946, Reprinted by Dover Publications, 1962.

Harrop, P., Lesartre, P. and Collet, A. GaAs integrated all FET front-end at 12 GHz. IEEE GaAs IC Symposium Research Abstracts, Las Vegas, USA, Nov. 1980, Paper 28.

Hobdell, J.L. Optimization of interdigital capacitors. IEEE Trans. Microwave Theory Tech., Vol. MTT-27, No. 9, Sept. 1979, pp.788-791.

Hornbuckle, D.P. and Van Tuyl, R.L. Monolithic GaAs direct-coupled amplifiers. IEEE Trans. on Electron Devices, Vol. ED-28, No. 2, February 1981, pp.175-182.

Hughes, R.S. Logarithmic video amplifiers. Artech House Inc.

Immorlica, A.A. and Eisen, F.H. Planar passivated GaAs hyperabrupt varactor diodes. Proc. Sixth Biennial Cornell Electrical Eng. Conf. Aug. 1977, pp.151-159.

Immorlica, A.A. and Wood, E.J. A novel technology for fabrication of beam-leaded GaAs Schottky barrier mixer diodes. IEEE Trans. on Electron Devices, Vol. ED-25, 1978, p.710.

Joshi, J.S., Cockrill, J.R. and Turner, J.A. Monolithic microwave gallium arsenide FET oscillators. IEEE Trans. on Electron Devices, Vol. ED-28, No. 2, Feb. 1981, pp.158-162.

Lawes, R.A. Electron beam lithography at the Rutherford Laboratory. Proceedings of the Technical Programme, Internepcon Semiconductor Symposium, Brighton, England. October 1979, pp.278-281.

Lee, F.S., Shen, E., Kaelin, G.R., Welch, B.M., Eden, R.C. and Long, S.I. High Speed LSI GaAs digital integrated circuits. 1980 GaAs IC Symposium Research Abstracts, Las Vegas, USA, November 1980, paper no. 3.

Liechti, C.A. GaAs FET logic. 1976 Int. GaAs Symposium Institute of Physics Conf. Series, Ser. No. 33a, 1977, Ch.5, pp.227-236.

Liechti, C.A. - private communication, 1981.

Lim, Y.C. and Moore, R.A. Properties of alternately charged coplanar strips by conformal mappings. IEEE Trans. on Electron Devices, Vol. ED-15, No. 3, March 1968.

Maissel, L.I. and Glang, R. Handbook of thin film technology. McGraw-Hill, New York, 1970, Section 18-18.

McLevige, W.V. and Sokolov, V. Resonated GaAs FET devices for microwave switching. IEEE Trans. on Electron Devices, Vol. ED-28, No. 2, Feb. 1981, pp.198-204.

Mehal, E.W. and Wacker, R.W. GaAs integrated microwave circuits. IEEE Trans. on Electron Devices, Vol. ED-15, 1968, p.513.

Mizutani, T., Kato, N., Ida, M. and Ohmori, M. High speed enhancement mode GaAs MESFET logic. IEEE Trans. on Microwave Theory Tech. Vol. MTT-28, No. 5, May 1980, p.479-483.

Mun, J. and Phillips, J.A. Improved circuit for normally-off GaAs FET logic. Colloquium on High Speed Logic, IEE, Savoy Place, London, October 1980, Digest No. 1980/44, pp.6/1-6/4.

Murphy, R.A. and Clifton, B.J. Surface oriented Schottky barrier diodes for millimeter and submillimeter wave applications. Proc. 1978 International Electron Devices Meeting, Washington, 1978, pp.124-128.

Niclas, K.B., Wilser, W.T., Gold, R.B. and Hitchens, W.R. A 350 MHz to 14 GHz GaAs MESFET amplifier using feedback. Digest of Technical Papers of 1980 ISSCC, San Francisco, 1980, pp.164-16 .

Nuzillat, G., Damay-Kovala, F., Bert, G. and Arnodo, C. Low pinch-off voltage FET logic; LSI oriented logic approach using quasi-normally off GaAs MESFETs. IEE Proc. Vol. 127, Pt. I, No. 5, October 1980, pp.287-295.

Ostrander, W. and Lewis, C., Trans. 8th Symp. Am. Vac. Soc., Vol. 12, Pergamon Press, Oxford, 1962, pp.881-888.

Pengelly, R.S. and Rickard, D.C. Design, measurement and application of lumped elements up to J-band. 7th European Microwave Conference Proceedings, Copenhagen, Denmark, pp.460-464, 1977.

Pengelly, R.S., Suffolk, J.R., Cockrill, J.R. and Turner, J.A. A comparison between actively and passively matched S-band GaAs monolithic FET amplifiers. To be presented at the 1981 International MMT-S Microwave Symposium, Los Angeles, USA, June 1981.

Pengelly, R.S., Suckling, C.W. and Cockrill, J.R. The application of feedback techniques to the realisation of broadband monolithic power amplifiers. To be published.

Pengelly, R.S. and Turner, J.A. Monolithic broadband GaAs FET amplifiers. Electronic Letters, Vol. 12, No. 10, pp.251-252, May 1976.

Pitt, K.E.G., Thin Solid Films, Vol. 1, 1967, pp.173-182.

Pucel, R.A., Masse, D.J. and Hartwig, C.P. Losses in microstrip.
IEEE Trans. Microwave Theory Tech., Vol. MTT-16, No. 6, pp.342-350,
June 1968.

Rickard, D.C. Private communication, 1978.

Sato, Y., Uchida, M., Ishibashi, Y. and Araki, T. Chip-type planar
Schottky barrier diodes fabricated from selectively grown GaAs.
Rev. Electro. Communication Lab. (Tokyo), Vol. 23, 1975, p.535.

Seely, S. Electron Tube circuits. McGraw-Hill, New York, 1958,
pp.519-520.

Shaw, D.W. Selective epitaxial deposition of gallium arsenide in
holes. J. Electrochem. Soc., Vol. 113, 1966, p.904.

Smith, J.L. The odd and even mode capacitance parameters for
coupled lines in suspended substrate. IEEE Trans. Microwave Theory
Tech., Vol. MTT-19, pp.424-431, May 1971.

Sokolov, V. and Williams, R.E. Development of GaAs monolithic power
amplifiers in X band. IEEE Trans. on Electron Devices, Vol. ED-27,
No. 6, June 1980, pp.1164-1171.

Stubbs, M.G. - private communication, 1980.

Suckling, C.W., Stubbs, M.G., Pengelly, R.S. and Cockrill, J.R.
Broadband S-band SPDT MESFET switches using monolithic lumped
element circuits. To be published.

Suffolk, J.R. - private communication, 1980.

Suffolk, J.R., Cockrill, J.R., Pengelly, R.S. and Turner, J.A.
An S-band image rejection receiver using monolithic GaAs circuits.
IEEE GaAs IC Symposium Research Abstracts, Las Vegas, USA,
Nov. 1980, Paper 27.

Terman, F.E. Radio Engineering Handbook, McGraw-Hill, New York,
1943, pp.48-60.

Tolliver, D.L. Plasma processing in microelectronics past, present
and future. Solid State Technology, November 1980, pp.99-105.

Tserng, H.Q., Macksey, H.M. and Nelson, S.R. Design, fabrication,
and characterization of monolithic microwave GaAs power FET
amplifiers.   IEEE Transactions on Electron Devices, Vol. ED-28,
No. 2, Feb. 1981, pp.183-190.

Tserng, H.Q. and Macksey, H.M. Performance of monolithic GaAs FET
oscillators at J-band. IEEE Trans. on Electron Devices, Vol.
ED-28, No. 2, Feb. 1981, pp.163-165.

Van Tuyl, R.L.   A monolithic GaAs IC for heterodyne generation of
  RF signals.   IEEE Trans. on Electron Devices, Vol. ED-28, No. 2,
  Feb. 1981, pp.166-170.

Van Tuyl, R.L., Liechti, C.A., Lee, R.E. and Gowen, E.   GaAs MESFET
  logic with 4 GHz clock rate.   IEEE J. Solid-State Circuits,
  Vol. SC-12, pp.485-496, Oct. 1977.

Vendelin, G.D.   A Ku-band integrated receiver front end.   IEEE J.
  Solid State Circuits, Vol. SC-3, No. 3, pp.255-257, Sept. 1968.

Vorhaus, J.L., Fabian, W., Ng, P.B. and Tajima, Y.   Dual-gate GaAs
  FET switches.   IEEE Trans. on Electron Devices, Vol. ED-28, No. 2,
  Feb. 1981, pp.204-211.

Vorhaus, J.L., Pucel, R.A., Tajima, Y. and Fabian, W.   A two-stage
  all monolithic X-band power amplifier.   1981 IEE Solid State
  Circuits Conf. Digest of Tech. Papers, pp.74-75.

Wilcox, D.J. - private communication, 1980.

Wilcox, D.J. - private communication, 1980.

Yokoyama, N.   GaAs MOSFET high speed logic.   IEEE Trans. on Microwave
  Theory Tech. MTT-28, No. 5, pp.483-486, May 1980.

Yokoyama, N., Mimura, T., Fukuta, M. and Ishikawa, H.   A self-aligned
  source/drain planar device for ultra high-speed GaAs MESFET VLSIs.
  1981 Int. Solid State Circuits Conf. Digest of Technical papers,
  Feb. 1981, pp.218-219.

Zucca, R., Welch, B.M., Asbeck, P.M., Eden, R.C. and Long, S.I.
  Semi-insulating III-V materials, Nottingham 1980, Shiva Publishing
  Ltd. pp.335-345.

Zuleeg, R., Notthoff, J.K. and Lehovec, K.   Femtojoule high speed
  planar GaAs EJFET logic.   IEEE Trans. on Electron Devices, Vol.
  ED-25, June 1978, pp.628-639.

Zuleeg, R., Notthoff, J.K. and Troeger, G.L.   GaAs junction FET LSI.
  Eascon 1980 Record, pp.146-150.

Other Papers of Interest

Belohoubek, E.F.   Advanced microwave circuits.   IEEE Spectrum,
  Feb. 1981, pp.44-47.

Bosch, B.G.   Device and circuit trends in gigabit logic.   IEEE Proc.
  Vol. 127, Pt.I, No. 5, October 1980, pp.254-265.

Curtice, W.R.   A MESFET model for use in the design of GaAs integrated
  circuits.   IEEE Trans. Microwave Tehory Tech., Vol. MTT-28, No. 5,
  May 1980, p.448-456.

Daly, D.A., Knight, S.P., Caulton, M. and Ekholdt, R. Lumped elements in microwave integrated circuits. IEEE Trans. Microwave Theory Tech. Vol. MTT-15, No. 12, Dec. 1967, pp.713-721.

Dobratz, B.E., Ho, N., Krumm, C.F. and Greiling, P.T. Gallium arsenide FET logic pseudorandom code generator. IEEE Trans. Microwave Theory Tech. Vol. MTT-28, No. 5, May 1980, pp.486-490.

Livingstone, A.W. and Mellor, P.J.T. Capacitor coupling of GaAs depletion mode FETs. IEE Proc. Vol. 127, Pt.I, No. 5, October 1980, pp.297-300.

Phillips, D.H. GaAs integrated circuits for military/space applications. Military Electronics/Countermeasures, March 1979, pp.24-30.

Taylor, D.M., Wilson, D.O. and Philips, D.H. Gallium arsenide review: past, present and future. IEE Proc. Vol. 127, Pt.I, No. 5, October 1980, pp.266-269.

Upadhyayula, L.C. GaAs MESFET comparators for gigabit-rate analog-to-digital converters. RCA Review, Vol. 41, No. 2, June 1980, pp.198-211.

Weidlich, H.P., Archer, J., Pettenpaul, E. et al. A GaAs monolithic broadband amplifier. IEEE Solid State circuits Conf. Digest Tech. Papers 1981, pp.192-193.

Yoder, M.N. Blazing speed monolithic integrated circuits. J. Vac. Sci. Technol. Vol. 16, No. 6, Nov/December 1979, pp.2041-2045.

CHAPTER 11
New Materials and New Structures

1. INTRODUCTION

The last twenty years has seen GaAs material being used increasingly
for many microwave devices including the field effect transistor.
There are, however, many other compound semiconductors which should
give considerable gain, noise figure and higher frequency advantages
over GaAs. These materials include indium phosphide (InP) and the
ternary and quaternary compounds such as $Ga_xIn_{1-x}As$ and $Ga_xIn_xAs_ySb_{1-y}$.
Of the ternary compounds InPAs appears to be most likely to give
substantial frequency of operation improvement and GaInAs to give
substantial gain and noise figure improvement, particularly at the
higher frequencies. However, in order to reach operating frequencies
approaching 50 GHz or so narrow gate length devices are still needed
and the technology and ingenuity of the device designer then limits
the device performance.

In order to substantially increase the maximum frequency of oscillation it is necessary to develop new transistor structures which,
although having the advantages of the FET (that is being majority
carrier devices), also have some of the advantages of the bipolar
transistor such as very high transconductance. One of the most promising structures receiving much attention at the present time is the
permeable base transistor which may ultimately have a maximum frequency of oscillation greater than 1000 GHz compared with typical
values of only 80 GHz for a conventional 0.5μm gate length GaAs FET.

2. THE InP MESFET

Indium phosphide (InP) has electronic properties very similar to
those of GaAs and can therefore be used, in principle, to produce
MESFETs. InP has some fundamental differences which suggest that the
material will produce FETs with superior performance over GaAs
devices. Table 11.1 lists the major properties of Si, GaAs and InP.
The primary performance advantage results from the higher peak
electron drift velocity which is the parameter of major importance
for short gate length FET's. Maloney and Frey (1976) have shown that
the frequency for unity gain, f_T, for InP FETs at room temperature

should be some 48% higher than GaAs devices having gate lengths of 1µm. Fig. 11.1 shows the theoretical f_T versus gate length, L_g curves for FETs using GaAs or InP having gate lengths varying from 0.5 to over 5 microns. The f_T advantage has, in fact, been observed experimentally (Barrera et al, 1975) but the advantages that were theoretically predicted initially for gain and noise figure have not been seen due to non-optimum material characteristics, buffer layer and substrate quality problems and technological problems associated with the low barrier characteristics of Schottky barrier gates to InP.

Gallium arsenide FET fabrication is highly developed, indeed as has been seen in the last chapter, many laboratories worldwide are currently working on sophisticated analogue and digital GaAs integrated circuits. Indium phosphide technology has been unable to benefit properly from this extensive GaAs work because of the low Schottky barrier height of metal/indium phosphide structures. Not only does the gate electrode of the device tend to have low breakdown voltage and considerable current leakage but non-destructive characterization of FET layers using capacitance voltage (C(V)) data has not been possible. The use of novel MOS techniques has been necessary (Clarke et al, 1979) to accurately profile the epitaxial layers grown for InP FETs.

Barrera and Archer (1975) have reported extensive work on InP MESFETs, the results of which indicated many of the intrinsic properties of this device. Relatively large drain current can flow when the device is 'pinched off' typically 0.5 mA for a 1µm gate length, 500µm gate width device. This is due to the reverse current that flows between the gate and the drain due to thermonic field emission at the perimeter of the gate. Such currents are some 1000 times greater than in the GaAs FET not only because of the lower Schottky barrier height of the gate for InP but also due to the surface properties of the InP.

D.C. transconductance is higher for InP FETs than for GaAs FETs and increases with the doping of the epitaxial layer. Typical values are 200 mS per mm gate width for $N_D = 1.4 \times 10^{17}$ cm^{-3} and 100 mS per mm for $N_D = 6 \times 10^{16}$ cm^{-3}. GaAs FETs have a typical value of 100 mS/mm for $N_D = 1 \times 10^{17}$ cm^{-3}.

Table 11.2 indicates the results of work conducted at Hewlett Packard in 1975 for 1µm gate length FETs. Even though the current-gain cut-off frequency, f_T is some 50% greater in the case of the InP device the combined influence of the much larger drain-to-gate feedback capacitance, C_{dg}, the smaller drain to source output resistance, R_{ds}, and the somewhat larger input resistance, $R_i + R_g + R_s$, in the InP case, give rise to lower available power gain and hence lower unilateral power gain cut-off frequency, f_{max} since

$$f_{max} = \frac{f_T}{2\left\{ \sqrt{(R_i + R_g + R_s)/R_{ds}} + 2\pi f_T R_g C_{dg} \right\}}$$

TABLE 11.1.   Comparison of Semiconductor Material Properties

Property	Units	Material		
		Silicon	Gallium Arsenide	Indium Phosphide
Low field mobility (450°K)	cm^2 V^{-1} s^{-1}	400	5000	3000
Peak (threshold) velocity (450°K)	cm s^{-1}	–	1.2×10^7	1.9×10^7
'Peak-to-valley' ratio (450°K)	–	–	2.2 – 2.4	3.0 – 3.1
Saturated drift velocity (450°K)	cm s^{-1}	8.5×10^6	5×10^6	6×10^6
Thermal conductivity (300°K)	W cm^{-1} $^{\circ}C^{-1}$	1.45	0.44	0.68
Energy gap (300°K)	eV	1.11	1.43	1.34

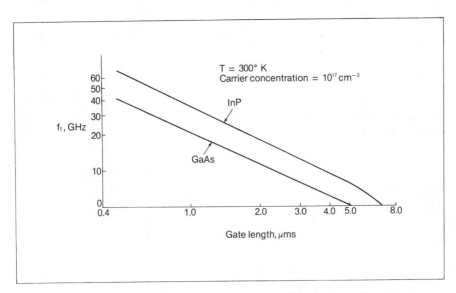

FIG. 11.1.   Current Gain Cut-off Frequency Versus Gate Length from a One-Dimensional Monte Carlo Analysis for GaAs and InP

TABLE 11.2.   Comparison of InP and GaAs MESFETs Fabricated at HP
Laboratories in 1975

Parameter	InP		GaAs	
Material	6×10^{16} cm^{-3}	10^{17} cm^{-3}	6×10^{16} cm^{-3}	10^{17} cm^{-3}
Current gain cut-off frequency, f_T (GHz)	20	20–24	11	13
Mason's unilateral power gain cut-off frequency, f_{max} (GHz)	33	32	40	40
Noise figure for MAG at 10 GHz dB	6	9.5	7.5	6.2
Minimum noise figure at 10 GHz dB	4.8	6.0	3.5	3.2
Associated gain at 10 GHz, dB	5.8	5.5	6.6	7.8
Transconductance g_{mo}, mS	50	84	33	53

TABLE 11.3.   Magnitudes of Equivalent Circuit Parameters for InP and
GaAs MESFETs with Similar Geometries

Parameter	InP	GaAs
Carrier Concentration	6×10^{16} cm^{-3}	6×10^{16} cm^{-3}
C_{gs}, pF	0.36	0.5
R_{gs}, Ω	10000	∞
R_i, Ω	6.0	3.5
C_{dg}, pF	0.056	0.01
R_{dg}, Ω	8500	∞
g_{mo}, mS	50	33
R_{ds}, Ω	260	660
R_g, Ω	5.5	3.0
L_g, nH	0.1	0
R_s, Ω	3.5	5.5
L_s, nH	0.04	0
R_d, Ω	5.0	7.0
C_{ds}, pF	0.06	0.06

Table 11.3 compares the equivalent circuit element values for GaAs and InP devices of the same geometry where the component designations are given in Fig. 11.2.

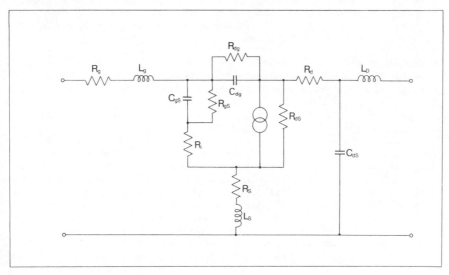

FIG. 11.2.   RF Equivalent Circuit for the InP MESFET

The largest contribution to the decrease in the power gain of the InP FET is due to the large value of capacitance, C_{dg} which increases degenerative feedback. The roughly five times higher feedback capacitance C_{dg} in current-saturated InP MESFETs as compared to GaAs ones is mainly due to the formation of weaker Gunn domains between the gate and drain at the drain bias levels needed for drain current saturation. This results in the depletion layer fringing capacitance not being so effectively decoupled at the drain.

The second most serious influence on power gain is that of the drain to source resistance, R_{ds} for InP which is half the GaAs value.

The smaller drain-to-source resistance, R_{ds}, in the InP device is correlated with the generally higher drain currents found in such devices. This is caused by the resistivity of the InP substrate being lower by a factor of up to 10^4 for a given Cr doping than for GaAs. It has been concluded that no electron trapping occurs at the epitaxial to substrate interface in InP FETs (Engelmann, et al, 1977).

As has been seen in Chapter 2 the formation of a Gunn domain at the drain side of the channel leads to the average carrier velocity in the channel decreasing. In the range of drain bias voltages employed for MESFETs (usually 4 to 5 volts) the effect is much stronger in GaAs than in InP and hence the InP FET appears to have a higher f_T

than GaAs which simple theories predict (Barrera et al, 1975). The difference in the Gunn domain formation mentioned above is related to the fundamental difference in the carrier velocity versus electric field characteristic of GaAs and InP. To achieve similar Gunn domain effects in InP as in GaAs one would have to increase the drain bias levels by a factor of at least three over those for GaAs. Such action is not, at present, possible because of the breakdown voltage limits of InP Schottky gate junctions.

Englemann and Liechti (1977) predicted that the use of Fe rather than Cr doped substrates might effect the process of Gunn domain formation and hence improve InP FET performance.

Gleason et al (1978) have produced ion-implanted InP FETs using Fe doped substrates having resistivities greater than 10^7 Ω cm grown by the liquid encapsulated Czochralski method. In contrast to the results mentioned earlier the gates of the implanted devices showed adequate breakdown voltages and low leakage currents. Typical d.c. leakage currents were 5 to 10 nA at a V_{DS} of 3V and with corresponding gate to drain breakdown values of 10 to 15 volts at I_{gs} equal to 1µA. Many of the breakdown and leakage current problems of early devices are thought to have been due to surface oxides forming prior to gate metallization. However, the r.f. results of the ion implanted devices were disappointing with 9.8 dB noise figure and 13.7 dB associated gain being achieved at 8 GHz.

The use of a plasma oxidation process prior to gate metal deposition giving reproducible low leakage gates has led to InP MESFETs being produced having the Plessey GAT4 geometry with a minimum noise figure of 3.8 dB and 7 dB associated gain at 12.75 GHz. This compares with a minimum noise figure of 2.6 dB with 6 dB associated gain for an ion implanted GaAs at the same frequency.

It has been concluded by many workers that the InP MESFET has no advantage over the GaAs device and activity is now being concentrated on the technology necessary to produce FET structures using the superior properties of certain of the ternary and quaternary alloys.

3. THE InP MISFET

InP has, however, not been completely forgotten as it is now forming the basis of much intense work on MISFETs (metal insulator semiconductor field effect transistor) specifically related to logic integrated circuit applications.

Unlike the situation with GaAs, it has been demonstrated that a layer of mobile electrons can be induced at the surface of both p-type (Messick et al, 1978; Meiners et al, 1979a) and semi-insulating InP (Meiners et al, 1979b) by the application of a positive potential to a metal gate electrode isolated from the semiconductor by a thin insulating film much as shown in Fig. 11.3. Normally-off enhancement mode metal-insulator field effect transistors on both p-type (Lile et al, 1979) and semi-insulating InP substrates have been fabricated.

The devices made on SI InP have shown power gain at microwave frequencies (Meiners et al, 1979b).

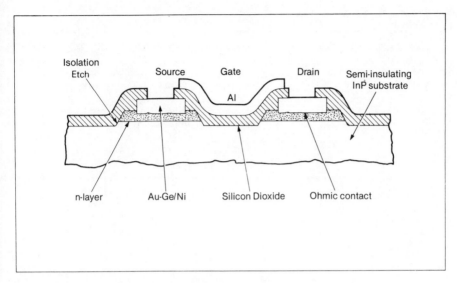

FIG. 11.3.   InP MISFET Structure

GaAs MISFETs or MOSFETs (Becker et al, 1965; Sugano et al, 1978) although exhibiting excellent microwave performance are limited in their operation at very low frequencies due to the high density of states at the GaAs to insulator interface, resulting in fixed charge densities. InP MISFETs exhibit much lower charge densities allowing the surface potential to be changed by over 1 volt compared to 0.4V for GaAs (Messick, 1979; Messick, 1976). MIS technology also offers the attractions of high dynamic range since the positive gate bias which is required to produce a surface layer of mobile electrons in a normally-off MESFET, is limited only by the dielectric breakdown voltage of the insulator in the MISFET. Thus InP MISFET technology apart from being a promising one for analogue microwave applications is also most useful for an integrated logic approach similar to NMOS on Si.

The procedures employed for the formation of the gate insulators (Hasegawa et al, 1975; Zeisse et al, 1977) as well as some of the properties of the insulator dielectrics used are outlined in Tables 11.4 and 11.5. It may be seen, for example, that the pyrolytic insulators produced using chemical vapour deposition have a higher breakdown voltage than those dielectrics produced using anodization of the GaAs.

The results of tuned power gain measurements for both GaAs and InP MISFETs have been reported by Messick, 1979) as a function of frequency. The gains of the MISFETs which had gate lengths of 4μm and gate

TABLE 11.4. Gate Insulation Processes

Type of FET	Gate Insulator Process
GaAs Schottky gate	No insulator present
GaAs/Anodic oxide insulated gate	Wet chemical anodization achieved at room temperature
GaAs/pyrolytic $Si_xO_yN_z$ insulated gate	Chemical vapour deposition at 600°C using N_2, NH_3 and SiH_4
InP/pyrolytic SiO_2 insulated gate	Chemical vapour deposition at 310°C using N_2, O_2 and SiH_4

TABLE 11.5. Gate Insulator Properties

Material	Properties
Anodic insulator on GaAs	0.1µm thickness, growth time: 2 to 4 mins; 20-40V breakdown voltage; leakage current at 10V is 10^{-12} to 10^{-10} A/mm^2. Surface state density = 10^{12} cm^2 eV^{-1}.
Pyrolytic $Si_xO_yN_z$ on GaAs	0.1µm thickness, growth time: 180-300 mins; 50-100V breakdown voltage; leakage current at 10V is 10^{-12} A/mm^2. Surface state density = 10^{-12} cm^2 eV^{-1}.
Pyrolytic SiO_2 on InP	0.12µm thickness, growth time: 20-80 mins; 50-100V breakdown voltage; leakage current at 10V is 10^{-12} to 10^{-11} A/mm^2; surface state density = 10^{11} cm^2 eV^{-1}.

widths of 260μm were very similar to those of commercially available
4μm gate length MESFETs. Fig. 11.4 shows the gain results where the
bias conditions for the GaAs FETs were individually optimized for
maximum power gain.

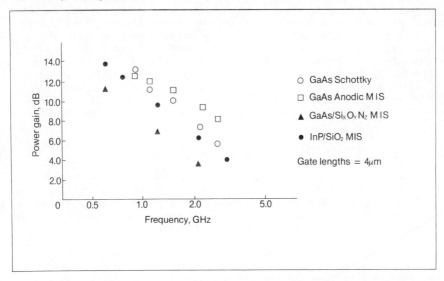

FIG. 11.4.   Power Gain Versus Frequency for MIS FETs (after Messick)

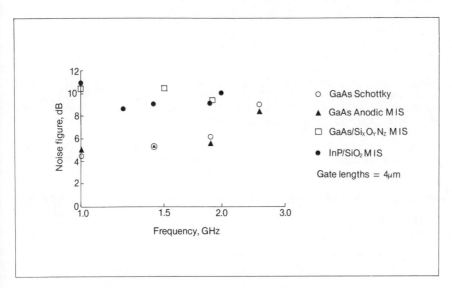

FIG. 11.5.   Noise Figure Versus Frequency for MISFETs (after Messick)

Noise figure results for the GaAs and InP MISFETs are compared with the commercially available 4μm gate length GaAs MESFET in Fig. 11.5, where it may be seen that the MESFET has a 2 dB lower noise figure, typically. The reason for the degradation in MISFET noise figure is probably due to a combination of ohmic contact value, epitaxial layer quality, gate insulator quality and the different gate biasing associated with each device. Certainly at microwave frequencies the surface states associated with the insulator are unable to follow the impressed gate signal. In this respect the InP MISFET appears to be preferable for applications requiring low as well as high frequency operation since this device has a lower surface state density and in devices fabricated to date using silicon dioxide (SiO_2) as the insulator full drain modulation by gate action down to 100 Hz has been demonstrated.

4. TERNARY AND QUATERNARY COMPOUNDS FOR MESFETs

As we have seen in this monograph the amount of effort afforded to GaAs FET device technology and circuit design has been considerable. This effort has shown that in order to justify the development costs involved any new material which is considered as a contender to GaAs must have some strong performance advantages.

Table 11.6 shows some of the important material parameters of a variety of ternary and quaternary materials which have considerable advantages over GaAs in terms of peak electron velocity. All the materials with the exception of InP have a higher mobility than GaAs leading to higher cut-off frequencies and shorter switching times since it may be shown that the cut-off frequency is given by

$$f_T = \frac{1}{2\pi} \frac{g_m}{C_{gs}} = \frac{\mu}{2\pi L_g} \left(1 - \frac{W}{a}\right) \sqrt{V_p(V_{bi} - V_g)} \qquad 11.1$$

where μ is the electron mobility in the channel, L_g is the gate length a is the thickness of the epitaxial layer, V_p is the pinch-off voltage, V_{bi} is the built-in voltage of the gate, V_g is the applied gate bias. W, the depletion width is given by

$$W = 2\varepsilon \sqrt{\frac{(V_{bi} - V_g)}{qN_D}} \qquad 11.2$$

where q is the electronic charge and N_D is the donor concentration.

The switching time, τ, is given by

$$\tau = \frac{Q}{I_D} = \frac{L_g^2}{\mu} \frac{\sqrt{(V_{bi} - V_g)}}{\left\{\sqrt{V_p} - \sqrt{(V_{bi} - V_g)}\right\} V_D} \qquad 11.3$$

where V_D is the applied drain voltage.

TABLE 11.6.   Important Ternary and Quaternary Compound Characteristics

Material	Low Field Mobility, $cm^2 V^{-1} s^{-1}$	Peak Velocity $10^7 cms^{-1}$	Saturated Velocity $10^7 cms^{-1}$	Dielectric Constant	Energy Gap eV	Saturation Field $kV cm^{-1}$
GaAs	4500	1.86	1.33	12.9	1.439	2.96
InP	3815	2.6	1.84	12.3	1.340	4.82
$Ga_{.47}In_{.53}As$	8875	2.2	1.43	13.73	0.717	1.61
$InP_{.8}As_{.2}$	5283	2.8	1.85	12.7	1.101	3.50
$Ga_{.27}In_{.73}P_{.4}As_{.6}$	7041	2.7	1.77	13.2	0.889	2.51
$Ga_{.5}In_{.5}As_{.96}Sb_{.09}$	9377	2.2	1.41	13.8	0.708	1.50

A value of 10^{17} donors per cm^3 for carrier concentration is assumed.

The properties of the materials, detailed in Table 11.6 have been used to predict the performance of MESFETs fabricated on the materials using the one dimensional FET model of Pucel et al (1975). Golio and Trew (1980) have shown, for example, that $Ga_{0.27}In_{0.73}P_{0.4}As_{0.6}$ should have a 58% higher available gain at minimum noise bias at 20 GHz than GaAs.

Referring to the basic small signal model of a FET (Fig. 11.2) Table 11.7 shows the predictions of the Pucel model for a GaAs FET (Golio et al, 1980). Parameters such as C_{gs} and g_m are in good agreement between measurement and theory indicating that the predicted f_T should be accurate.

TABLE 11.7. Comparison of Small Signal Equivalent Circuit Element Values Predicted and Obtained Experimentally

Parameter	Model	Measured
g_m (mS)	35.6	33
C_{gs} (pF)	0.468	0.5
C_{ds} (pF)	0.08	0.06
C_{dg} (pF)	0.05	–
R_{ds} (Ω)	3170	660
R_i (Ω)	3.24	3.5
f_T (GHz)	12.1	11

The values of L_S, L_g and L_D in Fig. 11.2 are assumed to be 0.05 nH, 0.1 nH and 0.1 nH respectively which are representative of values for chip FETs. The gate resistance, R_g, is expressed as:

$$R_g = \frac{\rho_g W_g}{3Nt_g L_g} \qquad 11.4$$

where ρ_g is the gate metal resistivity (Ωcm)

   W_g is the total gate width (cm)

   N is the number of gate fingers

   t_g is the gate metallization thickness (cm), and

   L_g is the gate length (cm).

The source resistance, R_s, is given by:

$$R_s = \left[\sqrt{\frac{R_c \rho_m}{a}} \coth \sqrt{\frac{\rho_m S^2}{R_c \cdot a}} + \frac{\rho_m L_{gs}}{a} \right] \frac{1}{W_g} \qquad 11.5$$

where a is the channel thickness (cm)

ρ_m is the bulk material resistivity (Ωcm)

S is the source length (cm)

L_{gs} is the gate-to-source spacing (cm)

R_c is the specific contact resistance in Ω cm^2 given by

L_{gs} is the gate-to-source spacing (cm) given by

$$R_C = \frac{100}{\sqrt{N_D}} \qquad 11.6$$

where N_D is the doping concentration (cm^{-3}).

The drain resistance, R_d, is expressed by:

$$R_d = \left[\sqrt{\frac{R_c \rho_m}{a}} \coth \sqrt{\frac{\rho_m S^2}{R_c \cdot a}} + \frac{\rho_m L_{gd}}{a} \right] \frac{1}{W_g} \qquad 11.7$$

where L_{gd} is the gate-to-drain spacing (in cm). Expressions for the other element values and the noise figure predictions come directly from Pucel's work (1975).

The built-in voltage of the Schottky barrier gate, V_{bi}, is given by

$$V_{bi} = \frac{2E_g}{3} - \frac{kT}{q} \cdot \ln \left(\frac{N_c}{N_D} \right) \qquad 11.8$$

where E_g is the energy gap

k is Boltzmann's constant (1.38 x 10^{-23} J/K)

T is temperature ($^{\circ}$K)

and N_c is the density of states (m^{-3}).

V_{bi} is therefore directly related to the technology and the material used.

Using the above information equivalent circuit models for 1μm gate length FETs have been predicted for the various alloys used as the active layers. Table 11.8 shows the predictions whilst Fig. 11.6(a) and (b) show graphically the calculated advantages in terms of available gain and minimum noise figure. The realisation of the noise figure predicted for GaAs itself has not yet been achieved partially due to substrate effects (Eastman et al, 1979) but noise

TABLE 11.8.   Pucel Model Predictions for some Ternary and Quaternary
Materials when the FETs are Biased for Minimum Noise Figure

Material	g_m, mS	C_{gs}, pF	R_{ds},	f_T, GHz	F_{min} at 20 GHz dB	G_{av} at 20 GHz dB
GaAs	21.4	0.207	11600	16.4	1.726	2.99
InP	24.5	0.186	8070	20.9	1.924	4.00
$Ga_{.47}In_{.53}As$	24.5	0.219	16900	17.9	0.977	4.03
$InP_{.8}As_{.2}$	27.8	0.200	7680	22.1	1.585	4.59
$Ga_{.27}In_{.73}P_{.4}As_{.6}$	27.8	0.207	10800	21.4	1.236	4.75
$Ga_{.5}In_{.5}As_{.96}Sb_{.09}$	24.5	0.220	17400	17.7	0.935	4.01

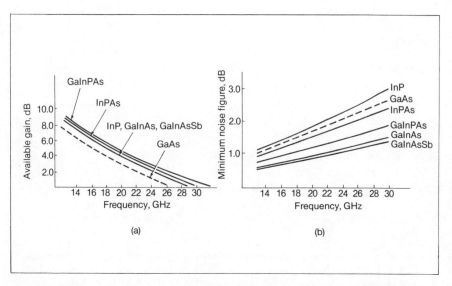

FIG. 11.6(a).   Predicted Available Gain as a Function of Frequency
    (b).   Predicted Minimum Noise Figure as a Function of
Frequency (after Golio)

figures close to 2 dB have been achieved at 18 GHz for 0.5μm gate
length GaAs FETs (Butlin, private communication). What Figures 11.6
(a) and (b) and Table 8 do show, however, is that there are substantial benefits to be gained at the higher frequencies by fabricating
MESFET structures on $Ga_{0.47}In_{0.53}As$ or on $Ga_{0.27}In_{0.73}P_{0.4}As_{0.6}$.
Interestingly the predicted performance of InP FETs shown in Fig.
11.6(a) and (b) is as found in practice – some small improvement in
gain but a degradation in noise figure over the GaAs device.

$Ga_{0.47}In_{0.53}As$ grown lattice matched onto InP has a measured room
temperature mobility of 11,000 cm^2 V^{-1} s^{-1} as compared to a value of
7000 cm^2 V^{-1} s^{-1} for GaAs of the same doping level of 10^{16} cm^{-3}.
Fig. 11.7 shows the mobility of $Ga_{0.47}In_{0.53}As$ over the electron
concentration range of 10^{14} to 10^{17} cm^{-3} representing data from
several laboratories.

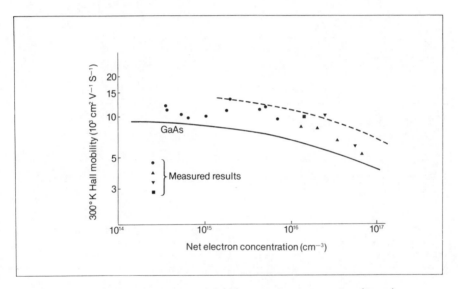

FIG. 11.7.   Room Temperature Mobility of $In_{.53}Ga_{.47}As$ (Lattice
   Matched to InP) and GaAs as a Function of Net Electron Concentration (after Morkoc et al)

The difficulty with fabricating Schottky barrier FETs on $Ga_{.47}In_{.53}$
As is associated with the low built-in voltage of the Schottky gate.
From equation 11.8 and Table 11.6 the built-in potential is calculated to be 0.3 eV for a carrier concentration of 1 x 10^{17} cm^{-3} at
room temperature for aluminium gated devices. The equivalent value
for GaAs is 0.7 to 0.8 eV. The small potential of $Ga_{0.47}In_{0.53}As$
results in a very thin barrier which gives rise to electron tunnelling
and thus to large leakage currents between gate and source. Methods
to reduce this leakage current have been proposed including the
growth of a very thin (500Å typically) layer of $In_{0.53}Ga_{0.47}$ having a

net concentration of 10^{15} cm^{-3} between the Schottky barrier and the channel region thus increasing the depletion width (from Equation 11.2). The barrier height can also be increased by making this thin layer n$^-$ InP (Morkoc et al, 1979). Such a FET structure is shown in Fig. 11.8 where the n$^-$ layer is shown to be etched away between the gate and source and drain areas.

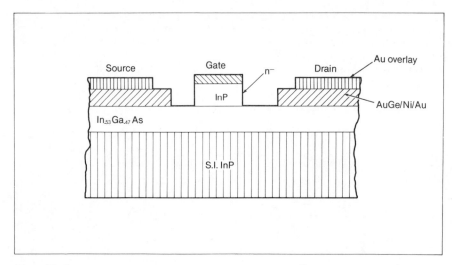

FIG. 11.8.   In$_{0.53}$Ga$_{0.47}$As FET Employing an n$^-$ InP Layer under
   Schottky Barrier Gate (after Morkoc et al)

Thus it may be appreciated from the foregoing that much interesting technological work is foreseen over the next few years in order to realize the potential advantages of higher electron mobility for normally-off FETs (particularly for logic applications) and peak electron velocity for normally-on FETs (for microwave applications).

5. PERMEABLE BASE TRANSISTOR

The permeable base transistor consists of a very thin grating embedded inside a single crystal of gallium arsenide. The grating forms a Schottky barrier with the GaAs and is used to increase or decrease the potential which occurs in the semiconductor which is between the grating lines. The permeable base transistor (PBT) has the advantages of the FET - in being a majority carrier device as well as having the higher current gain properties of barrier control-led devices.

A drawing of the PBT is shown in Fig. 11.9(a) together with a fabri-cated device in Fig. 11.9(b). The device consists of four layers,

the n[+] substrate, the n-type emitter layer, the metal (tungsten)
grating film and the n-type collector layer. Electrons flow from the
n[+] layer to the emitter where they are constrained by the proton
bombarded isolated region to flow vertically upwards through the
grating to the collector region. The metal base contact connected to
the grating is brought out to the top surface of the device.

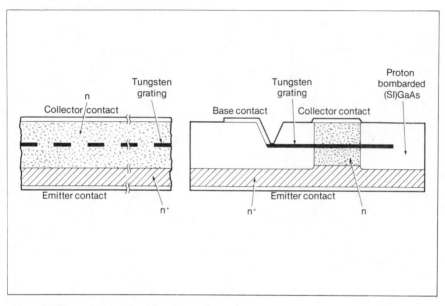

FIG. 11.9. The Permeable Base Transistor

The theory of operation of the PBT can be understood by considering
a small cross-section of the device which includes three fingers of
the tungsten grating (Fig. 11.10(a)). The grating is sandwiched
inside an n-layer whose carrier level is such that the zero biased
depletion width of the 'base' Schottky barrier is larger than the
openings in the grating. The electronic potential as one moves along
the line XX' in Fig. 11.10(a) will be that of back to back Schottky
diodes. As one moves along the line Y-Y' through the grating gaps
the potential will increase as the metal grating is approached and go
through a maximum in the plane of the metal grating as shown in Fig.
11.10(b), where the maximum potential will be less than the corresponding Schottky barrier height. This barrier height will depend on
the carrier concentration, the spacing between the grating fingers,
the base thickness and the Schottky built-in voltage. Applying a
small positive bias to the collector of the PBT as in Fig. 11.10(c)
results in a small collector current flow because of the barrier

between the gratings. The current can be considered to be due to
drift and diffusion components which are nearly equal when the barrier
is relatively large. If positive bias is now applied between base and
emitter as in Fig. 11.10(d) the drift component of the emitter current
is reduced and the collector current and transconductance increase
rapidly. At higher forward bias the barrier will be overcome in the
centre of the gaps and current flow will be space charge limited
(Fig. 11.10(e)). Because the device uses an extremely fine grating
embedded in a semiconductor of the correct carrier concentration the
high current densities that are present are barrier controlled resulting in large transconductance and maximum frequency of oscillation.

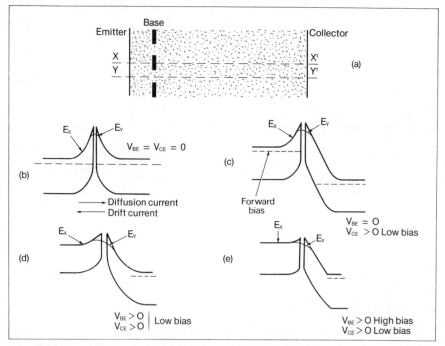

FIG. 11.10.   Theory of Permeable Base Transistor

The PBT has been simulated numerically by Alley et al (1979) using
finite difference methods to solve the simultaneous equations resulting from a two-dimensional numerical analysis.

If the device is designed so that the grating gaps are small compared to the zero-biased depletion width of the n-type layer, the
base metallization thickness is made approximately 5% of this depletion width at the largest operating voltage, high transconductance
and output resistance are achieved. Assuming that the grating fingers
have a low resistance leads to a high f_T and high f_{max}.

For example, a device with 2000 Angstrom (Å) grating gaps, 2000Å grating fingers, a 200Å thick grating and an n-region carrier concentration of 1×10^{16} cm^{-3} will have the calculated collector current density versus collector to emitter voltage characteristic of Fig. 11.11.

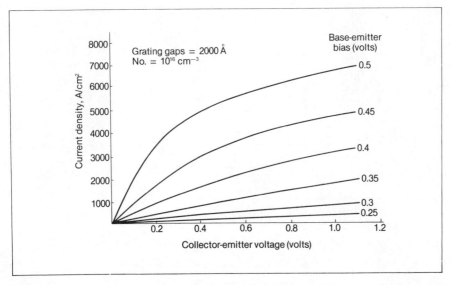

FIG. 11.11.   Permeable Base Transistor Collector Characteristics

It is also possible to calculate the current density within the grating gaps as a function of the base to emitter voltage.   In this respect there are two distinct regions.   Firstly as V_{BE} is increased from zero the current increases exponentially.   At larger values of V_{BE} the net negative charge in the middle of the gaps exceeds the positive charge forming a negative space charge.   The current density in the middle of the gaps is then proportional to $V_{BE}^{3/2}$. The current density close to the grating still continues to rise exponentially. The large non-linearity between base-emitter voltage and collector current results in values of transconductance normalized to collector current density of approximately 8 mS A^{-1} cm^2.   It is possible to determine the elements of a simple small signal equivalent circuit (neglecting parasitic components) - for example, the base to emitter resistance being the slope of the I_B to V_{BE} curve and the collector to emitter resistance being the slope of the I_C to V_{CE} curve (Bozler and Alley, 1980).

Results of calculations based on such an equivalent circuit for the geometry mentioned previously are shown in Fig. 11.12.

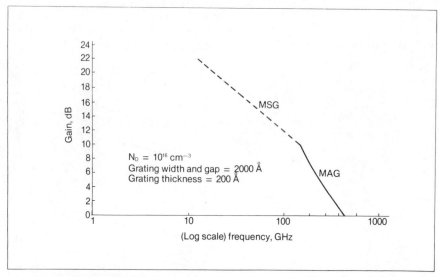

FIG. 11.12.   Theoretical Maximum Stable and Available Gains for a
    Permeable Base Transistor

The calculated value of the unit current gain frequency

$$(f_T = \frac{g_m}{2\pi C_{BE}})$$

is 88 GHz and from Fig. 11.12 an f_{max} of 400 GHz is predicted.
Devices made to date have not been optimized for frequency and have
used a simple structure to prove the concept (Ally et al, 1980). The
grating is produced using X-ray lithography and then the wafer put
back into the epitaxial reactor enabling GaAs to be grown over the
tungsten grating. Connection to this grating is achieved by ensuring
that the GaAs growth does not fully cover the metallisation used to
connect together the grating fingers. Devices to date have resulted
in f_{max} of 30 GHz with a noise figure of 2.5 dB at 8 GHz with 10 dB
associated gain.

6. BALLISTIC ELECTRON TRANSISTORS

Much theoretical work has been recently done on a new mode of
operation of FET structures (Eastman et al, 1980; Shur et al, 1979).
This mode of operation depends on the fact that under certain conditions the mean free path of carriers in the semiconductor becomes
larger than the length of the active region of the device itself.
Thus the velocity of the electrons can become considerably higher
than the peak electron velocity associated with normal 'collision-
dominated' GaAs devices. This electron motion is referred to as

'ballistic' since, at least to a first order approximation, the electron velocity is due to the acceleration caused by the applied electric field.

The mobility, for example, of electrons in high purity GaAs is approximately 150,000 cm^2 V^{-1} s^{-1} at a temperature of 77°K giving a mean free path of approximately 1.3μm. This distance is much larger than state-of-the-art high microwave frequency FET gate lengths which are less than 0.2μm.

Now, the switching time of a ballistic electron transistor τ is

$$\tau = \frac{L}{v_{eff}} \hspace{4cm} 11.9$$

where L is the gate length in the case of a normal FET and v_{eff}, the effective electron velocity is given by

$$v_{eff} = \frac{j}{qn_o} \hspace{4cm} 11.10$$

where n_o is the doping density, q is the electronic charge and j is the current density in the channel. The current density is given by

$$j = qn_o \sqrt{\frac{qU}{2m*}} \hspace{3cm} 11.11$$

where U is the electric potential between the electrodes and m* is the electronic effective mass.

The power consumed during the switching time τ is given by

$$P = jUS \hspace{4cm} 11.12$$

where S is the cross section of the active region. Thus from equations 11.9 and 11.11

$$\tau = L \sqrt{\frac{2m*}{qU}} \hspace{3.5cm} 11.13$$

and

$$P = qn_o \sqrt{\frac{qU^3}{2m*}} \hspace{1cm} \text{where } S = \frac{QU}{\tau} \hspace{1.5cm} 11.14$$

The power delay product $P_\tau = \dfrac{jUSL}{\tau}$

$$= qn_o ULS \hspace{3cm} 11.15$$

Equation 11.15 can be rewritten as

460

$$P_\tau = \frac{0.1602 \text{ U.L.Sn}_o}{10^{15}} \qquad\qquad 11.16$$

where P_τ is in femtojoules, U is in volts, L is in microns and n_o is in cm^{-3}.

Now a typical value of U is 0.1 volt, L is 1μm and S is 1μm^2 so that the switching time and power delay product for GaAs can be estimated from Equations 11.9 to 11.16 as $\tau \simeq 0.879$ picoseconds and $P_\tau \simeq 0.16$ femtojoules for $n_o = 10^{15}$ cm^{-3}. Such figures of merit are considerably better than those for conventional GaAs MESFET logic and rival the values expected for Josephson junction logic elements (Hernel, 1974).

The above theory based on the work of Shur and Eastman assumes strictly that the device is two terminal. However the agreement of this theory with the Monte Carlo simulations of low doped FETs by Rees et al (1977) is fair.

A simple one dimensional model for the FET can be used to calculate the current-voltage relationship under the ballistic regime (Shur et al, 1979). Consider Fig. 11.13 where the region under the gate is divided into a depletion region and the channel.

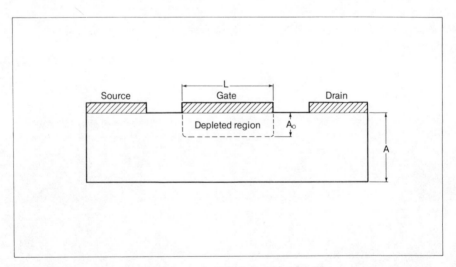

FIG. 11.13. Simple Model to Predict Ballistic Electron Transport Properties of GaAs MESFET.

The drain to source voltage V_{DS} is assumed to be smaller than the difference V_{bi}--V_G where V_{bi} is the built-in voltage of the Schottky gate and V_G is the applied gate voltage. Also V_{DS} is assumed to be less than the ideal pinch-off voltage V_{PO} which is the voltage necessary to fully deplete the region under the complete gate length L. In this context U_{PO} is defined as

$$U_{PO} = \frac{qn_o L^2}{2\epsilon_o \epsilon} \qquad\qquad 11.17$$

This expression may be seen to be analogous to the normal pinch-off voltage V_{PO} of a FET, which has already been derived in Chapter 2, and is given by

$$V_{PO} = \frac{qn_o A^2}{2\epsilon_o \epsilon} \qquad\qquad 11.18$$

where A is the thickness of the active layer.

In the ballistic case the width A_o of the depletion layer under the gate is independent of the distance along the gate length L as depicted in Fig. 11.13 and

$$A_o \simeq \left\{ \frac{2\epsilon_o \epsilon (V_{bi} - V_G)}{qn_o} \right\}^{\frac{1}{2}} \qquad\qquad 11.19$$

and the current density in the channel is given by

$$j = q^{3/2} n_o \sqrt{\frac{2V_{DS}}{m^*}} \qquad\qquad 11.20$$

As a result the drain to source current is

$$I_{ds} \simeq jA_o \cdot W$$

where W is the gate width, which yields

$$I_{ds} \simeq q^{3/2} n_o W \sqrt{\frac{2V_{ds}}{m^*}} \left\{ A - \frac{2\epsilon_o \epsilon (V_{bi} - V_G)}{qn_o} \right\}^{\frac{1}{2}} \qquad\qquad 11.21$$

This equation may be compared with equation 2.3 of Chapter 2.

Further work has shown that in n^+-n-n^+ and n^+-p-n^+ GaAs structures ballistic transport of electrons can occur at room temperature. For example, Fig. 11.14 shows the measured and calculated current-voltage characteristics of a 2 terminal n^+-p-n^+ structure fabricated by Eastman et al (1980) where the net acceptor level was 10^{15} cm^{-3} and the structure was 0.47µm long. There is no current until the applied

voltage exceeds the so-called punch-through voltage (which is given
by the same expression as equation 11.17 but where L is now the diode
length).

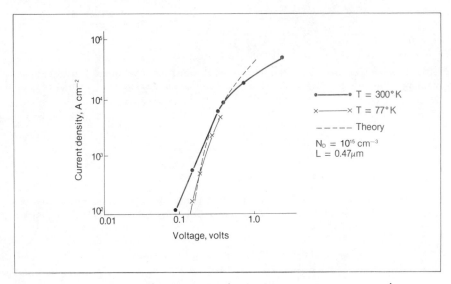

FIG. 11.14. Measured and Theoretical Electron Current Density as a
    Function of Applied Bias (after Eastman et al)

As soon as punch through occurs the current flow increases sharply.
Above approximately 0.5 volt applied voltage the measured current
densities deviate from the ballistic theory due to the ability of the
electrons to transfer, for example, to valleys in the (111) direction
if they were initially travelling in the (100) direction.

   Most of the theoretical assessment to date on ballistic electron
transport is related to the exploitation of the resultant devices in
high speed logic circuits. However, the periodic structure of the
permeable base transistor dealt with earlier in this Chapter is almost
exactly what is needed to induce ballistic transport due to the small
grating gaps and geometry. Thus the outcome of successful device
fabrication and the characterisation of these structures will be
followed with great interest in the next few years as it will lead to
the application of such devices in higher frequency and faster speed
circuits.

7. CONCLUSIONS

   This final chapter has attempted to give the reader a brief introduction to field effect transistor structures using either materials
other than GaAs or different operating principles to improve the

performance of or extend the useful frequency range of three terminal
solid state devices. Many of the results to date have shown the need
for a more advanced technology and understanding. Technologies such
as molecular beam epitaxy and ion implantation have already had considerable impact on the formation of reproducible homo- and heterostructures.

The state of most of the devices mentioned in this Chapter such as
the PBT and the BET are where the GaAs MESFET was fifteen or so years
ago - it will be interesting to see how many of them are as successful
and far reaching in their applications as that particular device has
been!

8. BIBLIOGRAPHY

Alley, G.D., Bozler, C.O. and Murphy, R.A., Lindley, W.T. Two dimen-
sional numerical simulation of the permeable base transistor.
Proceedings of the 7th Biennial Cornell Electrical Engineering
Conference, Aug. 1979, pp.43-51.

Alley, G.D., Bozler, C.O., Flanders, D.C., Murphy R.A. and Lindley,
W.T. Recent experimental results on permeable base transistors.
International Electron Devices Meeting Technical Digest, December
1980, pp.608.612.

Barrera, J.S. and Archer, R.J. InP Schottky-gate field-effect
transistors. IEEE Transactions on Electron Devices, Vol. ED-22,
No. 11, Nov. 1975, pp.1023-1030.

Becke, H., Hall, R. and White, J. Gallium arsenide MOS transistors.
Solid State Electronics, Vol. 8, 1965, pp.813-823.

Bozler, C.O. and Alley, G.D. Fabrication and numerical simulation of
the permeable base transistor. IEEE Trans. on Electron Devices,
Vol. ED-27, No. 6, June 1980, pp.1128-1141.

Butlin, R.S. - private communication.

Clarke, R.C. and Reed, W.D. Vapour phase epitaxy of indium phosphide
for FET fabrication. Proceedings of the Seventh Biennial Cornell
Electrical Engineering Conference, Cornell University, Ithaca, USA,
Aug. 1979, pp.81-92.

Eastman, L.F. and Shur, M.S. Substrate current in GaAs MESFETs.
IEEE Transactions on Electron Devices, Vol. ED-26, Sept. 1979,
pp.1356-1361.

Eastman, L.F. et al. Ballistic electron motion in GaAs at room
temperature. Electronics Letters, Vol. 16, No. 13, pp.524-525,
June 1980.

Engelmann, R.W.H. and Liechti, C.A. Bias dependence of GaAs and InP
MESFET parameters. IEEE Transactions on Electron Devices, Vol.

ED-24, No. 11, Nov. 1977, pp.1288-1296.

Gleason, I.R., Dietrich, H.B., Henry, R.L., Cohen, E.D. and Bark,M.L. Ion-implanted n-channel InP metal semiconductor field effect transistor. Applied Physics Letters, Vol. 32, No. 9, May 1978, pp.578-581.

Golio, J.M. and Trew, R.J. Compound semiconductors for low-noise microwave MESFET applications. IEEE Transactions on Electron Devices, Vol. ED-27, No. 7, July 1980, pp.1256-1262.

Hasegawa, H., Forward, K.E. and Hartnagel, H. Electronics Letters, Vol. 11, 1975, p.53.

Hernel, D.J. Femtojoule Josephson tunnelling logic gates. IEEE J. Solid State Circuits, Vol. SC-9, pp.277-282, 1974.

Lile, D.L., Collins, D.A., Meiners, L.G. and Messick, L. N-channel inversion-mode InP MISFET. Electronics Letters, Vol. 14, pp.657-659, Sept. 1978.

Maloney, T.J. and Frey, J. Frequency limits of GaAs and InP field effect transistors at $300^{\circ}K$ and $77^{\circ}K$ with typical active-layer doping. IEEE Transactions on Electron Devices, Vol. ED-23, No. 5, May 1976, p.519.

Meiners, L.G., Lile, D.L. and Collins, D.A. Inversion layers on InP. J. Vac. Sci. Technol. Vol. 16, pp.1458-1461, Sept./Oct. 1979.

Meiners, L.G., Lile, D.L. and Collins, D.A. Microwave gain from an n-channel enhancement mode InP MISFET. Electronics Letters, Vol. 15, Aug. 1979, p.578.

Messick, L., Lile, D.L. and Clawson, A.R. A microwave InP/SiO_2 MISFET. Applied Physics Letters, Vol. 32, April 1978, pp.494-495.

Messick, L. Power gain and noise of InP and GaAs insulated gate microwave FETs. Solid State Electronics, Vol. 22, pp.71-76, January 1979.

Messick, L. InP/SiO_2 MIS structure. J. Appl. Phys. Vol. 47, Nov. 1976, pp.4949-4951.

Morkoc, H., Oliver, J.D. and Eastman, L.F. High mobility $In_{.53}Ga_{.47}As$ for high performance MESFETs. Proceedings of the 7th Biennial Cornell Engineering Conference, Aug. 1979, pp.71-80.

Pucel, R.A., Haus, H.A., and Statz, H. Signal and noise properties of gallium ardenide microwave field effect transistors. Advances in Electronics and Electron Physics, Vol. 38, New York: Academic Press, 1975, pp.195-265.

Rees, H., Sanghera, G.S. and Warriner, R.A. Low temperature FET for low-power high speed logic. Electronics Letters, Vol. 13, No. 6, pp.156-158, 1977.

Shur, M.S. and Eastman, L.F. Near ballistic electron transport in GaAs devices at 77°K. Proceedings of the 7th Biennial Cornell Electrical Engineering Conference, Aug. 1979, pp.389-399.

Shur, M.S. and Eastman, L.F. Ballistic transport in semiconductor at low temperatures for low-power high-speed logic. IEEE Transactions on Electron Devices, Vol. ED-26, No. 11, pp.1677-1683, Nov. 1979.

Sugano, T., Koshiga, F., Yamasaki, K. and Takahashi, S. 30-40 GHz GaAs insulated gate field effect transistors. IEEE International Electron Devices Meeting Technical Digest (Washington DC, Dec. 1978), pp.148-151.

Zeisse, C.R., Messick, L.J. and Lile, D.L., Journal of Vac. Sci. Tech. Vol. 14, 1977, p.957.

Index

Absorption
  stabilisation, 264
Acceptor levels, 4
Active matching, 363
Airbridge, 78, 79, 387
AM noise spectra, 258
Amplifier, 165, 353
Annealing, 115
Arsenic trichloride process, 107
Auger analysis, 395
Avalanching, 88

Balanced amplifier, 180
Balanced mixer, 228
Ballistic motion, 459
Band structure, 1
Base, 455
Beam leads, 388
Bipolar transistor, 43
Bubbler system
  double, 108
  single, 107
Buffer layer, 99, 100, 103
Built-in voltage, 451
Burnout, 88

Capacitor, 343
Carrier concentration, 100, 115,
  116, 386
Cascode connection, 190
Cavity stabilisation, 264, 265
Cermet resistor, 389
Channel capacitance, 66, 315
Channel resistance, 25, 69, 315
Characteristic impedance, 341
Chirp, 290
Chloride transport, 106
Clock frequency, 367
Cofired ceramic package, 300
Collector, 455
Combining techniques, 201
Common gate connection, 187
Computer aided design, 172
Conduction band, 2
Constant gain circles, 149
Contact layer, 132
Conversion gain, 220, 235
Conversion transconductance, 220
Coplanar waveguide, 349
Correlation factor, 29-34

Cr-doping, 105, 126
Cut-off frequency, 24
CVD, 99
Czochralski process, 113

DCFL, 375
Depletion mode, 370
Depletion region, 15-17, 22, 456
Dielectric crossovers, 78
Dielectric resonators, 265
  mechanical tuning of, 275
  temperature coeff. of, 273
Diffused gate FET, 13
Diffusion noise, 29
Diffusion potential, 10
Digital circuit, 366
Diodes:-
  planar, 351
  Schottky, 350
  surface oriented, 350
Discriminator, 264, 329
Distributed elements, 161, 348
Divider, 426, 429
Doping density, 15
Dopper frequency, 290
Drain mixer, 225
Drain resistance, 25
Drift velocity, 13-14, 439
Dual-gate FET, 45-56, 315, 323
Dual-gate FET mixer, 233
Dynamic switching energy, 366

Effective dielectric constant,
  349
Effective inductance, 341
Electron beam lithography, 381
Electronic tuning, 280
Emitter, 455
Energy bands, 1,2
Energy gap, 2
Enhancement mode, 374, 444
Epitaxy, 99-113
Epoxies, 302
Equivalent circuit of FET, 153,
  442, 450

Feedback, 185, 246, 276, 353,357
Fermi level, 6
FET channel, 367
Flip chip, 81, 260

Float-off, 129-130
FM noise spectra, 258
Free-running oscillators, 256

GaInAs, 453
Gain control, 51
Gate, 15
Gate mixer, 216
Gate-to-channel capacitance, 25
Gate-to-drain avalanching, 88, 90
Gate-to-source resistance, 21
Glass-metal package, 300
Grating, 454
Gunn domain, 19, 69, 443
Gunn oscillator, 245

Hall mobility, 100, 116, 122
Hard fired ceramic package, 300
Heterojunction, 11
Holes, 3

IF, 216
Image rejection mixer, 237
Impatt oscillator, 245
Impedance mapping, 249, 250
Impurity semiconductors, 4
Inductors, 340, 393
Ingot, 113
InP, 439
Insertion loss mapping, 53
Interconnections, 398
Interdigital capacitor, 343
Intermetallic compounds, 5
Intermodulation products, 71-77,
  184
Internal matching, 199
Inverter, 368
Ionic crystals, 1
Ion implantation, 99, 114, 384
Ion implanted material proces-
  sing, 132, 381
Ion milling, 397
I-V characteristics, 62-63

Junction FET, 374

Large signal parameters, 194,195
Lattice phonons, 28
Level shifting, 370
Liquid encapsulated Czochralski
  process, 113
Liquid phase epitaxy (LPE), 99,
  109

Load line, 193
Local oscillator feedthrough, 355
Logic gate, 366
Logic swing, 368
Low pass filter prototypes, 197,
  315
LSI, 370
Lumped elements, 158, 197, 307,
  340

Matching, 156, 157, 160-172
Maximum available gain, 27
Maximum frequency of oscilla-
  tion, 27
Mercury probe, 105
Mesa process, 386, 398
Metal alkyl-hydride process, 109
Microstrip, 348
Millimetre wave mixer, 419
Minimum noise figure, 30, 40
MISFET, 444
Mixers, 213, 355, 418
Mixing products, 214
Mobility, 14, 115-119, 367, 449,
  459
Modelling, 65
Molecular beam epitaxy (MBE), 99,
  113
Monolithic circuit, 319
Monolithic:
  mixers, 418-422
  small signal amplifiers, 400-
    405
  power amplifiers, 407-415
  oscillators, 415-417
  switches, 417-418
MOSFET, 445
Multilevel logic, 373
Multiplication, 239
Multiplier, 425, 429
Multithrow switch, 321

NAND gate, 371
n-channel, 371
Negative feedback, 185
Negative resistance, 246, 287
Network synthesis, 174
Noise current generator, 28
Noise figure, 228, 447
  circles, 162
  modelling, 163
  optimum, 161
  single sideband, 227

Non linear circuit elements, 255
Non linear performance, 69
NOR gate, 371, 374
Normally-off, 374, 444
Normally-on, 370
n-type semiconductors, 4

Ohmic contact, 9, 129
Optimization, 173
Optimum reflection coefficient, 152
Organo-metallic chemistry, 99, 109
Oscillators, 245, 415
Oscillator noise, 262
Osciplier, 333
Out-diffusion, 103-104
Output power, 64
Output resistance, 69
Overlay capacitor, 344

Packages, 299
Packaged FET, 206
Package modelling, 303
Parasitics, 83, 154, 187
Peak velocity, 441, 448
Peripheral power density, 258
Permeable base transistor, 454
Phase locking, 264
Phase shifter, 323
Pinch off voltage, 20, 101, 461
Plasma etching, 397
Plasma enhanced CVD, 395
Polyimide, 386, 391
Power-delay product, 366, 459
Power FETs, 192
Power-gain saturation, 261
Prematching, 299, 306
Premature decline phenomenon, 285
Propagation delay, 366
P-type semiconductors, 4
Pulsed operation, 91, 203
Pulsed oscillators, 289, 334
Punch-through, 462
Push-pull amplifier, 407

Q factor, 342, 345
Qualification, 122
Quarter wavelength lines, 319
Quarter wave transformers, 196
Quasi-normally off logic, 378
Quaternary alloy, 448

Recessed channel, 130
Rectifying contact, 7, 9, 11
Reflection amplifier, 205
Resistor:
  active load, 347
  bulk, 346
  thin film, 389
Reverse channel oscillator, 260
Rieke diagram, 272
Ring oscillator, 377, 428

Satellite valleys, 28
Saturation velocity, 15
Scattering (S) parameters, 51, 66, 141, 155, 194, 246
Schottky barrier, 7, 9, 455
SDFL logic, 372
Self-aligned process, 129, 377
Sensitivity analysis 174, 359
Short gate-length, 19
Si implantation, 126
Silicon nitride, 393
Skin effect, 39
Small signal amplifiers, 141
Smith Chart, 145, 249
Source contact resistance, 21,25
Source-drain burnout, 88
Source follower connection, 190, 354
Source lead inductance, 84, 305
SPDT switch, 319
Specific contact resistance, 451
Speed-power product, 366
Spiral inductor, 192, 341
Spurious free dynamic range, 199
Stabilised oscillators, 263
Stability circles, 146, 288
Stability factor, 145
Steady-state oscillations, 252
Surface states, 445
Switches, 315, 417
Switching time, 24
Synthesis, 174, 182

Ternary alloy, 448
Thermal impedance, 87, 309
Thermal redistribution, 120
Thermal spreading, 310
Three-port device, 49, 249
Transfer characteristic, 101
Transconductance, 21, 68, 367
Transducer gain, 147
Transmission lines, 304

Transmission stabilisation, 264
Tunable oscillators, 253

Undoped substrates, 123-128
Unilateral figure of merit, 150
Unilateral power gain, 27
Up-conversion, 238

Valence bond, 2
Vapour phase epitaxy (VPE), 106
Varactor tuning, 281
Vector modulator, 327
Vias, 81, 82, 260, 387
VLSI, 370

Wrap-around ground, 78,79
Wire bonding, 80, 154, 303
Word generator, 425

Yield, 366
YIG sphere, 284
YIG tuning, 284